T0186984

WJEC
Chemistry
A2 Level

2nd Edition

David Ballard

Rhodri Thomas

Illuminate
Publishing

Published in 2021 by Illuminate Publishing Limited, an imprint of Hodder Education, an Hachette UK Company, Carmelite House, 50 Victoria Embankment, London EC4Y 0DZ

Orders: please contact Hachette UK Distribution, Hely Hutchinson Centre, Milton Road, Didcot, Oxfordshire, OX11 7HH.
Telephone: +44 (0)1235 827827. Email: education@hachette.co.uk.
Lines are open from 9 a.m. to 5 p.m., Monday to Friday. You can also order through our website: www.hoddereducation.co.uk

© David Ballard, Rhodri Thomas 2021

The moral rights of the authors have been asserted.

British Library Cataloguing in Publication Data

A catalogue record for this book is available from the British Library

ISBN 978-1-912820-7-33

Printed by Ashford Colour Press, UK

Impression 5
Year 2023

Hachette UK's policy is to use papers that are natural, renewable and recyclable products and made from wood grown in well-managed forests and other controlled sources. The logging and manufacturing processes are expected to conform to the environment regulations of the country of origin.

Every effort has been made to contact copyright holders of material reproduced in this book. If notified, the publishers will be pleased to rectify any errors or omissions at the earliest opportunity.

This material has been endorsed by WJEC and offers high quality support for the delivery of WJEC qualifications. While this material has been through a WJEC quality assurance process, all responsibility for the content remains with the publisher.

WJEC examination questions are used under licence from CBAC Ltd. WJEC bears no responsibility for example answers to questions taken from its past question papers that may be contained in this publication.

Editor: Geoff Tuttle
Design: Nigel Harriss
Layout: Neil Sutton, Cambridge Design Consultants

Cover image: © Sima/Shutterstock

Acknowledgements

One of the authors (DB) would like to thank Derby College for access to their science laboratories and also to his wife, Margaret for her infinite patience during the writing of this book.

MIX
Paper | Supporting responsible forestry
FSC™ C104740
www.fsc.org

Image credits:

p. 1 Sima/Shutterstock
p. 10 Jim West/Science Photo Library; **p. 11** (both bottom) ER Digginger/Science Photo Library; **p. 13** Science Photo Library; **p. 14** Martyn F Chillmaid/Science Photo Library; **p. 16** Andrew Lambert Photography/Science Photo Library; **p. 18** Takashi Images; **p. 21** Andrew Lambert Photography/Science Photo Library; **p. 24** Andrew Lambert Photography/Science Photo Library; **p. 27** (all) Andrew Lambert Photography/Science Photo Library; **p. 30** Turtle Rock Scientific/Science Source/Science Photo Library; **p. 31** isak55; **p. 36** (top) magnetix; **p. 36** (bottom) Boris15; **p. 38** Andrew Lambert Photography/Science Photo Library; **p. 39** Alexandre Dotta/Science Photo Library Alexandre; **p. 41** Andrew Lambert Photography/Science Photo Library; **p. 42** DB; **p. 43** GIPhotostock/Science Photo Library; **p. 48** GIPhotostock/Science Photo Library; **p. 49** GIPhotostock/Science Photo Library; **p. 50** (both) Science Photo Library; **p. 51** GIPhotostock/Science Photo Library; **p. 53** Charles D Winters/Science Photo Library; **p. 64** Power and Syres/Science Photo Library; **p. 69** (top) Public domain; **p. 69** (bottom) The Nobel Foundation/Public domain; **p. 73** Andrii Melnykov; **p. 74** (both) Phil Degginger/Science Photo Library; **p. 78** Stanislav/Adobe Photo Library; **p. 79** Billion Photos; **p. 87** Christina Richards; **p. 95** Martyn F Chillmaid/Science Photo Library; **p. 108** Juulijis/fotolia; **p. 111** Margarete Platzer; **p. 112** On The Run Photo; **p. 113** CEPTAP; **p. 114** magnetix; **p. 119** PiLensPhoto/fotolia; **p. 120** Ajamal; **p. 121** logos2012/fotolia; **p. 123** leisuretime70/fotolia; **p. 128** D&Jfoodstyling/fotolia; **p. 131** David Ballard/Illuminate; **p. 132** David Ballard/Illuminate; **p. 134** Evgeny Karandaev; **p. 139** (top) Pavel Drozda/fotolia; **p. 139** (bottom) HUIZENG; **p. 144** Swapen/fotolia; **p. 152** ExQuising/fotolia; **p. 156** spline_x; **p. 157** (top left) molekuul.be/fotolia; **p. 157** (top right) molekuul.be/fotolia; **p. 157** (middle) raimond14/fotolia; **p. 157** (bottom) FreeProd/fotolia; **p. 159** rueangwit/fotolia; **p. 163** smuay/fotolia; **p. 165** (top) alarts/fotolia; **p. 165** (bottom) Axel Bueckert/fotolia; **p. 167** (top) korta/fotolia; **p. 167** (bottom) D. Ballard/Illuminate; **p. 191** R. MACKAY PHOTOGRAPHY, LLC

Contents

What this book contains

The contents of this book match the specification for WJEC A2 Level Chemistry. It provides you with information and practice examination questions that will help you to prepare for the examinations at the end of the year.

This book covers all three of the Assessment Objectives required for your WJEC Chemistry course. The main text covers the three Assessment Objectives:

- AO1 Knowledge and Understanding
- AO2 Application of Knowledge and Understanding
- AO3 Analyse, interpret and evaluate information, ideas and evidence.

This book also addresses:

- The mathematics of chemistry, which will represent a minimum of 20% of your assessment
- Practical work. The assessment of your practical skills and understanding of experimental chemistry represents a minimum of 15% and will also be developed by your use of this book. Some practical details are mentioned in the chapters of Units 3 and 4. Further comments about the required experimental task and the practical methods and analysis task, are given in the details for Unit 5.

The book content is clearly divided into the units of this course. These are Unit 3 – Physical and Inorganic Chemistry, Unit 4 – Organic Chemistry and Analysis and Unit 5 – Practical. Each chapter covers one topic. The main text of each topic covers all the points from the specification that you need to learn. At the beginning of each chapter is a list of learning objectives highlighting what you need to know and understand and a maths checklist to help you know what maths skills will be required. At the end of each chapter is a set of Test Yourself questions, designed to help you practise for the examinations and to reinforce what you have learned. At the end of each unit, you will find questions selected from WJEC examination papers set over the past few years. Answers to all the questions are given at the end of the book. There is also a Periodic Table on page 232.

Marginal features

The margins of each page hold a variety of features to support your learning:

Key term

The **d-block** is the group of elements whose outer electrons are found in d-orbitals

▲ These are terms that you need to be able to explain. They are highlighted in blue in the body of the text. You will also find other terms in the text in **bold** type, which are explained in the text but have not been defined in the margin. The use of key terms is an important feature since examination papers may contain a number of terms that have to be explained.

10 Knowledge check

Write two chemical equations to show that zinc oxide is an amphoteric oxide.

▲ Knowledge checks are short questions to check your understanding of the subject, allowing you to apply the knowledge that you have acquired. These questions include filling in blanks in a passage, matching terms with phrases specific to the topic under study, and brief calculations. Answers are provided at the back of the book.

Study point

Nitric(III) acid is unstable and is made when required. It can be written as HNO_2 or HONO.

▲ As you progress through your studies, study points are provided to help you understand and use the knowledge content. In this feature, factual information may be emphasised, or restated to enhance your understanding.

Stretch & challenge

Energy calculations suggest that carbon tetrachloride should react easily with water to form carbon dioxide. The reaction does not occur however, as it would be too slow – we say that carbon tetrachloride shows **kinetic stability**.

▲ Stretch and challenge boxes may provide extra information not in the main text, but relevant. They may provide more examples, or questions, but do not contain information that will be tested in an examination.

Top tip ≫

It is not true to say that carbon has no d-orbitals, as all atoms have d-orbitals in higher shells; however, these are not available for use.

▲ Top tips provide general or specific advice to help you prepare for an examination. Read these very carefully.

Link ⟩

Reduction of nitriles page 142

▲ Links to other sections of the course are highlighted in the margin, near the relevant text. They are accompanied by a reference to any areas where sections relate to one another. It may be useful for you to use these Links to recap a topic, before beginning to study the current topic.

Practical check ≫

The construction of electrochemical half-cells and full cells, and the measurement of cell EMF values is a specified practical task. These can include a range of different types of half-cell. Although you do not need to perform practical tasks involving the standard hydrogen electrode, you will need to be able to recall how this is constructed and used.

▲ Occasionally a topic covers an experiment or a practical that is a **specified practical task**. This feature appears alongside in the margin to highlight its importance and to give you some extra information and hints on understanding it fully.

Maths tip ≫

When calculating rate it is important that the units of rate match the units of the measurements. For example, time can be given in alternative units such as minutes, which may give a rate unit of $mol\ dm^{-3}\ min^{-1}$.

▲ Maths skills are highlighted at the beginning of each chapter and relevant examples are included throughout the book. The mathematical skills feature gives you specific help with relevant maths you need to ensure you understand.

Assessment

Assessment objectives

Examinations test your subject knowledge and the skills associated with how you apply that knowledge. These skills are described in Assessment Objectives. Examinations are written to reflect these objectives, with marks allocated in the proportions shown:

	AO1	AO2	AO3
A2 Level	27.3%	43.7%	29.0%

You must meet these Assessment Objectives in the context of the subject content, which is given in detail in the specification. Your ability to select and communicate information and ideas clearly, using appropriate scientific terminology, will be tested within each Assessment Objective. The Assessment Objectives are explained below, with suitable examples. Example answers to these questions are given on page 219.

Assessment Objective 1 (AO1)

Demonstrate knowledge and understanding of scientific ideas, processes, techniques and procedures.

This AO tests what you know, understand and remember. It is a test of how well you can recall and what is relevant. That is why it is essential that you know the content of the specification.

The questions that test this AO are often short-answer questions, using words such as 'state', 'give', 'explain' or 'describe'.

Here are two examples:

AO1 Demonstrate knowledge

Explain why amino acids are amphoteric compounds. [1]

This question tests AO1 because it asks you to recall factual information

AO1 Demonstrate understanding of scientific ideas

State how the method implies that glucose pentaethanoate is insoluble in water. [1]

This question tests AO1 because it asks for an explanation based on your factual knowledge.

Assessment Objective 2 (AO2)

Apply knowledge and understanding of scientific ideas, processes, techniques and procedures:

- **in a theoretical context**
- **in a practical context**
- **when handling qualitative data**
- **when handling quantitative data.**

AO2 tests how you use your knowledge and apply it to different situations, in the four ways outlined. A question may present you with a situation that you have not met before, but it will give you enough information so that you can use what you already know to provide an answer.

Make sure that you understand the methods of all the experiments that you have done; and that you understand how to do all the calculations needed to process the results.

In testing AO2, a question may use words such as 'use your knowledge of …', 'calculate' or 'explain …'

Here are four examples:

AO2 In a theoretical context

Draw a dot and cross diagram of the ammonium ion and use it to explain the difference between covalent and co-ordinate bonds. [2]

Here you are demonstrating your theoretical knowledge of the structure of the ammonium ion and then using this to explain the difference between the two types of bonding present.

AO2 In a practical context

When copper(II) compounds are dissolved in water they appear pale blue. Give the formula of the complex ion present in aqueous solutions of copper(II) compounds. [1]

This is AO2 because you are demonstrating your knowledge and understanding in the context of your practical work.

AO2 When handling qualitative data

A particular oxidation can be carried out using either benzene or water as the solvent. Give a reason why the use of water is seen as a 'greener' process. [1]

This question focuses on qualitative data that is used in a context where knowledge is being applied to a new situation.

AO2 When handling quantitative data

Explain why the reaction between chlorine and sodium hydroxide to give $NaCl$ and $NaClO_3$ is a disproportionation reaction. [2]

This requires you to deduce the oxidation numbers of the chlorine species present and then to apply the numbers obtained.

Assessment Objective 3 (AO3)

Analyse, interpret and evaluate scientific information, ideas and evidence, including in relation to issues, to:

- **make judgements and reach conclusions**
- **develop and refine practical design and procedures.**

The AO3 marks on a paper are awarded for developing and refining practical design and procedures, and for making judgments and drawing conclusions. You may be asked to criticise a method or analysis, or be asked how to improve aspects of it. You may be asked to design a method to test a particular hypothesis or why you might apply a particular test. You may be asked to interpret a test or draw a conclusion from evidence presented to you.

Some examination questions will give information about novel situations. As with AO2 questions, you may be presented with unfamiliar scenarios, but you will be tested on how well you use your knowledge to understand and interpret them.

For all the experiments that you have done, make sure that you know how to improve the accuracy of your method and the repeatability of readings.

AO3 questions tend to use words such as 'evaluate', 'suggest', 'justify', 'improve' or 'design' and often ask you to manipulate experimental data.

Here are two examples:

AO3 Make judgements and reach conclusions

Assuming that the amino acid with M_r 131 is a straight chain aliphatic α-amino acid, deduce its structure. [2]

You have been given the relative molecular mass of the amino acid and need to use your judgement to deduce a structure that satisfies the question.

AO3 Develop and refine practical design and procedures

A student plans a method to distinguish between solid samples of lead carbonate, magnesium carbonate, magnesium chloride and magnesium hydroxide. She plans to add hydrochloric acid to each sample to form a solution. To separate samples of each solution she then plans to add silver nitrate solution and excess sodium hydroxide solution.

Her teacher says that hydrochloric acid is not the correct reagent to use in this method. Give two reasons why hydrochloric acid is not appropriate and suggest an alternative reagent that could be used to avoid these problems. [3]

This question is AO3 because it asks you to modify the method to avoid conflicting results.

Examinations

Three units will be examined in the second year of your A-level. The assessment is summarised in the table below.

Theory units

Paper	Unit 3 Physical and Inorganic Chemistry	Unit 4 Organic Chemistry and Analysis
Topics covered	Topic 3.1 to 3.9	Topic 4.1 to 4.8
% of the overall A level	25%	25%
Length of exam	1 hour 45 minutes	1 hour 45 minutes
Marks available	80	80
Question types	Short answer Structured Calculation Extended prose	Short answer Structured Calculation Extended prose

The theory papers may include a synoptic element where questions, although focused on A2 material, might also contain topics contained in the AS material. Questions focusing solely on AS material will not be set. These theory papers will not include multiple choice style questions.

Practical Unit 5

Practical work is the foundation of scientific study – each idea or concept you have studied has been formed based on the evidence collected in practical work. Your practical skills and your understanding of practical techniques and their results are assessed throughout the course.

Your practical work is assessed in four ways:

- In every theory paper, a proportion of marks will be associated with practical skills, and one type of AO3 question may ask you to evaluate a practical method and make improvements.

- You must show that you have undertaken a set of practical tasks over the two years of the course and the lab book showing this can be requested by a visiting examiner.

- You will undertake a practical examination that contributes 5% of your final mark for the course.

- You will sit a written exam based on the practical skills you have developed, called the Practical Methods and Analysis Task and this also contributes 5% of your final mark.

- The final two assessments are classed as the two parts of Unit 5, and are usually examined in the same week.

Suggested practical exercises

There are many opportunities for practical work during the A2 year and the specification lists some that should be done. These are:

Unit	Topic	Practical
3	3.1	Construction of electrochemical cells and measurement of E_{cell}
	3.2	Simple redox titration Estimation of copper in copper(II) salts
	3.5	Determination of the order of a reaction; for example, the oxidation of iodide ions by hydrogen peroxide in acid solution
	3.8	Determination of an equilibrium constant; for example, for the equilibrium established when ethanol reacts with ethanoic acid
	3.9	Titration using a pH probe; for example, titration of a weak acid against a weak base
4	4.8	Synthesis of a liquid organic product, including separation using a separating funnel Synthesis of a solid organic product, including recrystallisation and determination of melting temperature Two-step organic synthesis, including purification and determination of melting temperature of product Planning a sequence of tests to identify organic compounds from a given list Paper chromatography separation, including two-way separation

Examination questions

As well as being able to recall facts, write equations, name structures and describe their functions, you need to appreciate the underlying principles of the subject and understand associated concepts and ideas. You will need to develop skills where you can apply your knowledge and understanding to new and novel situations.

You will be expected to answer different styles of question, for example:

- Short-answer questions – these require a brief answer, perhaps a formula, an equation, an observation or a definition.

- Structured questions – these are in several parts, usually based on a common theme. They generally become more difficult as you work your way through the question. These structured questions may ask for a brief response or require you to answer using extended writing. The number of line spaces and the mark allocation for each part of the question are there to help you gain an indication of the length of answer required. For example, if there are three marks for a particular part of the question, then you should be looking to identify at least three separate points.

- Extended prose question – each paper will contain one six-mark question marked 'QER' which requires extended prose for its answer. Some candidates rush into this type of question but you should take time to read the question carefully to understand what the examiner is requiring in your response. If the question includes bullet points then make sure you cover each one. It is helpful if you construct a plan for this type of question, you should then be able to organise your thoughts in a logical way and produce a response which clearly emphasises your ideas, without repetition. In this type of question examiners do not award marks for individual points of information but use a more holistic approach, using banded levels of assessment.

5–6 marks will be awarded if you include most of the relevant information with clear and logical reasoning in your answer. The examiners will be looking for a clear response to what is required, using well-constructed sentences and suitable chemical terminology.

3–4 marks will be awarded if the answer is clear but there are some factual omissions.

1–2 marks will be awarded if only a few valid points are made, and there is less use of scientific vocabulary and the response does not always follow a logical sequence.

0 marks will be gained if the question has not been attempted or no relevant points have been made.

Practical work is an essential part of any chemistry course and questions in theory papers may well assess your practical knowledge and understanding. Questions partly focused on practical procedures could be found in any of the three question types outlined above.

Command words

Command words generally start the question and give an indication of what is required in your response. The most common command words are summarised in the table below.

Command words	What to do
Give / name / state	Give a one-word answer or a short sentence with no explanation.
State what you see / Describe an observation	Write about what you expect to see in a reaction, e.g. effervescence occurs, an orange precipitate forms and the reaction gets warmer.
Describe or outline an experiment	Give a step-by-step account of what is taking place. You should include any equipment that you would use, the reagents required and any particular reaction conditions.
Explain	Give chemical reasons for something.
Suggest / predict	There may not be a definite answer to this type of question. You should use your chemical knowledge to put forward a sensible answer as to what the answer might be.
Compare	Give the similarities and/or differences between two things. Make an explicit comparison in each case.
Deduce	Use the information provided to answer the question.
Calculate	Use the information provided and your mathematical knowledge to work out the answer. It is always important to show your working in this type of question.
Justify	You will be given a statement for which you should use your chemical knowledge as evidence. You are advised to give a conclusion as to whether the initial statement can be accepted.
Write an equation	You will need to know the formula of the reactants and products. It is essential to ensure that the equation is balanced.

Sometimes questions do not use the command words outlined above. These may be questions that require you to use information that is included in the question. This type of question could include the drawing of a graph or constructing a table.

Exam data booklet

When you sit the examination, you will be given a data booklet. This contains useful information to help you in answering the questions. Copies of this are available to you throughout the course and it is important that you familiarise yourself with it. Many calculations will require you to use values listed on the data sheet, and you will be expected to refer to the sheet without being told to do so. This booklet includes:

- Avogadro's constant (N_A) and the molar gas constant (R)
- Molar gas volumes at 273 K / 298 K and 1 atm
- Planck's constant (h) and the speed of light (c)
- The specific heat capacity of water (c). Do not confuse the two constants using the symbol c – their values are very different.
- The relationship between cm^3 and m^3, K and °C, atm and Pa
- The fundamental electronic charge (e)
- Infrared absorption values of some functional groups
- ^{13}C NMR chemical shifts of some types of carbon
- ^1H NMR chemical shifts of some types of proton
- A copy of the Periodic table

Make sure that you use the information from the Periodic table in your answer, even if it is different from that seen elsewhere. The examiners use the information in the data booklet to mark the examination papers.

Exam preparation and technique

We all vary in speed and natural ability but by attacking the challenge of A2 Chemistry in the right way, the best possible outcome can be achieved. The authors have had many years experience in teaching and preparing candidates for exams. Every year examiners publish reports that outline the areas that impact the performance of candidates and based on these and our experience here are some of the best tips:

1. Give yourself time. Take each topic slowly, clear up any uncertainties then try the exam practice questions. If you need to, then make sure you return to each topic after an interval of time to ensure that you still have it mastered. This may take more than one return trip. It is known that the unconscious mind continues to work and sort learned material so you must give it time. Last-minute cramming is of little use.

2. Be careful to understand what the question is really asking for. Candidates sometimes rush ahead down the wrong track and lose both time and marks. Questions on AO1 will ask you to show that you know and/or understand something; those on AO2 will ask you to apply these and AO3 to analyse, interpret and evaluate something. The actual lead words in the questions may include, 'state', 'describe', 'draw', 'name' and 'explain' for AO1, 'calculate', for AO2 and 'suggest' and 'analyse data' for AO3.

3. There is no substitute for work and concentration. Chris Froome did many very long training runs on his bicycle. Practice trains the mind and produces understanding and enjoyment in mastery of the subject.

3.1

Redox and standard electrode potential

Chemists organise the properties and reactions of materials to see patterns and make predictions. One way to classify reactions is as oxidation, with the name linked to the original definitions as processes that gained oxygen. The opposite reaction, reduction, referred to the loss of mass when oxygen was removed from a compound.

Over time, chemists have adjusted their ideas of oxidation and reduction as their understanding of the processes have developed, so the focus is now on electrons in the definitions of these terms. The harnessing of flows of electrons between compounds is key to the function of the batteries used to power all the mobile devices used today.

Topic contents

You should be able to demonstrate and apply knowledge and understanding of:

- The definitions of redox reactions in terms of electron transfer.

- Using ion/electron half-equations to represent redox systems.

- Building half-cells into cells and showing these as cell diagrams.

- The concept of standard electrode potentials and the role of the standard hydrogen electrode in finding these.

- Half-cells based on metal/metal ion electrodes and electrodes based on different oxidation states of the same element.

- How simple electrochemical cells are formed by combining electrodes.

- The concept of cell EMF and how it can be used to deduce the feasibility of reactions.

- The principles of the hydrogen fuel cell and its benefits and drawbacks.

Maths skills ≫

- Carry out simple EMF calculations.

Redox reactions

Redox reactions are reactions where electrons are transferred from one chemical species to another. They are combinations of **oxidation** and **reduction**:

Oxidation is a process where *electrons are lost.*   **OIL** (**O**xidation **i**s **L**oss of electrons)

Reduction is a process where *electrons are gained.*   **RIG** (**R**eduction **i**s **G**ain of electrons)

These two processes always occur together, giving a redox reaction (**Red**uction-**Ox**idation). They must occur at the same time, as the electrons gained during reduction must have come from a different material. We call the material that donates the electrons the reducing agent, and as it loses electrons in the process, the reducing agent is oxidised.

- An **oxidising agent** is a species that oxidises another species, and gains electrons in the process, so an oxidising agent is being reduced.

- A **reducing agent** is a species that reduces another species, and loses electrons in the process, so a reducing agent is being oxidised.

Oxidation and reduction half-equations

A full equation describes what happens to all the substances in a chemical reaction; however, it can be useful to focus on what happens to each substance separately. To do this we can split a full ionic equation into two ion-electron half-equations which separately describe what happens to each of the substances present. One half-equation represents electrons being lost (oxidation) and one half-equation represents electrons being gained (reduction).

Worked example

The ionic equation for the reduction of Cu^{2+} by metallic zinc is:

$$Zn(s) + Cu^{2+}(aq) \rightarrow Cu(s) + Zn^{2+}(aq)$$

In this reaction we can split the equation into what happens to the zinc and what happens to the copper:

The zinc changes from $Zn(s)$ to $Zn^{2+}(aq)$. To do this it must lose two electrons.

$$Zn(s) \rightarrow Zn^{2+}(aq) + 2e^- \qquad \text{OXIDATION}$$

The copper changes from $Cu^{2+}(aq)$ to $Cu(s)$. To do this it must gain two electrons.

$$Cu^{2+}(aq) + 2e^- \rightarrow Cu(s) \qquad \text{REDUCTION}$$

The half-equations must be balanced in terms of atoms and charge, so it is important to remember that each electron has a negative charge even if this is not shown.

◄ The zinc placed into copper sulfate solution rapidly changes from a silvery surface to brown as copper metal covers the zinc.

Cells and half-cells

Writing redox reactions as two half-equations is not a theoretical process – it is possible to physically separate the two processes so that oxidation happens in one place and reduction happens somewhere else. To do this we need to set up two half-cells, one for the oxidation and one for the reduction. We join them together to complete the circuit:

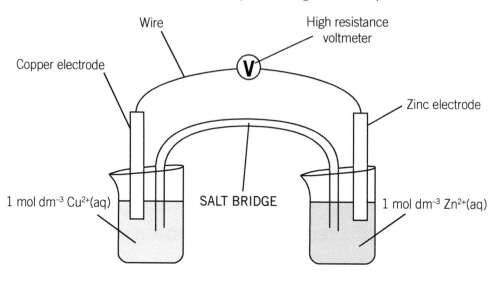

 Key term

A **salt bridge** is a piece of apparatus that connects the solutions in two half-cells so that the circuit is complete and the current can flow without the solutions mixing.

- The wire allows electrons to flow from the half-cell where oxidation occurs to the half-cell where reduction occurs. We often include a high resistance voltmeter if we are measuring the potential difference in the cell.

- The **salt bridge** completes the circuit and allows ions to flow without the solutions mixing. Current is a flow of charge and in the salt bridge the current is a flow of ions. A simple salt bridge that you may use in your experimental work is made of filter paper soaked in potassium nitrate solution.

- The entire apparatus is called a cell, with the two parts called half-cells.

Half-cells

Each cell is made up of two half-cells, each of which must contain both the reactants and products of the half-equation. This means the half-cell for the reduction of Cu^{2+} ions to copper atoms must contain both Cu^{2+} ions and copper atoms. It must also contain a metal to allow electrons to flow into or out of the half-cell. There are various types of half-cell depending on the physical states of each substance in the reaction.

There are three key types of electrochemical half-cell that you need to be aware of:

1. Metal / metal ions

The half-cells described so far have consisted of metals in contact with solutions containing metal ions, with two key examples being $Zn(s)$ with $Zn^{2+}(aq)$ and $Cu(s)$ with $Cu^{2+}(aq)$. In both cases we have a piece of metal as the electrode, with a solution containing a 1 mol dm^{-3} solution of the metal ions. In the case of zinc, there is no apparent colour change, but for copper the blue solution may lose colour as the copper ions are reduced.

2. A gas in contact with a solution of non-metal ions, with an inert metal electrode

Since non-metals are not conductors, we must use an <u>inert</u> platinum electrode to allow electrons to flow in or out of the half-cell. This is typically used for hydrogen ($H_2(g)$ I $H^+(aq)$) or oxygen ($O_2(g)$ I $OH^-(aq)$) half-cells. The gas is bubbled over the inert electrode which is dipping into a solution of the ions. These changes do not cause any apparent colour change.

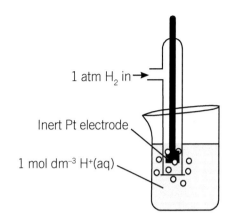

1 atm H_2 in →

Inert Pt electrode

1 mol dm^{-3} $H^+(aq)$

Top tip

When drawing labelled diagrams of standard half-cells, the conditions must be stated clearly in the labels, with all solutions having a concentration of 1.00 mol dm^{-3} and all gases being at a pressure of 1 atm.

3. A solution containing ions of a metal in two different oxidation states, again using an inert metal electrode

Again there is no conductor in the system so we must use an inert platinum electrode to allow electrons to flow in or out of the half-cell. This is typically used for the transition metals, where the metal may have several oxidation states. Good examples are Fe^{2+}/Fe^{3+} and Mn^{2+}/MnO_4^-. Both these half-cells cause colour changes when oxidation or reduction occurs. Fe^{2+} is pale green and Fe^{3+} is yellow/orange; Mn^{2+} is colourless, MnO_4^- is purple.

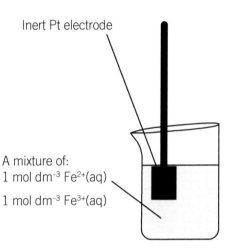

Inert Pt electrode

A mixture of:
1 mol dm^{-3} $Fe^{2+}(aq)$
1 mol dm^{-3} $Fe^{3+}(aq)$

Representing half-cells

We can show half-cells using cell diagrams. These list all the essential substances present in a half-cell, starting with the metal that is used to conduct electrons into or out of the half-cell. Each substance is then listed, with vertical lines separating substances in different physical states (solid, liquid, gas or solution) and commas separating substances in the same physical state.

$Mg(s)$ I $Mg^{2+}(aq)$ This is the metal/metal ion half-cell for magnesium.

$Pt(s)$ I $H_2(g)$ I $H^+(aq)$ This is a gas/solution half-cell containing an inert platinum electrode. As this is for hydrogen, it is the standard hydrogen electrode if it is under standard conditions.

$Pt(s)$ I $Mn^{2+}(aq)$, $MnO_4^-(aq)$ This is a mixed solution half-cell containing an inert platinum electrode. In this case a comma separates the two manganese-containing ions in aqueous solution as they are in the same physical state.

Top tip

A cell diagram is a list of substances written on one line, whilst a diagram of a cell is a labelled picture showing all the parts of a cell.

Knowledge check 3

Draw a labelled diagram that shows the cell represented by:
$Pt(s)$ I $Fe^{2+}(aq)$, $Fe^{3+}(aq)$ II $Mg^{2+}(aq)$ I $Mg(s)$.

Top tip

Don't draw a vertical line between two different ions that are present in the same solution, use a comma to separate them instead.

◀ The cell involving copper and zinc electrodes is called the Daniell cell after its inventor, John Daniell.

Standard electrode potentials

Although the chemical reactions that occur in electrochemical cells can be performed directly, using these cells can allow us to obtain detailed information about the reaction. We can measure the EMF of the reaction, and this tells us a lot about how easy it is to oxidise or reduce the substances in each half-cell. The EMF measured is the difference in the redox power of the two half-cells, with the largest EMF values occurring when a half-cell containing a species that is easy to oxidise, such as magnesium, is connected to a half-cell with a species that is easy to reduce, such as manganate(VII).

The ability of a half-cell to gain or lose electrons is measured using the **standard electrode potential, $E^\ominus$**. The scale of the standard electrode potential uses hydrogen as zero, and any species that is easier to reduce has a negative $E^\ominus$ value, with ones that are easier to oxidise having a positive $E^\ominus$ value. The hydrogen half-cell is called the **standard hydrogen electrode**.

Key term

The **standard electrode potential** is the potential difference between any half-cell under standard conditions and the standard hydrogen electrode. Standard conditions are a temperature of 298K, a pressure of 1 atm and concentrations of 1 mol dm^{-3}.

The standard hydrogen electrode

The standard hydrogen electrode consists of a platinum electrode coated with fine platinum grains, called platinum black. This is dipped into a 1.0 mol dm^{-3} solution of H$^+$(aq), usually as hydrochloric acid, and hydrogen gas is slowly passed over the electrode at a pressure of 1 atmosphere and a temperature of 298K .

Top tip

When describing any standard cell the standard conditions must be stated clearly: 1 atm pressure for any gas; 1 mol dm^{-3} for any solution; 298K for temperature.

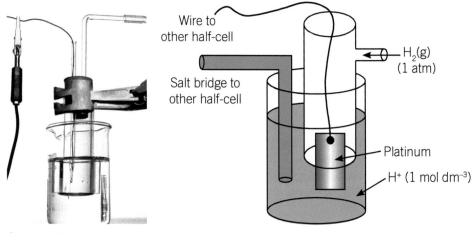

▲ Standard hydrogen electrode

Standard electrode potential

The standard electrode potential is the potential difference between the standard hydrogen electrode and any other system in which the concentrations of all the active ions in solution are 1.0 mol dm^{-3} and all gases are at 1 atm pressure at 298K (25°C). Its symbol is $E^\ominus$.

Knowledge check 4

(a) Draw a cell diagram for the cell that measures the standard electrode potential of the following process:

Fe^{3+}(aq) + e$^-$ ⟶ Fe^{2+}(aq)

(b) Draw a labelled diagram of the cell that measures the standard electrode potential of the following process:

Fe^{3+}(aq) + e$^-$ ⟶ Fe^{2+}(aq)

To measure a standard electrode potential, we must set up a half-cell under standard conditions, and connect it to the standard hydrogen electrode, e.g. Zn^{2+}(aq) I Zn(s) connected to H$^+$ I H$_2$ I Pt. The lines in each of these systems represent a change of physical state. If combining two half-cells together we use a double line between them to represent a salt bridge and place the metals at both ends, to produce a cell diagram such as:

Pt(s) I H$_2$(g) I H$^+$(aq) II Zn^{2+}(aq) I Zn(s)

The standard electrode potential for a half-cell is measured by joining it to the standard hydrogen electrode. This creates a system like the one shown opposite:

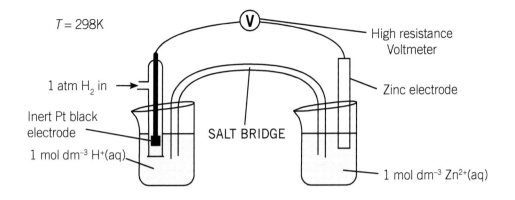

Using standard electrode potentials

The $E^\ominus$ for the standard hydrogen electrode is taken as 0 V and all other values are stated relative to this. The half-cell with the most positive (or least negative) electrode potential will act as the positive electrode. The half-cell with the most negative electrode potential will act as the negative electrode. Electrons will flow from the negative to the positive electrode. The $E^\ominus$ value for the zinc half-cell is -0.76 V, which is more negative than the potential of the standard hydrogen electrode. This means that the negative electrode will be the zinc half-cell, and the standard hydrogen electrode will be the positive electrode. Electrons flow from the negative to the positive electrode (i.e. Zn loses electrons more readily than H_2).

- The positive electrode has the more positive $E^\ominus$.
- The electrons flow to the half-cell with the more positive $E^\ominus$.

The $E^\ominus$ value for the copper half-cell (Cu^{2+} I Cu) is $+0.34$ V. The plus sign signifies that the standard hydrogen electrode has the more negative potential. Electrons therefore flow along the wire from the standard hydrogen electrode to the copper and the hydrogen electrode becomes the negative electrode. (i.e. H_2 loses electrons more readily than Cu).

We can use this method to put the reducing power of any set of half-cells in order. These are always quoted as reduction potentials, and it is noticeable that the order of these potentials for metals reflects the reactivity series, so this is sometimes called the electrochemical series. The most reactive metals are the strongest reducing agents and have the most negative $E^\ominus$ so Na(s) is the strongest reducing agent in the list. The most reactive non-metals are the strongest oxidising agents and have the most positive $E^\ominus$, so MnO_4^- is the strongest oxidising agent. The electrochemical series has the same order as the reactivity series.

Reaction			$E^\ominus$ / Volt
$Na^+(aq) + e^-$	$\rightleftharpoons$	Na(s)	-2.71
$Mg^{2+}(aq) + 2e^-$	$\rightleftharpoons$	Mg(s)	-2.36
$Zn^{2+}(aq) + 2e^-$	$\rightleftharpoons$	Zn(s)	-0.76
$2H^+(aq) + 2e^-$	$\rightleftharpoons$	$H_2(g)$	0.00
$Cu^{2+}(aq) + 2e^-$	$\rightleftharpoons$	Cu(s)	$+0.34$
$I_2(s) + 2e^-$	$\rightleftharpoons$	$2I^-(aq)$	$+0.54$
$Fe^{3+}(aq) + e^-$	$\rightleftharpoons$	$Fe^{2+}(aq)$	$+0.77$
$Br_2(l) + 2e^-$	$\rightleftharpoons$	$2Br^-(aq)$	$+1.09$
$Cr_2O_7^{2-}(aq) + 14H^+(aq) + 6e^-$	$\rightleftharpoons$	$2Cr^{3+}(aq) + 7H_2O(l)$	$+1.33$
$Cl_2(g) + 2e^-$	$\rightleftharpoons$	$2Cl^-(aq)$	$+1.36$
$MnO_4^-(aq) + 8H^+(aq) + 5e^-$	$\rightleftharpoons$	$Mn^{2+}(aq) + 4H_2O(l)$	$+1.51$

In these equations, the oxidising agents are the substances that are being reduced such as Cl_2, I_2 or Zn^{2+}. The strongest oxidising agents are those that have the most positive standard electrode potentials. The strongest reducing agents are present in the half-equations with the most negative standard electrode potentials, such as Mg, Zn or H_2.

◀Stretch & challenge

If electrons are allowed to flow, this will affect the concentrations of each species. It is possible to use Le Chatelier's principle to predict the effect of the changing concentration on the standard electrode potential. If we consider the reaction $Fe^{3+}(aq) + e^- \rightleftharpoons Fe^{2+}(aq)$, this has a standard electrode potential of $+0.77$ V, which means it will have a greater tendency to gain electrons than the standard hydrogen electrode. If electrons flow then reduction will occur in this half-cell, increasing the concentration of Fe^{2+} and decreasing the concentration of Fe^{3+}. Le Chatelier's principle would suggest that this would shift the equilibrium to the left, decreasing the tendency of the half-cell to gain electrons and hence making the electrode potential less positive. Note that this is not the standard electrode potential any more as when the concentration changes it is not under standard conditions.

Knowledge check 5◀

(a) Identify the positive electrode in the following cells:

   (i) $Pt(s)|H_2(g)|H^+(aq)\|Cu^{2+}(aq)|Cu(s)$

   (ii) $Pt(s)|Fe^{2+}(aq), Fe^{3+}(aq)\|Mg^{2+}(aq)|Mg(s)$

(b) Identify the direction of electron flow in the wire in the cells in (i) and (ii).

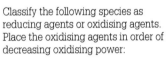

Knowledge check 6◀

Classify the following species as reducing agents or oxidising agents. Place the oxidising agents in order of decreasing oxidising power:

Na^+, Cu, I_2, Cl^-, H_2, MnO_4^-, Mg.

Calculating the EMF of an electrochemical cell

7 Knowledge check

Draw a labelled diagram that shows the cell represented by: $Mg(s) \mid Mg^{2+}(aq) \parallel Zn^{2+}(aq) \mid Zn(s)$ labelling the positive and negative electrodes and the direction of flow of electrons in the wire. Calculate the EMF of this cell.

8 Knowledge check

Calculate the EMF values of the following cells:

(a) $Pt(s)\mid H_2(g)\mid H^+(aq) \parallel Cu^{2+}(aq) \mid Cu(s)$

(b) $Pt (s)\mid Fe^{3+}(aq), Fe^{2+}(aq) \parallel Mg^{2+}(aq) \mid Mg(s)$

(c) $Pt(s) \mid MnO_4^-(aq), Mn^{2+} (aq) \parallel Zn^{2+}(aq) \mid Zn(s)$

When two half-cells are connected, the high resistance voltmeter will give a reading showing the EMF of the cell. The value of the EMF is given by the difference between the standard electrode potentials of the two half-cells. The EMF for an electrochemical cell is calculated as a positive value.

Worked example

Calculate the EMF for a cell represented by the cell diagram below:

$$Zn(s) \mid Zn^{2+}(aq) \parallel Cu^{2+}(aq) \mid Cu(s)$$

The two standard electrode potentials are −0.76 V and +0.34 V. Subtract the most negative value from the most positive to find the difference.

$$EMF = +0.34 - (-0.76) = +1.10V$$

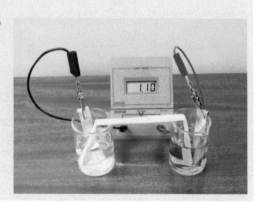

Feasibility of reactions

9 Knowledge check

Write the equation for the reaction that occurs when the following pairs of half-cells are connected together:

(a) $Zn \mid Zn^{2+}$ with $Fe^{2+}, Fe^{3+} \mid Pt$

(b) $Cl_2 \mid Cl^- \mid Pt$ and $Cu^{2+} \mid Cu$

You will need to refer to the table of standard electrode potentials.

When an oxidising agent and reducing agent are mixed, a redox reaction may occur. For a successful reaction, the oxidising agent needs to be strong enough to oxidise the reducing agent. The standard electrode potentials give information about the strength of oxidising agents and reducing agents and allow you to work out if a reaction is feasible. A feasible reaction is one that can occur spontaneously. There are two ways to show this – by calculating the EMF or by comparing standard electrode potentials. Most of these reactions will happen easily, but the standard electrode potentials do not give any information about the rate of reaction so for a few reactions the rate may be too slow to be effective.

Calculating the EMF

A chemical reaction is feasible if the EMF for the redox process is positive. The EMF can be calculated from the standard electrode potentials for two half-cells. To assess the feasibility of the reaction between $Zn^{2+}(aq)$ and $Cu(s)$ we need the two standard electrode potentials below. This situation is different from an electrochemical cell where all reactants and products are present so reactions can go either way. In this case we have only substances from the start of one reaction and the end of the other.

Study point

When combining two half-equations to get a full equation, you need to ensure that there are the same numbers of electrons in both halves before they are combined. If they don't have the same electrons already, each equation needs to be multiplied by an appropriate number.

Example: If we have two electrons in one half-equation and one in the other then we double everything in the equation with one electron.

			$E^\ominus$ / Volt
$Zn^{2+}(aq) + 2e^-$	$\rightleftharpoons$	$Zn(s)$	−0.76
$Cu^{2+}(aq) + 2e^-$	$\rightleftharpoons$	$Cu(s)$	+0.34

The reaction between $Zn^{2+}(aq)$ and $Cu(s)$ would need the first reaction to occur as written and the second to occur in reverse. This makes the first equation a reduction and the second an oxidation. The EMF for the redox process will be $E^\ominus$ for the reduction half-equation minus $E^\ominus$ for the oxidation half-equation.

Zn²⁺(aq) + 2e⁻	$\rightleftharpoons$	Zn(s)	REDUCTION
Cu(s)	$\rightleftharpoons$	Cu²⁺(aq) + 2e⁻	OXIDATION

EMF for the reaction $= E^{\ominus}_{\text{REDUCTION}} - E^{\ominus}_{\text{OXIDATION}} = -0.76 - (+0.34) = -1.10\text{V}$

As this value is negative the reaction is not feasible.

Comparing standard electrode potentials

If a piece of magnesium metal is placed in a solution containing Zn^{2+} ions, a redox reaction occurs with the magnesium forming Mg^{2+} ions and the Zn^{2+} being reduced to zinc metal. This can be proved using standard electrode potential values.

			$E^{\ominus}$ / V
Mg²⁺(aq) + 2e⁻	$\rightleftharpoons$	Mg(s)	−2.36
Zn²⁺(aq) + 2e⁻	$\rightleftharpoons$	Zn(s)	−0.76

In this case, we have one reactant from the left-hand side of a half-equation (the Zn^{2+} ions) and one from the right-hand side of a half-equation (the magnesium metal). For the overall reaction to occur, the Zn^{2+} must be a stronger oxidising agent than Mg^{2+} so it can oxidise magnesium metal to form Mg^{2+} ions. In this case the Zn^{2+} is a stronger oxidising agent as it has a more positive (less negative) standard electrode potential than the Mg^{2+} half-cell.

It can be easier to remember this as the 'anticlockwise rule':

1. Write the two half-equations in order of increasing standard electrode potentials, with the most negative value first.

2. Start in the top right-hand corner, and move anticlockwise around the equations. This means that in the feasible reaction the top reaction goes in reverse and the bottom goes in the direction is written.

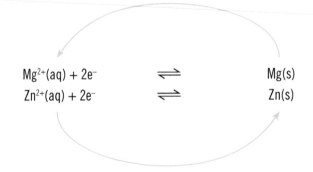

The feasible reaction is therefore $Mg(s) + Zn^{2+}(aq) + 2e^- \rightarrow Mg^{2+}(aq) + 2e^- + Zn(s)$

Cancelling out the electrons that are present on both sides of the equation gives the following feasible reaction.

$$Mg(s) + Zn^{2+}(aq) \rightarrow Mg^{2+}(aq) + Zn(s)$$

In terms of standard potentials, the calculation here would be

$\text{EMF} = E^{\ominus}_{\text{REDUCTION}} - E^{\ominus}_{\text{OXIDATION}} = -0.76 - (-2.36) = 1.60 \text{ V}.$

As the value is positive, the reaction is feasible.

<< **Top tip**

When discussing and comparing standard electrode potentials many values are negative. To make any comparison clear use 'more positive' or 'more negative' rather than 'larger' or 'smaller'.

<< **Top tip**

When identifying or discussing oxidising or reducing agents you need to be clear which substances are being discussed. It is not good enough to write 'zinc' when you mean 'zinc ions', 'zinc metal' or 'zinc atoms'. At times it is easier to write a formula such as Zn^{2+} rather than words to ensure your answer is clear.

Knowledge check 10

Are the following reactions feasible?

(a) $2\,H^+(aq) + Zn(s) \rightarrow Zn^{2+}(aq) + H_2(g)$

(b) $Cu(s) + Mg^{2+}(aq) \rightarrow Cu^{2+}(aq) + Mg(s)$

You will need to refer to the table of standard electrode potentials.

Fuel cells

Fuel cells have been developed as a method of releasing energy very efficiently from fuels such as hydrogen, methane or methanol. Traditional engines burn the fuel, followed by using this heat to cause gas expansion which is used to move a motor. Energy is lost at each stage of the this process, with much of the energy released from the fuel being lost as heat through the exhaust or radiator. Fuel cells used in some car engines are an electrochemical method of releasing energy and using this directly.

▲ Hydrogen fuel cell in a car

The fuel cell system passes the fuel over platinum metal which acts as a catalyst, but also as an electrode for the electrochemical system. Electrons are removed from hydrogen atoms at one electrode:

$$H_2 \rightarrow 2H^+ + 2e^-$$

The protons (H^+) diffuse through a semi-permeable membrane to the other electrode where they receive electrons and oxygen molecules to form water molecules:

$$O_2 + 4H^+ + 4e^- \rightarrow 2H_2O$$

The overall reaction is $2H_2 + O_2 \rightarrow 2H_2O$ and the maximum theoretical voltage that can be produced is 1.23 V; however, in most cases the voltage is between 0.6 and 0.7 V.

 **Knowledge check**

Most fuel cells produce a voltage of 0.6–0.7 V rather than the theoretical value of 1.23 V. Suggest reason(s) for this.

Top tip 〉〉

When you are discussing the advantages and drawbacks you should link these to chemical concepts and specific reasons. Avoid overgeneralisations such as 'cheaper' or 'better for the environment'.

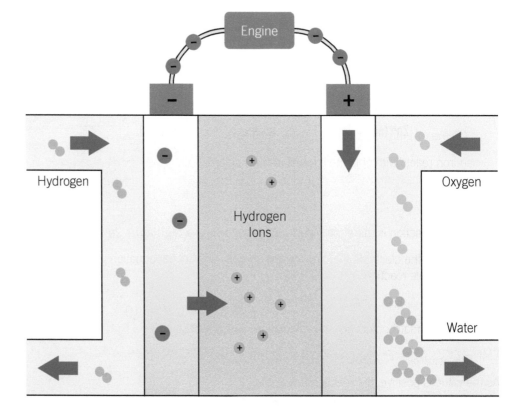

These cells have been particularly studied as ways of storing energy to be released as electricity or heat, for instance in the development of 'zero emissions' cars which use hydrogen fuel and release no carbon dioxide.

Benefits	Drawbacks
• They are a convenient way of storing and releasing energy.	• The hydrogen fuel must be generated elsewhere and this is likely to use fossil fuel energy sources which will cause their own carbon dioxide emissions. There is also an energy loss here as the conversion process is not 100% efficient.
• The energy efficiency is much higher than standard fuel systems (e.g. 36–45% for fuel cell compared to 22% for diesel).	• The gases needed are difficult to store compared to liquid fuels.
• Emissions from fuel cells are less damaging than the carbon dioxide from traditional engines.	• The fuel cells operate at lower temperatures (about 80°C) so need very efficient catalysts which use expensive metals.

Test yourself

1. State what is meant by an oxidising agent. [1]

2. Show that the following reaction is a redox reaction: [2]
$$NbCl_5 + H_2 \longrightarrow 2HCl + NbCl_3$$

3. Write the half-equations for oxidation and reduction processes occurring in the following reaction: [2]
$$UO_2{}^{2+}(aq) + 4H^+(aq) + Zn(s) \longrightarrow U^{4+}(aq) + 2H_2O(l) + Zn^{2+}(aq)$$

4. Explain what is meant by the standard hydrogen electrode. [2]

5. (a) Draw the apparatus used to measure the standard electrode potential for the following process: [3]
$$Cu^{2+}(aq) + 2e^- \rightleftharpoons Cu(s)$$

   (b) Identify the positive electrode in the reaction. [1]

   (c) On the diagram from part (a), draw the direction of electron flow in the wire. [1]

6. When iron reacts with chlorine it forms iron(III) chloride, $FeCl_3$. When iron reacts with iodine it forms iron(II) iodide, FeI_2.

   Use the standard electrode potential values given in the table to explain this difference. [2]

	Standard electrode potential, $E^\ominus$ /V
$I_2 + 2e^- \rightleftharpoons 2I^-$	+0.54
$Fe^{3+} + e^- \rightleftharpoons Fe^{2+}$	+0.77
$Cl_2 + 2e^- \rightleftharpoons 2Cl^-$	+1.36

7. Chromium forms a range of oxidation states and the standard electrode potentials involving several of these are shown in the table.

	Standard electrode potential, $E^\ominus$ /V
$Cr^{2+} + 2e^- \rightleftharpoons Cr$	−0.91
$Cr^{3+} + e^- \rightleftharpoons Cr^{2+}$	−0.42
$Cr_2O_7^{2-} + 14H^+ + 6e^- \rightleftharpoons 2Cr^{3+} + 7H_2O$	+1.33
$H_3PO_3 + 2H^+ + 2e^- \rightleftharpoons H_3PO_2 + H_2O$	−0.50

Phosphinic acid, H_3PO_2, can be used as a reducing agent. Identify the main chromium-containing product when a solution containing dichromate ions, $Cr_2O_7^{2-}$, is treated with phosphinic acid. [2]

8. The sale of new cars powered by petrol or diesel engines is to be banned in the UK from 2030. One possible alternative power source is fuel cells. Give two advantages of fuel cells over petrol engines. [2]

9. Use the table of standard electrode potentials below to answer the questions that follow.

	Standard electrode potential, $E^\ominus$ /V
$K^+ + e^- \rightleftharpoons K$	-2.92
$Na^+ + e^- \rightleftharpoons Na$	-2.71
$Mg^{2+} + 2e^- \rightleftharpoons Mg$	-2.38
$Mn^{2+} + 2e^- \rightleftharpoons Mn$	-1.18
$Cr^{2+} + 2e^- \rightleftharpoons Cr$	-0.91
$Zn^{2+} + 2e^- \rightleftharpoons Zn$	-0.76
$2H^+ + 2e^- \rightleftharpoons H_2$	
$Cu^{2+} + 2e^- \rightleftharpoons Cu$	+0.34
$Ag^+ + e^- \rightleftharpoons Ag$	+0.80

(a) Complete the table by inserting the missing value. [1]

(b) Give the formula of the strongest reducing agent. [1]

(c) Find the EMF of the cell $Mg | Mg^{2+} || Mn^{2+} | Mn$. [1]

(d) A piece of chromium metal is placed in a solution containing $Mn^{2+}(aq)$ ions. State, giving your reasons, whether a reaction will occur. [1]

Redox reactions

Redox reactions can be used in the synthesis of a range of organic and inorganic substances. Once any substance is produced it is important to ensure that it is pure, and that the levels of any impurity can be measured to make sure they do not affect the use of the material. The labels on many chemical substances that are sold will show the amounts of impurities – you will see similar labels on bottles of mineral water.

The measurement of purity for a range of substances was an early use of chemical reactions, with methods of qualitatively ensuring the purity of precious metals being particularly important. We can now use redox reactions to analyse metals and their compounds quantitatively using redox titrations. In this chapter you will see how the copper content of compounds can be derived, but similar approaches can be used for a range of metals due to the versatility of redox reactions. These methods allow very accurate values to be obtained, which is very important when the value of the metal involved is high.

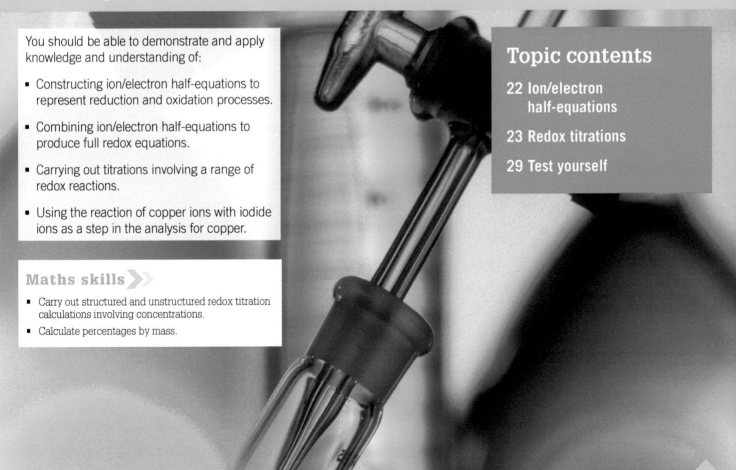

You should be able to demonstrate and apply knowledge and understanding of:

- Constructing ion/electron half-equations to represent reduction and oxidation processes.

- Combining ion/electron half-equations to produce full redox equations.

- Carrying out titrations involving a range of redox reactions.

- Using the reaction of copper ions with iodide ions as a step in the analysis for copper.

Maths skills ❯❯

- Carry out structured and unstructured redox titration calculations involving concentrations.
- Calculate percentages by mass.

Topic contents

Ion/electron half-equations

The use of simple ion/electron half-equations has been discussed in the last topic; however, it is important to be able to construct these for a range of simple and compound ions. In the case of elements where the oxidation state of a single atom is changing, such as a magnesium atom becoming an Mg^{2+} ion or an Fe^{3+} ion becoming an Fe^{2+} ion, then a half-equation will only need electrons to be included to balance. The charge in each equation must balance so the number of electrons must provide the negative charges needed for this balancing.

$$Mg(s) \longrightarrow Mg^{2+}(aq) + 2e^-$$

$$Fe^{3+}(aq) + e^- \longrightarrow Fe^{2+}(aq)$$

Link

Half-equations and half-cells pages 12-13.

When compound ions are involved in an ion/electron half-equation it can be more complex. Ions such as dichromate(VI) and manganate(VII) include oxygen atoms, and each oxygen requires two hydrogen ions to contribute to the formation of water in the products:

$$MnO_4^- + 8H^+ + 5e^- \longrightarrow Mn^{2+} + 4H_2O$$

$$Cr_2O_7^{2-} + 14H^+ + 6e^- \longrightarrow 2Cr^{3+} + 7H_2O$$

The electrons are needed to ensure each half-equation is balanced in terms of charge, and so the numbers of electrons can be calculated by counting the charges on each side of the half-equation.

Worked example

The bromate(V) ion, BrO_3^-, can be reduced to bromine, Br_2, in acid solution. Write the ion/electron half-equation for this process.

STEP 1: Write the reagents and products:

$$BrO_3^- \longrightarrow Br_2$$

STEP 2: Balance the atoms present:

$$2BrO_3^- \longrightarrow Br_2$$

STEP 3: Add two hydrogen ions to combine with each oxygen to form water.

$$2BrO_3^- + 12H^+ \longrightarrow Br_2 + 6H_2O$$

STEP 4: Find the total charges on both sides to find the number of electrons needed:

Total charge at the start $= 2- + 12+ = 10+$

Total charge at the end $= 0 + 0 = 0$

Change in charge = 10, so there need to be 10 electrons in the ion/electron half-equation.

$$\textbf{2BrO}_3^- + \textbf{12H}^+ + \textbf{10e}^- \longrightarrow \textbf{Br}_2 + \textbf{6H}_2\textbf{O}$$

1 Knowledge check

Write ion/electron half-equations for the processes below:

(a) The reduction of an acidified solution of perchlorate(VII), ClO_4^-, to chlorine gas.

(b) The reduction of acidified manganate(VII) to manganese dioxide, MnO_2.

(c) The oxidation of ethandioate ions ($C_2O_4^{2-}$) to form carbon dioxide gas.

(d) The conversion of tantalum(V) oxide (Ta_2O_5) to metallic tantalum.

Combining half-equations

When undertaking a redox reaction, the overall reaction combines the ion/electron half-equation for the oxidation process and the one for the reduction process. These must be combined in the correct ratio to have the same numbers of electrons in each half-equation.

Worked example

Bromate(V) ions can act as an oxidising agent, with the half-equation for this process given below. This can be used to oxidise bromide ions to form bromine molecules, Br_2.

$$2BrO_3^- + 12H^+ + 10e^- \rightarrow Br_2 + 6H_2O$$
$$2Br^- \rightarrow Br_2 + 2e^-$$

To combine the two half-equations above, we need to multiply the second by five to give ten electrons.

$$2BrO_3^- + 12H^+ + 10e^- \rightarrow Br_2 + 6H_2O$$
$$10Br^- \rightarrow 5Br_2 + 10e^-$$

These two half-equations are now added together:

$$2BrO_3^- + 12H^+ + 10e^- + 10Br^- \rightarrow Br_2 + 6H_2O + 5Br_2 + 10e^-$$

We now cancel out the electrons on both sides of the equation, and add the two sets of bromine molecules in the product together:

$$2BrO_3^- + 12H^+ + 10Br^- \rightarrow 6Br_2 + 6H_2O$$

As there are even numbers of each species we can halve the equation to give the final answer below.

$$BrO_3^- + 6H^+ + 5Br^- \rightarrow 3Br_2 + 3H_2O$$

Knowledge check 2

Combine the following pairs of half-equations to form an overall equation:

(a) $2BrO_3^- + 12H^+ + 10e^- \rightarrow Br_2 + 6H_2O$ and $Fe^{2+} \rightarrow Fe^{3+} + e^-$

(b) $Cr_2O_7^{2-} + 14H^+ + 6e^- \rightarrow 2Cr^{3+} + 7H_2O$ and $2HI \rightarrow I_2 + 2e^- + 2H^+$

(c) $MnO_4^- + 8H^+ + 5e^- \rightarrow Mn^{2+} + 4H_2O$ and $H_2O_2 \rightarrow O_2 + 2H^+ + 2e^-$

Study point

When combining half-equations then any species appearing on both sides of the equation should be cancelled out, not just electrons. Other species that may be affected are H^+ ions and H_2O molecules.

Redox titrations

Redox titrations are carried out in the same way as acid–base titrations.

A titration measures the exact volumes of two solutions that react together. One solution is usually a standard solution, which is a solution where we know the exact concentration. These are made in a volumetric flask and this method was studied when you performed acid–base titrations. The exact concentration of the other solution is usually unknown.

A titration involves a known volume of a solution being measured using a volumetric pipette and placed in a conical flask. A second solution is added a little at a time from a burette, swirling the mixture during addition. This is continued until the desired colour change is seen. Many redox titrations do not need an indicator as the colours of the reactants frequently allow the end point to be seen. The volume of solution added is measured using initial and final burette readings. These readings are always recorded to two decimal places – the first decimal place is read from the burette and the second is either a zero (meniscus is on the line on the burette) or 5 (the meniscus lies between the burette lines).

As in all titrations, the procedure is repeated until the readings taken are sufficiently close together. When results are in agreement they are said to be 'concordant'. Although the best outcome is when results match exactly, in most cases there will be slight differences. Results that agree with each other within 0.20 cm³ are acceptable but a smaller variation range suggests better results.

Practical check

When planning a titration, concentrations are chosen so that the volume added from the burette is usually between 20 and 30 cm³. If the volume is too small the percentage error in the measurement may be too large, whilst if the volume is too large you may empty the burette before reaching the equivalence point.

Link

The redox titrations use the same techniques as acid–base titrations found in AS Topic 1.7 described on pages 85–89 of the AS book.

Examples of redox titrations

Acidified manganate(VII) ions with iron(II) ions

In the case of a titration involving potassium manganate(VII), the purple solution of this oxidising agent is added from the burette. When it reacts with the species to be oxidised, it forms Mn^{2+}, which is almost colourless. The end point is reached when the solution goes pale pink. This is because a very small amount of the purple MnO_4^- remains, which appears pink when dilute in the conical flask. You can get a guide to the end point by judging how long it takes the purple of the manganate(VII) to disappear. At the start of the titration it disappears quickly, but as the end point gets closer it takes longer to go colourless. The half-equation for the reduction of the manganate(VII) is:

$$MnO_4^-(aq) + 8H^+(aq) + 5e^- \longrightarrow Mn^{2+}(aq) + 4H_2O(l) \qquad [1]$$

The colour change of the solution as Fe^{2+} (pale green) is oxidised to Fe^{3+} (yellow) is usually difficult to see as the solutions are generally dilute, and so the colour change associated with the manganate is used for the end point. The iron ions are oxidised according to the half-equation:

$$Fe^{2+}(aq) \longrightarrow Fe^{3+}(aq) + e^- \qquad [2]$$

To obtain the overall equation we need to combine these two half-equations to balance the electrons. In this case, we combine equation [1] + 5 × equation [2] giving:

$$MnO_4^-(aq) + 8H^+(aq) + 5Fe^{2+}(aq) \longrightarrow Mn^{2+}(aq) + 4H_2O(l) + 5Fe^{3+}(aq)$$

The 5 electrons on each side of the equation have balanced out and so do not appear in the overall equation. The acid **must** be present for this reaction to occur, and this is usually added as H_2SO_4. Sometimes one of the two solutions is made using dilute acid instead of deionised water. If there is insufficient acid, a different reaction occurs and the manganate(VII) forms a brown precipitate of MnO_2. If this happens, the results of that titration are not valid and must not be used.

Calculations

There are two ways to approach a titration calculation. You can either perform the calculation step-by-step or use the short method shown here. If you need to perform the calculation step-by-step then you would need to find the number of moles of MnO_4^- ions then use this to find the number of moles of Fe^{2+}. In this method the most common error is to forget the factor of 1000 needed to convert between cm^3 and dm^3. For the shortened method we can use the chemical equation to see that 1 MnO_4^- reacts with 5 Fe^{2+}. The calculations are therefore:

$$\frac{\text{Number of moles of } Fe^{2+}}{\text{Number of moles of } MnO_4^-} = \frac{5}{1}$$

Giving:

$$\frac{C_{Fe^{2+}} \times V_{Fe^{2+}}}{C_{MnO_4^-} \times V_{MnO_4^-}} = \frac{5}{1}$$

This can be rearranged to calculate the concentration of one specific solution, so for the concentration of Fe^{2+} this would be:

$$C_{Fe^{2+}} = \frac{5 \times C_{MnO_4^-} \times V_{MnO_4^-}}{1 \times V_{Fe^{2+}}}$$

---

3 Knowledge check

(a) Write an ionic equation for the oxidation of oxalic acid (ethanedioic acid), $C_2O_4H_2$, to carbon dioxide by acidified potassium manganate(VII).

(b) A 25.0 cm^3 sample of a solution of oxalic acid required 27.20 cm^3 of acidified potassium manganate(VII) solution of concentration 2.00×10^{-3} mol dm^{-3} for complete reaction. Find the concentration of the oxalic acid solution.

Maths tip »

amount in moles = concentration in mol dm^{-3} × volume in dm^3

Worked example

Samples of 25.0 cm³ of a solution of acidified iron(II) sulfate were titrated using a solution of potassium manganate(VII) of concentration 0.0200 mol dm⁻³. The results obtained are shown in the table below. Calculate the concentration of the iron(II) sulfate solution, giving your answer to an appropriate number of significant figures.

	1	2	3	4
Initial reading / cm³	0.00	0.55	0.20	0.75
Final reading / cm³	30.40	30.35	30.10	30.60

First calculate the volumes used in each titration:

Volume used / cm³	30.40	29.80	29.90	29.85

The first volume is anomalous as it is 0.50 cm³ away from the nearest other value. We use the remaining values in our calculations so the average volume is $(29.80 + 29.90 + 29.85) \div 3 = 29.85$ cm³.

The reacting ratio is $5Fe^{2+}$ reacting with $1MnO_4^-$, so the relationships can be written as:

$$\frac{c_{Fe^{2+}} \times V_{Fe^{2+}}}{c_{MnO_4^-} \times V_{MnO_4^-}} = \frac{5}{1}$$

Rearranging gives:

$$c_{Fe^{2+}} = \frac{5 \times c_{MnO_4^-} \times V_{MnO_4^-}}{1 \times V_{Fe^{2+}}}$$

So

$$c_{Fe^{2+}} = \frac{5 \times 0.0200 \times 29.85}{1 \times 25.0}$$

$$c_{Fe^{2+}} = 0.1194 \text{ mol dm}^{-3}$$

As the lowest numbers of significant figures in the measurements are three (25.0 and 0.0200) then the answer should be given to three significant figures:

$$c_{Fe^{2+}} = 0.119 \text{ mol dm}^{-3}$$

It is possible to perform the same calculation step-by-step, calculating the moles of MnO_4^- first, multiplying by 5 to get the moles of Fe^{2+} then dividing by the volume used.

Acidified dichromate(VI) ions with iron(II) ions

Potassium dichromate(VI) in acid solution will oxidise Fe^{2+} to Fe^{3+}, with a colour change from dark orange $Cr_2O_7^{2-}$ to a green solution of Cr^{3+} according to the half-equation:

$$Cr_2O_7^{2-}(aq) + 14H^+(aq) + 6e^- \longrightarrow 2Cr^{3+}(aq) + 7H_2O(l) \qquad [1]$$

The iron is again oxidised according to the half-equation:

$$Fe^{2+}(aq) \longrightarrow Fe^{3+}(aq) + e^- \qquad [2]$$

To obtain the overall equation we need to combine these two half-equations to balance the electrons. In this case, we combine equation [1] + 6 × equation [2] giving:

$$Cr_2O_7^{2-}(aq) + 14H^+(aq) + 6Fe^{2+}(aq) \longrightarrow 2Cr^{3+}(aq) + 7H_2O(l) + 6Fe^{3+}(aq)$$

The 6 electrons on each side of the equation have balanced out and so do not appear in the overall equation.

◀ Practical check

A redox titration is a **specified practical task**. These can include a range of oxidising agents and reducing agents, and you should be familiar with the reactions of MnO_4^- with Fe^{2+} and $S_2O_3^{2-}$ with I_2 (as part of the analysis of copper). You should be aware that an indicator is not needed in most cases.

Knowledge check 4 ◀

A metal nail is made of an alloy containing iron. A sample of 1.740 g of the alloy is dissolved in acid which converts all the iron to Fe^{2+}. The mixture is diluted to form a 250 cm³ solution and samples of 25.0 cm³ are taken for titration using 0.0200 mol dm⁻³ acidified potassium manganate(VII). 23.30 cm³ of the potassium manganate(VII) is needed for complete reaction. Calculate the percentage of iron in the original alloy.

Study point

For each common redox titration, you need to learn the colours of the species involved. This is because the colour changes take the place of indicators.

Top tip

When using oxidation states to identify whether a reaction is a redox reaction, you should calculate the oxidation states of each atom at the start and end. If any of these oxidation states changes then the reaction is a redox reaction – one oxidation state must become more negative (reduction) and one more positive (oxidation).

Link

Copper(II) ions in solution are blue whilst copper(I) ions in CuI are colourless. The reasons for these differences in colour are discussed on pages 49 and 50.

Calculations

From the equation above,

$$\frac{\text{Number of moles of } Fe^{2+}}{\text{Number of moles of } Cr_2O_7^{2-}} = \frac{6}{1}$$

Giving:

$$\frac{c_{Fe^{2+}} \times V_{Fe^{2+}}}{c_{Cr_2O_7^{2-}} \times V_{Cr_2O_7^{2-}}} = \frac{6}{1}$$

NOTE: The acid **must** be present for this reaction to occur, and this is usually added as H_2SO_4. If the pH of the solution rises too high, then the dichromate(VI) ion is broken up into two chromate(VI) ions.

$$Cr_2O_7^{2-}(aq) + H_2O(l) \rightleftharpoons 2CrO_4^{2-}(aq) + 2H^+(aq)$$

Dark orange — Yellow

The chromium always has an oxidation state of +6, so there is no change in oxidation state: **this is not a redox reaction.**

By Le Chatelier's principle, this equilibrium will shift to the left-hand side if acid is added, and to the right-hand side if base is added. This can be seen as a change of the colour of the solution.

Redox titration for copper(II) ions

It is not straightforward to analyse a solution for copper(II) ions directly, so an indirect route is used. Addition of a colourless solution containing iodide ions, such as potassium iodide, to a blue solution containing copper(II) ions leads to the formation of a cloudy brown solution. Cu^{2+} ions in solution react with iodide ions to generate a brown solution of iodine, and are reduced to copper(I) in a white precipitate of CuI. The equation for this process is:

$$2Cu^{2+}(aq) + 4I^-(aq) \rightarrow 2CuI(s) + I_2(aq)$$

We can titrate this iodine with sodium thiosulfate, to work out how much copper was present originally. Sodium thiosulfate is a common reducing agent, working according to the half-equation:

$$2S_2O_3^{2-}(aq) \rightarrow S_4O_6^{2-}(aq) + 2e^-$$

A common reaction of thiosulfate ions is with iodine molecules (I_2), to form iodide ions (I^-):

$$I_2(aq) + 2e^- \rightarrow 2I^-(aq)$$

Brown — Colourless

This gives an overall equation of:

$$2S_2O_3^{2-}(aq) + I_2(aq) \rightarrow S_4O_6^{2-}(aq) + 2I^-(aq)$$

Experimental details

- Measure 25.0 cm³ of $Cu^{2+}(aq)$ solution into a conical flask using a standard pipette. Add excess iodide ions (e.g. KI) to the $Cu^{2+}(aq)$ to ensure **all** Cu^{2+} reacts to produce I_2. The mixture formed is a cloudy brown solution.

Top tip

The observations upon mixing solutions containing Cu^{2+} and I^- can also be used as a qualitative chemical test. If you see a 'white precipitate in a brown solution' upon mixing two solutions you should know that one contains Cu^{2+} and one contains I^-. Other observations, such as a blue coloured solution, may help you identify which solution is which.

▲ Copper(II) ions in solution

▲ Mixture after addition of potassium iodide

- Titrate the brown $I_2(aq)$ against sodium thiosulfate ($Na_2S_2O_3$), until the mixture is straw-coloured. This means that most of the iodine has reacted, leaving only a little remaining. This tells us we are approaching the equivalence point.

- Add starch indicator which goes blue-black, and continue until the colour vanishes. The mixture is often described as flesh-coloured. The starch indicator is not added at the start of the titration as this would make it harder to spot when the equivalence point is near and so it would be more likely to overshoot.

Practical check

The insoluble CuI precipitate formed when KI is added to the copper(II) ions in solution remains throughout the experiment. This makes the mixture cloudy throughout the experiment. In some cases, it is better to let the solid settle to see the colour of the solution.

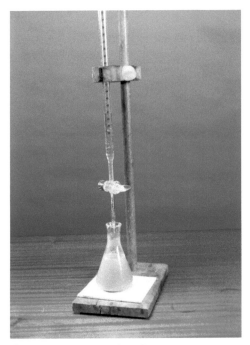

▲ Titration until straw coloured

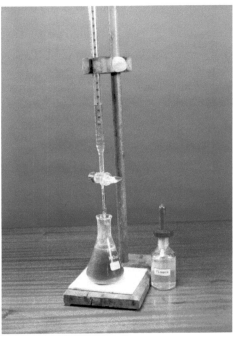

▲ Mixture after addition of starch indicator

< Link >

Titration calculations on page 48 of the AS book.

Calculations

Looking at the equations above we see that two copper(II) ions make one iodine molecule: $2Cu^{2+} \equiv 1I_2$

We can also see that one iodine molecule reacts with two thiosulfate ions: $1I_2 \equiv 2S_2O_3^{2-}$

Combining this gives $2Cu^{2+} \equiv I_2 \equiv 2S_2O_3^{2-}$ so $1Cu^{2+} \equiv 1S_2O_3^{2-}$

The titration calculations therefore use:

$$\frac{c_{Cu^{2+}} \times V_{Cu^{2+}}}{c_{S_2O_3^{2-}} \times V_{S_2O_3^{2-}}} = \frac{1}{1}$$

Alternatively the calculation can be performed step-by-step. This method is used in the worked example below. The step-by-step method requires you to convert everything to moles, use the 1:1 ratio to show that the number of moles of Cu^{2+} and $S_2O_3^{2-}$ would be the same and then calculate the answer required. A typical example is:

Moles of $S_2O_3^{2-}$ = Concentration of $S_2O_3^{2-}$ × Volume of $S_2O_3^{2-}$ in cm³ ÷ 1000

Reaction ratio is 1:1 so:

$$\text{Moles of copper ions} = \text{moles of thiosulfate}$$

$$\text{Concentration of } Cu^{2+}(aq) = \text{Moles } Cu^{2+} \div (\text{Volume in cm}^3 \div 1000)$$

$$= \text{Moles } Cu^{2+} \div \text{Volume in cm}^3 \times 1000$$

Worked example

A transition metal complex contains copper(II) ions. A sample of 4.242 g of the complex was dissolved in 250 cm³ of deionised water, samples of 25.0 cm³ were taken and excess potassium iodide was added to each. The samples were titrated against sodium thiosulfate solution of concentration 0.0500 mol dm⁻³ and 28.35 cm³ was needed for complete reaction. Find the M_r of the complex.

The number of moles of sodium thiosulfate used in the reaction is given by:

Moles thiosulfate = $0.0500 \times (28.35 \div 1000) = 14.175 \times 10^{-4}$ moles

Since 1 thiosulfate $\equiv 1Cu^{2+}$ then the number of moles of $Cu^{2+} = 14.175 \times 10^{-4}$ moles in 25.0 cm³.

The full volume was 250 cm³, which would contain ten times the number of moles = 0.014175 moles.

Since moles = mass ÷ M_r then M_r = mass ÷ moles

$$M_r = 4.242 \div 0.014175$$

$$= 299.3$$

Test yourself

1. In a solution containing a limited amount of acid, manganate(VII) ions (MnO_4^-) are reduced to manganese(IV) oxide, MnO_2. Write the ion–electron half-equation for this process. [1]

2. Cerium(IV) ions (Ce^{4+}) are strong oxidising agents, forming Ce^{3+} ions as products. The concentration of a solution containing Ce^{4+} ions can be found by titration against a standardised solution containing Fe^{2+} ions.

   (a) Write an equation for the oxidation of Fe^{2+} ions by Ce^{4+} ions. [1]

   (b) In an experiment, a 25.0 cm^3 sample of the Ce^{4+} solution required 18.40 cm^3 of Fe^{2+} solution of concentration 0.220 mol dm^{-3} for complete reaction. Calculate the concentration of the Ce^{4+} solution. [2]

3. Oxalic acid, $(COOH)_2$, can be oxidised to form carbon dioxide gas and H^+ ions. The half-equation for this process is:

$$(COOH)_2 \rightarrow 2CO_2 + 2H^+ + 2e^-$$

   One oxidising agent that can be used for the process is acidified potassium manganate(VII).

   (a) Write the half-equation for the reaction that occurs when manganate(VII) ions act as oxidising agents in the presence of excess acid. [1]

   (b) Find the ionic equation for the oxidation of oxalic acid by acidified manganate(VII) ions. [1]

   (c) Oxalic acid is present in spinach leaves, and the amount varies significantly between samples. In an experiment the oxalic acid present in 100 g of spinach leaves was extracted and used to make 250 cm^3 of a solution. A student decides to measure the oxalic acid present by taking 25.0 cm^3 samples of the solution and titrating these using acidified potassium manganate(VII) of concentration 1.00×10^{-2} mol dm^{-3}.
   The results of the titrations are given in the table:

Experiment	1	2	3	4
Volume of $KMnO_4$(aq) / cm^3	29.85	29.35	29.75	29.65

   (i) Find the volume of $KMnO_4$(aq) that should be used in calculations. [1]

   (ii) Calculate the mass of oxalic acid ($M_r = 90.02$) that is present in 100 g of the spinach leaves. [2]

   (iii) The student also had access to acidified potassium manganate(VII) solutions of concentration 0.50×10^{-2} mol dm^{-3} and 2.00×10^{-2} mol dm^{-3}. Suggest why these concentrations would be less suitable for use in the experiment. [2]

4. A commercial aquarium additive, used to destroy parasites without harming fish, contains copper(II) hydroxide, $Cu(OH)_2$. A sample of 1.944 g of the additive is dissolved in a small volume of nitric acid and used to make 250 cm³ of a standard solution.

   25.0 cm³ samples of the standard solution are placed in conical flasks and excess potassium iodide is added. The samples are then titrated against sodium thiosulfate solution of concentration 0.100 mol dm⁻³ and the mean volume of thiosulfate solution needed was 17.85 cm³.

   (a) Find the concentration of Cu^{2+}(aq) ions in the standard solution. [2]

   (b) Find the percentage $Cu(OH)_2$ by mass in the additive. [2]

5. Addition of NaOH (aq) and Fe^{2+}(aq) separately to two beakers of containing acidified aqueous potassium dichromate causes colour changes as seen below.

   (a) Which solution causes the acidified aqueous potassium dichromate to turn yellow? Explain why this change occurs. [4]

   (b) Which solution causes the acidified aqueous potassium dichromate to turn dark green? Explain why this change occurs. [3]

   (c) Use oxidation states to identify which, if any, of these changes are due to redox reactions. [2]

Chemistry of the p-block

The periodic table is the greatest tool available to a chemist. Patterns and similarities in behaviour across periods and down groups allow chemists to make predictions. The explanations for these patterns have led to many of the significant ideas in chemistry.

The p-block makes up a significant part of the periodic table, from Group 3 to Group 0. Like all groups in the periodic table, there are similarities between the elements in a group; however, the p-block also shows patterns with significant differences. The differences we see in some fundamental properties of these elements, such as the metal/non-metal behaviour or their common oxidation states, can make elements in the same group behave very differently.

You should be able to demonstrate and apply knowledge and understanding of:

- Acid/base properties of the elements and oxides, including amphoteric behaviour.

- The variations in oxidation states on going down groups, including octet expansion and the inert pair effect.

- Electron deficiency in Group 3 compounds, and the formation of coordinate bonds involving these.

- The bonding and structure in hexagonal and cubic boron nitride and how these relate to their properties and uses.

- The bonding, physical properties and redox behaviour of lead and carbon oxides.

- Changes in the types of bonding down Group 4 as shown by the chlorides CCl_4, $SiCl_4$ and $PbCl_2$ and their reactions with water.

- The reactions of $Pb^{2+}(aq)$ with aqueous NaOH, Cl^- and I^-.

- The different reactions of Cl_2 with both cold and warm aqueous NaOH and the various disproportionation reactions involved, and the uses of the products of the reactions.

- The differences in behaviour of NaCl, NaBr and NaI with concentrated sulfuric acid.

Topic contents

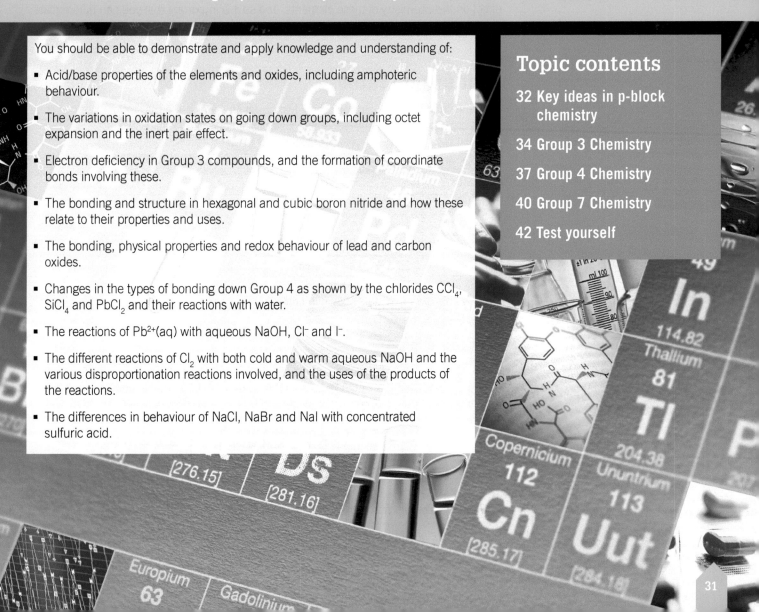

Link

The chemistry of Group 7, part of the p-block, was discussed in AS Topic 1.6 on pages 70–72. The chemistry of the s-block studied at AS can also appear as part of A2 questions as synoptic content.

Key ideas in p-block chemistry

The p-block elements are called this as their outer electrons are located in p sub-shells (2p, 3p, 4p, etc.). You need to be able to write the electronic configurations for these p-block elements in the second and third periods, i.e. those with outer electrons in the 2p or 3p sub-shells.

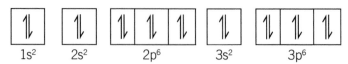

Every p-block element has a full s sub-shell in their outer shell, with between 1 and 6 further electrons in their p sub-shell.

- Group 3 atoms have 1 electron in their p sub-shells, i.e. outer electrons are s^2p^1.

- Group 4 atoms have 2 electrons in their p sub-shells, i.e. outer electrons are s^2p^2.

- Group 5 atoms have 3 electrons in their p sub-shells, i.e. outer electrons are s^2p^3.

- Group 6 atoms have 4 electrons in their p sub-shells, i.e. outer electrons are s^2p^4.

- Group 7 atoms have 5 electrons in their p sub-shells, i.e. outer electrons are s^2p^5.

The division of the outer electrons into s-electrons and p-electrons has a substantial effect on the chemistry of these elements, and so it is important that you can differentiate between these sets of electrons.

Oxidation states

Study point

It is easier to learn the patterns in oxidation states rather than learn each element separately. For all elements, the highest oxidation state equals the group number, with a second oxidation state two lower in many cases.

Elements in the p-block typically show two oxidation states – the maximum oxidation state which equals the group number, and a lower oxidation state which is two less.

Group 3		Group 4		Group 5	
B	3	C	2, **4**	N	3, 5
Al	3	Si	4	P	3, 5
Ga	1, **3**	Ge	2, **4**	As	**3**, 5
In	**1**, 3	Sn	2, **4**	Sb	**3**, 5
Tl	**1**, 3	Pb	**2**, 4	Bi	**3**, 5

(Most stable oxidation state shown in **bold**)

Inert pair effect

Key term

Inert pair effect is the tendency of the s^2 pair of electrons in an atom to stay paired, leading to a lower oxidation state.

The stability of the lower oxidation states becomes greater down the group, as is shown in the table. This tendency of the heavier elements to form the lower oxidation state is called the **inert pair effect**. For an element in Group 4, the outer electronic configuration is:

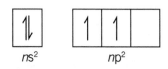

- When the element has an oxidation state of 4, it involves all four electrons.

- When the element has an oxidation state of 2, the inner two electrons do not become involved, and this ns^2 pair is called the *inert pair*.

The trend for the ns^2 electron pair to become an inert pair occurs in Groups 3, 4 and 5 of the periodic table. The inert pair effect increases down the group so it is only the lower members of these groups that favour the lower oxidation states: +1 for Group 3; +2 for Group 4 and +3 for Group 5. In Groups 3 and 4, the lower oxidation state is not observed in the first two members of each group (with the exception of +2 in carbon monoxide), but in Group 5 the upper members (nitrogen and phosphorus) exhibit the lower oxidation state more frequently.

Octet expansion

There are significant differences seen between the first members of the p-block groups and the lower members of these groups. The maximum number of outer shell electrons that can surround the atoms in the first members of each group (boron to neon, the second period elements) is eight – four pairs of electrons. This limits the numbers of bonds that can be formed in these elements:

- Boron: can form 3 covalent bonds and is electron deficient, e.g. BCl_3.

- Carbon: can form 4 covalent bonds, e.g. CH_4.

- Nitrogen: can form 3 covalent bonds and one lone pair, e.g. NH_3.

- Oxygen: can form 2 covalent bonds and two lone pairs, e.g. H_2O.

The other members of each group (third period and below) have access to d-orbitals that are not present in the second shell. This allows them to 'expand their octet' which means every electron in the outer shell can be used to form a covalent bond as there is no longer a limit of 8 electrons in this shell. This affects the numbers of bonds that can be formed for elements in Groups 5, 6 and 7:

- Phosphorus: can form 5 covalent bonds, e.g. PCl_5.

- Sulfur: can form 6 covalent bonds, e.g. SF_6.

- Chlorine: can form up to 7 covalent bonds, e.g. ClO_4^-.

Key term

Octet expansion is the ability of some atoms to use available d-orbitals to have more than 8 electrons in their valence shell.

Knowledge check 1

Nitrogen and bismuth are both Group 5 elements that favour a +3 oxidation state over a +5 oxidation state. Explain why each favours the +3 oxidation state.

Metallic properties

In the p-block, the elements at the top of each group are non-metals, such as boron, carbon or nitrogen, whilst the elements at the bottom of each group are metals such as thallium, lead or bismuth. This change in properties leads to the characteristic zig-zag line between metals and non-metals.

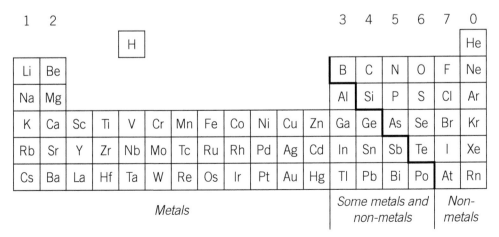

This change in metallic character has a significant effect on the bonding and properties of the p-block compounds.

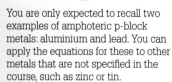

Top tip

You are only expected to recall two examples of amphoteric p-block metals: aluminium and lead. You can apply the equations for these to other metals that are not specified in the course, such as zinc or tin.

2 Knowledge check

Zinc is an amphoteric metal that forms +2 ions. Its reactions are similar to those of lead. Write two chemical equations to show that zinc oxide is an amphoteric oxide.

Amphoteric behaviour

Many p-block elements form **amphoteric** oxides, and these are typically metals close to the line separating metals and non-metals. They show both acidic and basic properties. To show amphoteric behaviour, we must demonstrate that an element or its compounds reacts with both acids and bases. Typically this would be to show the material reacting with an acid such as hydrochloric or nitric acid and a base such as sodium hydroxide.

For aluminium oxide or hydroxide:

$$Al_2O_3 + 6HCl \rightarrow 2AlCl_3 + 3H_2O \quad \text{or} \quad Al(OH)_3 + 3H^+ \rightarrow Al^{3+} + 3H_2O$$

$$Al_2O_3 + 2NaOH + 3H_2O \rightarrow 2Na[Al(OH)_4] \quad \text{or} \quad Al(OH)_3 + OH^- \rightarrow [Al(OH)_4]^-$$

For lead(II) oxide or hydroxide:

$$PbO + 2HNO_3 \rightarrow Pb(NO_3)_2 + H_2O \quad \text{or} \quad Pb(OH)_2 + 2H^+ \rightarrow Pb^{2+} + 2H_2O$$

$$PbO + 2NaOH + H_2O \rightarrow Na_2[Pb(OH)_4] \quad \text{or} \quad Pb(OH)_2 + 2OH^- \rightarrow [Pb(OH)_4]^{2-}$$

Solutions containing amphoteric metal compounds form precipitates when sodium hydroxide is added to their solutions. These precipitates are metal hydroxides. Since the hydroxides can react with more sodium hydroxide, these precipitates will then redissolve.

For aluminium:

$$Al^{3+}(aq) + 3OH^-(aq) \rightarrow \underset{\text{white precipitate}}{Al(OH)_3(s)} \quad \text{then } Al(OH)_3(s) + OH^-(aq) \rightarrow \underset{\text{colourless solution}}{[Al(OH)_4^-(aq)]}$$

For lead:

$$Pb^{2+}(aq) + 2OH^-(aq) \rightarrow \underset{\text{white precipitate}}{Pb(OH)_2(s)} \quad \text{then } Pb(OH)_2(s) + 2OH^-(aq) \rightarrow \underset{\text{colourless solution}}{[Pb(OH_4)]^{2-}(aq)}$$

Group 3 chemistry

The two most common elements in Group 3 are the first two in this group: boron and aluminium. These elements have very different physical properties as one is a non-metal and the other is a metal. Although aluminium is a metal, it has a relatively high electronegativity, so some of its compounds are covalent and show some similarities to analogous compounds of boron.

Electron deficiency

An electron-deficient atom is one that does not have a full outer shell, i.e. has fewer than eight electrons in its outer shell. When elements in Group 3 form compounds, they commonly form three covalent bonds such as in BF_3, BCl_3 and $AlCl_3$. In each of these cases, the three electrons from the Group 3 atom (shown as dots in the diagrams opposite) form covalent bonds with halogen atoms.

Top tip

When discussing electron deficiency in Group 3 compounds, always be specific about the atom that is electron deficient. Writing 'Aluminium chloride is electron deficient' does not identify the aluminium atom as the source of the electron deficiency.

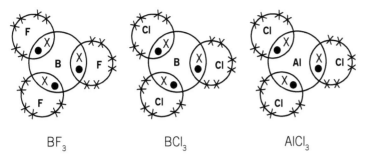

BF$_3$ BCl$_3$ AlCl$_3$

⟨ Link ⟩

The formation of coordinate bonds in compounds such as these is included in AS Topic 1.4 on page 51.

To fill their outer shell, these atoms will often form coordinate bonds to gain extra electron pairs (they are ***electron acceptors***). This may be done by reacting with other compounds, or by forming dimers, as in the case of AlCl$_3$ in the gas phase, which forms Al$_2$Cl$_6$ dimers:

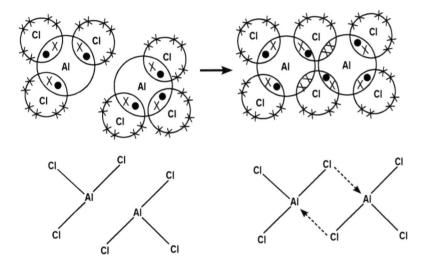

Knowledge check 3

Draw a dot-and-cross diagram to show the bonding in the compound formed when boron trichloride (BCl$_3$) is mixed with phosphine (PH$_3$).

In this case, each electron-deficient aluminium atom uses a lone pair on a chlorine atom to form a coordinate bond. The aluminium chloride dimer no longer has any electron-deficient atoms as each aluminium atom has eight electrons in its outer shell.

Other molecules can also form coordinate bonds to remove their electron deficiency. These compounds are classified as donor–acceptor compounds, where one molecule donates a lone pair and the other accepts it. A typical example is the compound formed between electron-deficient BF$_3$ and the lone pair on NH$_3$. Once again the compound formed no longer has an electron-deficient boron atom.

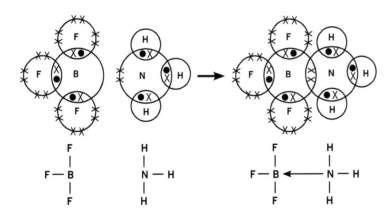

Top tip

When discussing the structures of BN and the allotropes of carbon, focus on the requirements of the question. You may be asked to discuss similarities or differences in structures or properties, and if you try to simply repeat everything you know about these substances, you may not address the specific points required.

Link

The structures of diamond and graphite are included in AS Topic 1.5 on page 61 of the AS book.

Boron nitride

Boron forms a large variety of compounds with nitrogen, and these have been of great interest due to the analogy between the B–N bond and the C–C bond. In each case there are a total of 12 electrons on the two atoms. The atomic radii of all the atoms are similar, with carbon being almost exactly the average of the radii of boron and nitrogen, and the electronegativity of carbon lies midway between boron and nitrogen. This leads to boron nitride, BN, having several forms which are similar to the several different forms of carbon.

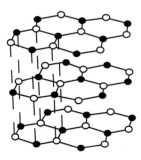

Hexagonal boron nitride
Layers where nitrogen and boron atoms combined in a hexagonal network are superimposed and have a structure similar to graphite.

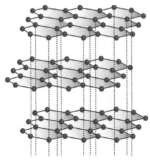

Graphite

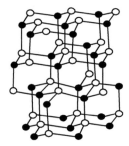

Cubic boron nitride
Boron and nitrogen atoms combine three-dimensionally replacing carbon atoms in diamond.

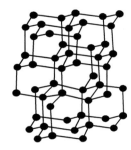

Diamond

Hexagonal boron nitride ('graphite structure')

Boron nitride can form hexagonal sheets similar to those found in graphite, but in this case the atoms in different layers lie directly above one another, with each boron having a nitrogen atom directly above and below it. This differs from graphite as the layers in graphite are arranged so that atoms on adjoining layers are not directly above one another. The forces between layers are weak so boron nitride shares the ability for layers to slip over one another with graphite, making it soft so it can be used as a lubricant.

The electrical properties of boron nitride are very different from graphite as there are no delocalised electrons present, with electrons localised as lone pairs on nitrogen atoms. The B–N bonds are polar due to the different electronegativities of the two atoms. This makes BN an insulator and leads to its use in electronics as a substrate for semiconductors.

Cubic boron nitride ('diamond structure')

Like diamond, cubic boron nitride is extremely hard with a high melting point as covalent bonds must be broken to break or melt the solid. This leads to its use as a wear-resistant coating or an industrial abrasive.

Group 4 chemistry

Group 4 contains a range of metals and non-metals, with most being familiar names to chemists and non-chemists alike. Indeed most of the elements in this group have been known for a very long time, with carbon, tin and lead known for thousands of years. The changes from the top to the bottom of the group are amongst the most significant of all the groups in the periodic table. The first element, carbon, is a non-metal which forms a huge range of covalent compounds with the carbon in the +4 oxidation state, whilst the heaviest stable member of the group, lead, is a metal which generally forms ionic compounds, with the +2 oxidation state being most stable.

The oxides of carbon and lead: redox properties

The oxidation states shown in Group 4 are +2 and +4, with the stability of the +2 oxidation state increasing down the group as the inert pair effect becomes more significant. The most stable oxidation state for all the elements in the group is +4 apart from lead where the +2 oxidation state is most stable. The stability of these oxidation states governs the redox properties of the compounds, and the oxides of carbon and lead exemplify this.

Carbon

Carbon dioxide, CO_2, is the most stable oxide of carbon. Carbon monoxide, CO, is the only stable compound to contain carbon in the +2 oxidation state. CO will act as a reducing agent as it easily becomes oxidised from +2 to +4. It is used for extracting metals from their oxides. For example:

| Iron | $Fe_2O_3(s) + 3CO(g) \longrightarrow 2Fe(l) + 3CO_2(g)$ |
| Copper | $CuO(s) + CO(g) \longrightarrow Cu(s) + CO_2(g)$ |

This method can only be used for the oxides of the less reactive metals. The oxides of the more reactive metals (anything above zinc in the reactivity series) are too stable, and so will not react.

Lead

Lead(II) oxide, PbO, is the most stable oxide of lead. All lead(IV) compounds are oxidising agents, as they are easily reduced from an oxidation state of +4 to +2. An example is PbO_2:

$$PbO_2(s) + 4HCl(conc) \longrightarrow PbCl_2(s) + Cl_2(g) + 2H_2O(l)$$
Lead(IV) oxide   Lead(II) chloride

The oxides of carbon and lead: acid–base properties

In general we can classify metal oxides as basic oxides and non-metal oxides as acidic oxides. Some metals form amphoteric oxides that can show both acidic and basic properties. The change from non-metal at the top of Group 4 to metals at the bottom of the group is reflected in the acid–base properties of the oxides. This means the non-metal oxide CO_2 is acidic whilst the metal oxide PbO is amphoteric.

≪ Top tip

When discussing any of the oxides of carbon or lead you must make it totally clear which one you mean. Any reference to lead oxide without stating an oxidation state of +2 or +4 is unlikely to be specific enough.

Knowledge check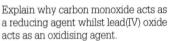

Explain why carbon monoxide acts as a reducing agent whilst lead(IV) oxide acts as an oxidising agent.

≪ Top tip

It is not enough to say that an oxide is acidic as it forms acidic solutions, you need to show it reacts with bases. Similarly basic oxides react with acids. You will need to be able to write equations for these reactions. For amphoteric oxides you need to write equations for reactions with acids and bases.

◀ Stretch & challenge

Although we focus on carbon and lead here, the patterns are still relevant for the elements in between. The stability of the +2 oxidation state increases down the group; however, it is only lead that has +2 as the most stable oxidation state.

Carbon

Carbon dioxide is a colourless gas made up of small covalent molecules. This is an acidic oxide and the oxide is soluble in water to give the very weak acid, carbonic acid:

$$CO_2(g) + H_2O(l) \rightleftharpoons H^+(aq) + HCO_3^-(aq)$$

Like all acidic oxides, carbon dioxide will react with alkalis to form a salt. All salts produced in this way are carbonates or hydrogencarbonates:

$$CO_2(g) + 2NaOH(aq) \rightarrow Na_2CO_3(aq) + H_2O(l)$$

$$CO_2(g) + NaOH(aq) \rightarrow NaHCO_3(aq)$$

>>> **Study point**

It is not true to say that carbon has no d-orbitals, as all atoms have d-orbitals in higher shells; however, in carbon, these are not available for use.

Lead

Lead(II) oxide, PbO, is an orange solid which contains bonding which is mainly ionic. Lead(II) oxide is an amphoteric oxide, so it reacts with acids and bases:

$$PbO(s) + 2HNO_3(aq) \rightarrow Pb(NO_3)_2(aq) + H_2O(l) \qquad \text{Acting as base}$$

$$PbO(s) + 2NaOH(aq) + H_2O(l) \rightarrow Na_2[Pb(OH)_4](aq) \qquad \text{Acting as an acid}$$

The chlorides of carbon, silicon and lead

Carbon and silicon

▲ Silicon chloride fumes in moist air as moisture reacts with it to release acidic HCl fumes.

The stable chlorides of carbon and silicon are the tetrachlorides, CCl_4 and $SiCl_4$. These are both colourless liquids containing individual covalent molecules. The molecules found in both are tetrahedral, due to the 8 electrons in the valence shell. For CCl_4, the outer electronic configuration is as shown:

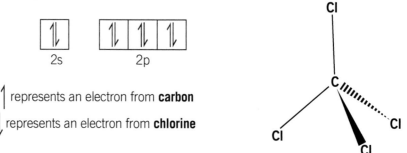

Reactions with water

CCl_4 does not react with water, instead it forms a separate layer under the water. The carbon atom cannot combine easily with water molecules. This lack of reactivity is due to the absence of available d-orbitals in the valence shell meaning that it cannot undergo octet expansion to allow the water molecules to combine with the carbon atom.

Silicon tetrachloride, $SiCl_4$, reacts very quickly with water in a hydrolysis reaction. This reaction produces fumes of hydrogen chloride gas, and silicon dioxide, SiO_2, as a solid precipitate. The reaction becomes more vigorous down the group as the bonds in the compounds become weaker.

$$SiCl_4(l) + 2H_2O(l) \rightarrow SiO_2(s) + 4HCl(g)$$

The reason for the increased reactivity is that silicon possesses available 3d-orbitals in addition to the 3s and 3p orbitals involved in bonding to the chlorine atoms. The lone pairs of the water can form coordinate bonds with these empty d-orbitals, giving a complex intermediate that eliminates two HCl molecules to give $Si(OH)_2Cl_2$: .

This molecule can then eliminate two more molecules of HCl to leave SiO_2.

You may also see the product of the reaction written as $Si(OH)_4$, with the overall reaction written as:

$$SiCl_4(l) + 4H_2O(l) \rightarrow Si(OH)_4(s) + 4HCl(g)$$

Both of these alternatives are acceptable, as the nature of the product formed is not well defined. Hydrated silicon dioxide ($SiO_2.2H_2O$) has the same overall composition as silicon hydroxide ($Si(OH)_4$) and spectroscopic analysis does not distinguish between these.

Lead

Lead(II) chloride is the most stable chloride of lead. It is a white ionic solid made up of Pb^{2+} and Cl^- ions. As it is an ionic compound, it does not react with water. It can be dissolved in hot water but not cold water. This is in common with most lead(II) compounds, which are insoluble in cold water.

Reactions of solutions of lead(II) compounds, $Pb^{2+}(aq)$

Lead(II) compounds are ionic compounds and practically all of them are insoluble in water. The only two common compounds which dissolve readily in cold water are lead nitrate, $Pb(NO_3)_2$, and lead ethanoate, $Pb(CH_3COO)_2$. The reactions of solutions of these salts with various anions produce a range of precipitates:

Solution added	Anions present	Observation/explanation
NaOH(aq)	OH⁻(aq)	An initial white precipitate of $Pb(OH)_2$ is formed: $Pb^{2+}(aq) + 2OH^-(aq) \rightarrow Pb(OH)_2(s)$
excess NaOH(aq)	OH⁻(aq)	The white precipitate redissolves in excess OH⁻ (aq) to form the tetrahydroxoplumbate(II) ion: $Pb(OH)_2(s) + 2OH^-(aq) \rightarrow [Pb(OH)_4]^{2-}(aq)$
HCl(aq)	Cl⁻(aq)	A dense white precipitate of lead(II) chloride, $PbCl_2$, is formed: $Pb^{2+}(aq) + 2Cl^-(aq) \rightarrow PbCl_2(s)$
KI(aq)	I⁻(aq)	A dense bright yellow precipitate of lead(II) iodide, PbI_2, is formed: $Pb^{2+}(aq) + 2I^-(aq) \rightarrow PbI_2(s)$ **THIS IS A KEY OBSERVATION TO IDENTIFY Pb^{2+} IONS**

⟨ **Link** ⟩

The reasons for the solubility, or lack of solubility, of ionic compounds are covered in Topic 3.6 on page 67.

Top tip

There are two yellow precipitates that are commonly seen in qualitative analysis. These are silver iodide and lead iodide; however, they are very different colours. The precipitate of lead iodide is often described as bright yellow or canary yellow whilst the precipitate for silver iodide is pale yellow. Whenever a bright yellow precipitate is mentioned in qualitative analysis, this is a good starting point for identifying all solutions present.

▲ Lead iodide precipitate is bright yellow.

Group 7 chemistry

The Group 7 elements are familiar to chemists as a set of diatomic molecules, containing elements in all three physical states. The chemical properties of the elements are very similar, with all the elements being non-metals although the range and stability of oxidation states shows clear patterns on descending the group.

The oxidising power of halogens

The halogens have different strengths as oxidising agents, with their oxidising power decreasing down the group. This ability to remove electrons from other species can be measured using the standard electrode potentials for the halogens. The values for chlorine, bromine and iodine are given below:

Reaction	$E^{\ominus}$ / Volts
$Cl_2(g) + 2e^- \rightleftharpoons 2Cl^-(aq)$	+1.36
$Br_2(l) + 2e^- \rightleftharpoons 2Br^-(aq)$	+1.09
$I_2(s) + 2e^- \rightleftharpoons 2I^-(aq)$	+0.54

5 Knowledge check

Use the standard electrode potentials to explain why chlorine is able to oxidise iodide ions and write an equation for this process.

- As the value for chlorine is the most positive, it readily gains electrons to form chloride ions, Cl^-. This also shows that it is difficult to oxidise chloride ions to chlorine molecules.
- As the value for iodine is the least positive, it gains electrons less readily than bromine and chlorine. This also shows that it is easier to oxidise iodide ions than it is to oxidise bromide or chloride ions.

Example

If we bubble chlorine gas into a solution containing bromide ions, the solution goes orange, showing bromine is being formed.

$$Cl_2(g) + 2Br^-(aq) \rightarrow Br_2(aq) + 2Cl^-(aq)$$

This occurs because chlorine is a stronger oxidising agent than bromine (it has a more positive standard electrode potential). This means that chlorine is able to oxidise bromide ions to form bromine molecules.

 Top tip

When discussing the chemistry of the halogens, it is essential to distinguish between a halogen and a halide. Students frequently use these terms interchangeably but they mean different things – a halide is always a negative ion whilst a halogen ion is a positive ion such as I^+. If in doubt, use the formula each time to make your answer clear.

 Link

Standard electrode potentials are covered in Topic 3.1 on page 14.

Reactions of concentrated sulfuric acid with sodium halides

Concentrated sulfuric acid is a strong acid and an oxidising agent. As an acid, it can react with all sodium halides to form hydrogen halides (HCl, HBr and HI). These all appear as steamy gases.

The halides become easier to oxidise as you go down the group. This means that sulfuric acid is a strong enough oxidising agent to oxidise the lower halides but not to oxidise chloride. The additional products formed by oxidation change as you go down the group.

Sodium chloride, NaCl

Addition of sulfuric acid to sodium chloride produces HCl gas. The hydrogen chloride is difficult to oxidise ($E^{\ominus} = +1.36$ V), and so the sulfuric acid does not cause any redox reaction.

$$NaCl(s) + H_2SO_4(conc.) \rightarrow NaHSO_4(s) + HCl(g)$$

Observations: Steamy fumes of HCl

Sodium bromide, NaBr

Addition of sulfuric acid to sodium bromide produces HBr gas.

$$NaBr(s) + H_2SO_4(conc.) \longrightarrow NaHSO_4(s) + HBr(g)$$

The sulfuric acid oxidises some of the HBr to form brown fumes of Br_2, and SO_2 gas. The HBr is slightly easier to oxidise ($E^\ominus = +1.09$ V) than HCl, and so the sulfuric acid causes the redox reaction below:

$$2HBr(s) + H_2SO_4(conc.) \longrightarrow SO_2(g) + Br_2(g) + 2H_2O(l)$$

S is reduced from +6 to +4 in SO_2; Br is oxidised from −1 to 0 in Br_2.

Observations: Steamy fumes of HBr; orange fumes of Br_2.

Sodium iodide, NaI

Addition of sulfuric acid to sodium iodide initially produces HI gas.

$$NaI(s) + H_2SO_4(conc.) \longrightarrow NaHSO_4(s) + HI(g)$$

The sulfuric acid easily oxidises the HI ($E^\ominus = +0.54$ V) to form a complex mixture of products including $I_2(s)$, $SO_2(g)$ and $H_2S(g)$. The reaction below occurs:

$$2HI(s) + H_2SO_4(conc.) \longrightarrow SO_2(g) + I_2(s) + 2H_2O(l)$$

S is reduced from +6 to +4 in SO_2; I is oxidised from −1 to 0 in I_2.

Further reduction of the sulfuric acid to S (O.S. = 0) and H_2S (O.S. = −2) can occur as the HI is a better reducing agent than HBr or HCl.

Observations: Steamy fumes of HI, purple fumes of I_2 or black solid/brown solution; smell of rotten eggs (H_2S), yellow solid (S).

Reactions of chlorine with sodium hydroxide

We have already seen the most common oxidation states of the halogens, which are 0 in the element and −1 in the halide ions, but there are many other oxidation states possible including +1 and +5. The higher oxidation states become more stable as you go down the group. We will examine the chemistry of the ions ClO^- (chlorate(I), +1 oxidation state) and ClO_3^- (chlorate(V), +5 oxidation state). These ions are formed by reaction of chlorine with an alkali.

When chlorine is bubbled through water, a reversible reaction occurs:

$$Cl_2(g) + H_2O(l) \rightleftharpoons HCl(aq) + HOCl(aq)$$

In this reaction, the chlorine is being both oxidised and reduced: the element at the start has an oxidation state of 0. The chlorine at the end has an oxidation state of −1 in HCl, and +1 in HOCl. A process where an element ends up in two different compounds, one with a higher oxidation state and one with a lower oxidation state is called a **disproportionation reaction**.

This is an equilibrium reaction, and the products are both acids. If we use the alkali sodium hydroxide instead of water, this will force the equilibrium over to the right-hand side:

$$Cl_2(g) + 2OH^-(aq) \longrightarrow Cl^-(aq) + OCl^-(aq) + H_2O(l)$$

The ClO^- ion is stable in the solution at room temperature, but on heating with concentrated sodium hydroxide a further disproportionation reaction occurs:

$$3Cl_2(g) + 6OH^-(aq) \longrightarrow 5Cl^-(aq) + ClO_3^-(aq) + 3H_2O(l)$$

This gives the chlorate(V) ion, where chlorine is in the +5 oxidation state, and to balance this change from 0 $\longrightarrow$ +5 there must be five chlorine atoms undergoing the 0 $\longrightarrow$ −1 change in oxidation state.

▲ The addition of concentrated sulfuric acid to sodium bromide.

Link

The use of chlorine in the sterilisation of water p72 in AS book.

▲ Bleach is usually a solution of sodium chlorate(I) in water.

Uses of chlorine and chlorate(I) ions

The chlorate(I) ion is an oxidising agent:

$$ClO^-(aq) + 2H^+(aq) + 2e^- \longrightarrow Cl^-(aq) + H_2O(l)$$

Similarly, elemental chlorine is an oxidising agent:

$$Cl_2 + 2e^- \longrightarrow 2Cl^-$$

The oxidising power of chlorine and of chlorate ions is the basis of their use in bleaches, which are often labelled with the old name of this chemical, sodium hypochlorite. Bleaching is an oxidation reaction, with the oxidised form of the coloured material or dye being colourless. Similarly the oxidising ability of ClO^- leads to its ability to kill bacteria, as the microbe cells are oxidised. This is the basis of chlorination of water supplies to disinfect them.

Test yourself

1. Explain why phosphorus forms PCl_3 and PCl_5 but nitrogen only forms NCl_3. [1]

2. Explain why PbO_2 is used as an oxidising agent but SnO_2 is not. [2]

3. Aluminium is classed as an amphoteric element.

    (a) Give the meaning of the term amphoteric. [1]

    (b) Give the equations for reactions that show that aluminium hydroxide is amphoteric. [2]

4. Explain why aluminium chloride exists as dimers containing coordinate bonds. [2]

5. A student adds a few drops of CCl_4 and $SiCl_4$ separately to samples of water. Give the expected observations and explain any differences. [2]

6. Concentrated sulfuric acid is added to a sample of a white solid. Coloured fumes are released but there is no smell of rotten eggs. A flame test on the solid gives a brick-red colour. Identify the solid, giving your reasoning. [2]

7. Chlorine reacts with aqueous sodium hydroxide in two different disproportionation reactions, depending on the temperature. Write the equations for both reactions, using oxidation states to show that they are disproportionation reactions. [3]

8. A student is provided with three solutions labelled A, B and C, each containing a compound formed from familiar ions. [4]

    ▪ She notes that A and B are colourless solutions and C is a pale blue solution.

    ▪ Flame tests on all three give colours, but the only one familiar to her from the A-level course is a lilac colour from solution B.

    ▪ Her teacher tells her that one substance contains sulfate ions.

    ▪ Mixing pairs of the solutions gave the following observations.

Pair	Observation
A+B	Bright yellow precipitate
B+C	White solid in a brown solution
A+C	White precipitate in a blue solution

Use **all** the information to identify A, B and C, explaining your reasoning.

3.4

Chemistry of the d-block transition metals

The periodic table can be divided into the main group elements (s-block and p-block) and the transition elements found in the d-block and f-block. The transition elements show key similarities with other elements in the same block, as well as the expected similarities with those in the same group. The similarity is greatest in the lanthanides, a set of f-block transition elements, but this chapter focuses on the first period of the d-block. These elements exemplify some key ideas such as variable oxidation states, coloured compounds and the formation of complexes which help us understand the chemistry of the entire d-block.

You should be able to demonstrate and apply knowledge and understanding of:

- Variable oxidation states in the d-block elements, including the important oxidation states of Cr, Mn, Fe, Co and Cu and the colours of key species containing these ions.

- The bonding in tetrahedral and octahedral complexes.

- The origin of colour in octahedral complexes such as $[Cu(H_2O)_6]^{2+}$ and $[Fe(H_2O)_6]^{3+}$.

- Examples of tetrahedral and octahedral complexes containing copper(II) and cobalt(II) and how ligand exchange can interconvert these.

- The origin of the catalytic properties of transition metals and their compounds, including both homogeneous and heterogeneous examples.

- The aqueous reactions of sodium hydroxide with Cr^{3+}, Fe^{2+}, Fe^{3+} and Cu^{2+}.

Topic contents

d-block transition elements

The **d-block** consists of the elements scandium to zinc, along with similar elements in the next two periods; however, this topic focuses on the first row of the d-block alone, showing how the chemistry of these elements is very different from the metals of the s-block. The elements include some of the most familiar metals, such as iron, copper and zinc. These elements are hard, dense metals with some of the highest melting and boiling points out of all the elements.

Atomic No.	21	22	23	24	25	26	27	28	29	30
Symbol	Sc	Ti	V	Cr	Mn	Fe	Co	Ni	Cu	Zn
Name	Scandium	Titanium	Vanadium	Chromium	Manganese	Iron	Cobalt	Nickel	Copper	Zinc

>>> **Key terms**

The **d-block** is the set of elements whose highest energy electrons are found in d-orbitals.

A **transition element** is a metal that possesses a partially filled d sub-shell in its atom or stable ions.

Transition elements are considered to be the metals with partially filled d-orbitals. These include all the elements from scandium to nickel as they have partially filled d-orbitals in the unreacted metals. Copper has a full set of d-orbitals as a metal but is considered to be a transition metal as it has partially filled d-orbitals in most of its compounds. Zinc has a filled d sub-shell in a zinc atom, and maintains this in its compounds. As its d sub-shell is never partially filled, it is not a transition element; however, all the other elements in the d-block can be classified as transition elements.

Electronic configurations for the elements

When working out the electronic configurations of elements beyond argon, we need to fill 3d and 4s orbitals. As studied in the AS units, the 3d-orbitals are filled after the 4s-orbitals. Despite this, the 'arrows in boxes' are usually written with the 3d-orbitals first. The order of filling orbitals for these elements is: 4s then 3d then 4p. So we have the following electronic configurations, with 4s filled first:

Potassium (Atomic Number 19)

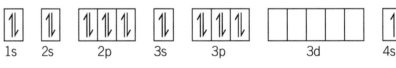

Calcium (Atomic Number 20)

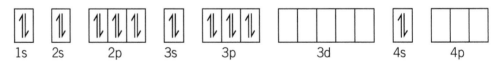

 Knowledge check

Write the electronic configurations of the transition elements Ti, V, and Fe.

< **Link** >

Electronic configurations in AS Topic 1.2 on pages 22–24 of the AS book.

Then the d-orbitals are filled, with one electron going into each:

Scandium (Atomic Number 21)

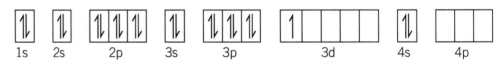

Manganese (Atomic Number 25)

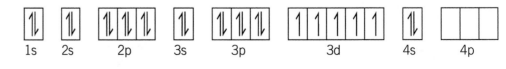

Then the electrons pair up in the d-orbitals:

Zinc (Atomic Number 30)

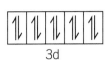

1s
2s
2p
3s
3p
3d
4s
4p

And finally, we fill the 4p orbitals:

Gallium (Atomic Number 31)

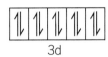

1s
2s
2p
3s
3p
3d
4s
4p

Exceptions to the rules

Two transition elements do not obey the rules given above. *You must know these exceptions.* They are chromium (Cr) and copper (Cu), whose electronic configurations are given below:

Chromium, Atomic Number 24, $1s^2\ 2s^2\ 2p^6\ 3s^2\ 3p^6\ 3d^5\ 4s^1$.

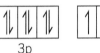

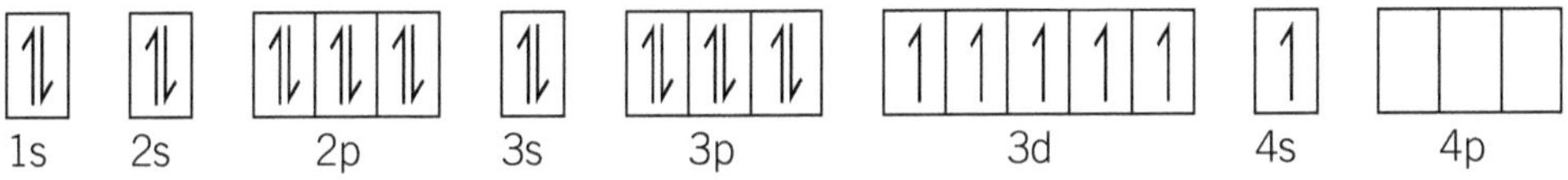

1s
2s
2p
3s
3p
3d
4s
4p

Copper, Atomic Number 29, $1s^2\ 2s^2\ 2p^6\ 3s^2\ 3p^6\ 3d^{10}\ 4s^1$.

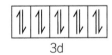

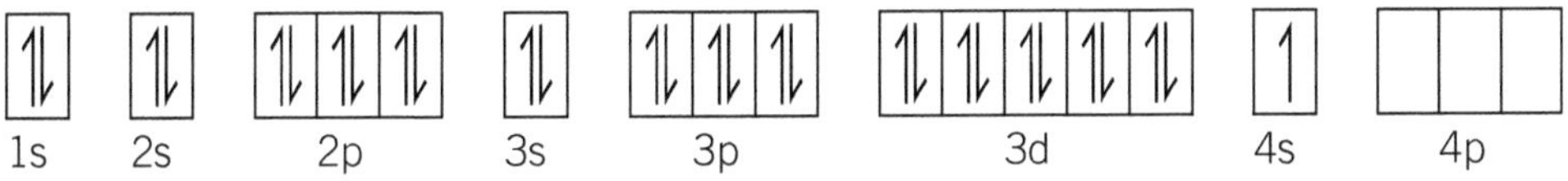

1s
2s
2p
3s
3p
3d
4s
4p

In these cases, we see that we have one electron in the 4s orbital, even though it is slightly lower in energy than the 3d orbitals. The small difference between the energy of the 3d and 4s orbitals, and the extra energy required to pair up electrons, leads to these configurations being more stable than the alternative $3d^4\ 4s^2$ and $3d^9\ 4s^2$ configurations. **You can remember these exceptions by recalling that sub-shells are more stable if they are filled or half-filled.**

Stretch & challenge

The idea that chromium and copper adopt unexpected electronic configurations due to the stability of filled or half-filled d sub-shells is a simplification that reminds us of the pattern, but does not truly explain it. Due to the similar energy of the 3d and 4s sub-shells, it does not take much to move electrons between them and so factors such as repulsion between electrons in the same sub-shell and pairing energy for electrons in the same orbital are enough to change the stable electronic configuration.

Electronic configurations for the ions

When the electronic configurations of the transition metal atoms are found, they suggest that the 4*s* orbital is filled before the 3d orbitals. When electrons are removed to form positive ions, the **4s electrons are lost first**. This is because the 4s and 3d orbitals are very close together in energy, so on balance it is more energetically favourable to lose these 4s electrons before the 3d electrons. If we examine an atom of iron, the electronic structure is:

[Ar]
3d
4s
4p

Study point

When an atom contains many electrons, the electronic configuration can be shortened by writing [Ne], [Ar] or [Kr] followed by the rest of the electronic configuration. These represent full inner shells equivalent to the electronic configurations of these atoms. For example, [Ar] would represent 18 electrons in $1s^2 2s^2 2p^6 3s^2 3p^6$.

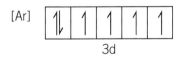

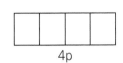

Common iron ions are Fe^{2+} and Fe^{3+}. To work out the electronic configurations of these ions, we need to remember to remove the 4s electrons first:

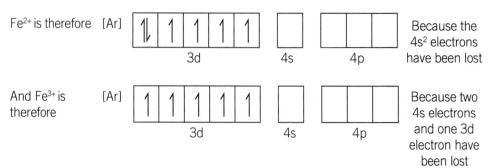

Fe^{2+} is therefore [Ar] 3d 4s 4p — Because the $4s^2$ electrons have been lost

And Fe^{3+} is therefore [Ar] 3d 4s 4p — Because two 4s electrons and one 3d electron have been lost

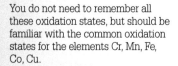
Oxidation states for the transition elements

Although we have already observed metals showing more than one oxidation state when we studied the p-block, the range of oxidation states seen in the transition metals is much greater. Manganese shows the most, with 7 different positive oxidation states, ranging from +1 to +7. The table below lists all the possible oxidation states of the first row of transition metals, with the most common stable oxidation states shown in **bold**.

Sc	Ti	V	Cr	Mn	Fe	Co	Ni	Cu	Zn
	+1	+1	+1	+1	+1	+1	+1	**+1**	
	+2	**+2**	+2	**+2**	**+2**	**+2**	+2	+2	+2
+3	**+3**	+3	**+3**	**+3**	**+3**	**+3**	+3	+3	
	+4	+4	+4	**+4**	+4	+4	+4		
		+5	+5	+5	+5	+5			
			+6	**+6**	+6				
				+7					

These elements can form these different oxidation states because the energies of the 4s and 3d orbitals are very similar so the energy required to remove any of these electrons is similar. As the elements form compounds, energy is released, either through the formation of covalent bonds or when the ionic lattice forms. The energy needed to reach higher oxidation states and the energy released in compound formation is finely balanced, allowing a range of oxidation states to form.

The oxidation state favoured by each metal depends on many factors. The oxidising power of the other atoms in the compound is one factor, so when iron metal reacts with chlorine gas the product is iron(III) chloride but when it reacts with iodine vapour the product is iron(II) iodide. Iodine is a much weaker oxidising agent than chlorine and so it cannot oxidise the iron to the +3 oxidation state.

Transition metal complexes

Transition metal ions are small and can have large positive charges. They have many orbitals available for bonding, some of which are empty. Electron-rich molecules have lone pairs, so these can form coordinate bonds with the empty orbitals on the transition metal ions:

 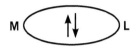

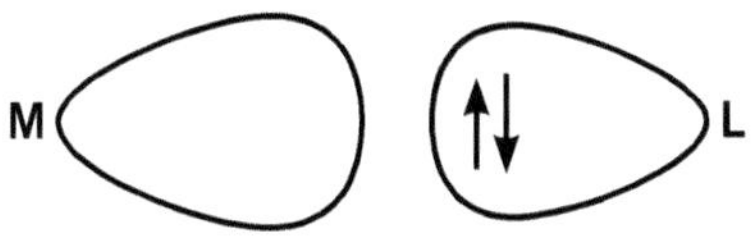

The metal (M) has an empty orbital, and the ligand (L) has a lone pair of electrons.

The two atomic orbitals overlap to form a molecular orbital

A coordinate bond is formed.

A *ligand* is a small molecule with a lone pair that can form a bond to a transition metal, e.g. H_2O, NH_3, Cl^-, CN^-.

The combination of the transition metal ion with its ligands is called a *complex*.

Most of the ions we have written as simple ions are actually complexes with water molecules as ligands around the transition metal ion.

Typically the transition metal complexes have either:

6 ligands arranged octahedrally around the metal ion. [MOST COMMON]   OR   4 ligands arranged tetrahedrally around the metal ion [LESS COMMON]

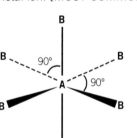 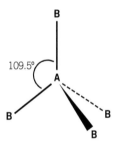

Examples in solution:

$[Fe(H_2O)_6]^{2+}$, a pale green complex

$[Fe(H_2O)_6]^{3+}$, a yellow complex

$[Cu(H_2O)_6]^{2+}$, a blue complex

$[Cr(H_2O)_6]^{3+}$, a dark green complex

$[Co(H_2O)_6]^{2+}$, a pink complex

Examples:

$[CuCl_4]^{2-}$, a yellow or green complex

$[CoCl_4]^{2-}$, a blue complex

Both shapes can be seen for the same transition metal ion with different ligands. The shape found is dependent on the metal, the oxidation state of the metal, and the ligands, and these factors often favour the octahedral complex with six ligands around the metal atom.

Top tip

If you are drawing the structures of complexes you need to represent them in three dimensions to show their full structures. This involves drawing wedge and dashed lines. The bonds must clearly show how the atoms are connected so bonds between water molecules and a metal ion must be attached to the oxygen atom of the water.

Top tip

You need to know the colours of the complexes formed by the following ions in aqueous solution: Fe^{2+}, Fe^{3+}, Co^{2+}, Cu^{2+} or Cr^{3+}. These may be written as the ions themselves or as their complexes containing six water molecules as ligands. If the question specifies there are no ligands then the compounds are often colourless.

Typical transition metal complexes

The complexes of copper can be used to demonstrate the variation in transition metal complexes. The three complexes $[Cu(H_2O)_6]^{2+}$, $[Cu(NH_3)_4(H_2O)_2]^{2+}$ and $[CuCl_4]^{2-}$ all contain Cu^{2+} ions, but their different structures and properties are due to their different ligands. Similarly cobalt can form $[Co(H_2O)_6]^{2+}$ and $[CoCl_4]^{2-}$ both containing Co^{2+} ions.

$[Cu(H_2O)_6]^{2+}$ and $[Co(H_2O)_6]^{2+}$

These are the complexes present in most aqueous solutions of Cu^{2+} and Co^{2+} giving the familiar colours of these solutions and many copper(II) and cobalt(II) compounds. The complexes are octahedral, with one lone pair from each oxygen atom of the water molecules used for bonding to the metal ion.

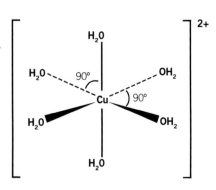

$[Cu(NH_3)_4(H_2O)_2]^{2+}$

Addition of ammonia to a solution containing $[Cu(H_2O)_6]^{2+}$ causes four ammonia molecules to replace water molecules, forming a royal blue solution containing $[Cu(NH_3)_4(H_2O)_2]^{2+}$ ions. This complex is octahedral, but because it contains two different ligands, there could be two different arrangements:

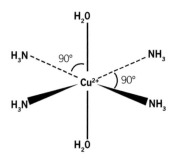

Trans *isomer*

Two water molecules opposite each other.
THIS IS THE COMMON ISOMER

Cis *isomer*

Two water molecules next to each other.

$[CuCl_4]^{2-}$ and $[CoCl_4]^{2-}$

These are tetrahedral complexes, with all four chlorides at 109.5° to each other. The complexes are formed when copper(II) or cobalt(II) ions react with concentrated hydrochloric acid, which displaces the water molecules. There are distinct colour changes as the change in ligands and coordination geometry both contribute to changes in the light absorbed. The colour changes are:

- Copper(II) goes from pale blue to yellow/green.
- Cobalt(II) goes from pink to blue.

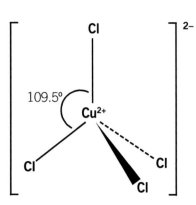

Ligand exchange

When a transition metal ion is exposed to a mixture of ligands, such as an aqueous solution containing chloride ions, ligands can be exchanged to form a new complex. This is an equilibrium process, so the concentrations of the metal ions and any possible ligands are key to identifying the species that will be present in solution.

$$[Cu(H_2O)_6]^{2+} + 4NH_3 \rightleftharpoons [Cu(H_2O)_2(NH_3)_4]^{2+} + 4H_2O$$

▲ From left to right: Solutions of zinc sulfate ($ZnSO_4$), cobalt(II) chloride ($CoCl_2$), iron(II) sulfate ($FeSO_4$), iron(III) chloride ($FeCl_3$), copper(II) sulfate ($CuSO_4$), copper(II) chloride ($CuCl_2$).

‹ Link ›

In Topic 3.2 on pages 26 the colours of the oxoanions MnO_4^- (dark purple), $Cr_2O_7^{2-}$ (orange) and CrO_4^{2-} (yellow) were discussed. You should be able to recall these colours as well as the colours discussed in this topic.

According to Le Chatelier's principle, addition of extra ammonia forces the equilibrium to the right, producing more of the $[Cu(H_2O)_2(NH_3)_4]^{2+}$ Addition of extra water forces the equilibrium to the left, producing more of the $[Cu(H_2O)_6]^{2+}$ complex. This is associated with a colour change, with the ammonia-containing complex being royal blue compared to the pale blue of the original complex.

The equilibrium between complexes can lead to a change in geometry depending on the ligands used. The equilibrium below shows the interconversion between two complexes of cobalt. If a large amount of chloride is used, such as by adding concentrated hydrochloric acid, then the equilibrium shifts from the octahedral complex to the tetrahedral chloro-complex.

$$[Co(H_2O)_6]^{2+} + 4Cl^- \rightleftharpoons [CoCl_4]^{2-} + 6H_2O$$

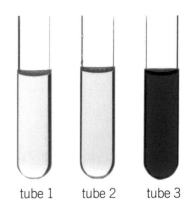

tube 1   tube 2   tube 3

▲ Copper(II) ions (tube 1) change colour when concentrated hydrochloric acid (tube 2) or ammonia (tube 3) are added.

Colour in transition metal ions and complexes

Transition metal complexes are almost always coloured, with almost every colour being seen in the wide range of transition metal complexes. Although we are familiar with these colours, it is important to remember that *transition metal atoms are only coloured in* **complexes**. In the absence of any ligands around the metal ion, the compound would be colourless.

When ligands are introduced around a transition metal ion, they have a significant effect on the orbitals in the atom. Without the ligands, the transition metal atom has 5 degenerate d-orbitals, that is 5 d-orbitals with the same energy. The shapes of these orbitals are shown below.

Knowledge check **3**

A blue solution turns pink when more water is added, but then turns blue again when concentrated hydrochloric acid is added. Identify the initial solution and explain the observations.

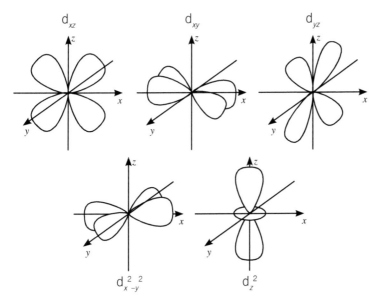

◀ **Stretch & challenge**

The equilibrium constant, K_c for any ligand exchange equilibrium gives a guide to the relative stability of the complex and the free ligands and metal ions. If the value of K_c is large then the complex is more stable than the free ion and ligands.

The first three orbitals, d_{xz}, d_{xy} and d_{yz}, point between a pair of axes: the first orbital points between the x and z axes.

The last two orbitals $d_{x^2-y^2}$ and d_{z^2} point along the axes: the first along the x and y axes, the second along the z axis.

In an octahedral complex, six negatively charged ligands approach the transition metal ion along the directions of the three axes. These negative charges repel the electrons in the orbitals that point along these axes, which makes these orbitals less stable. The orbitals that do not point along the axes are not made less stable. This means that the energies of the orbitals are no longer the same, i.e. they are no longer degenerate.

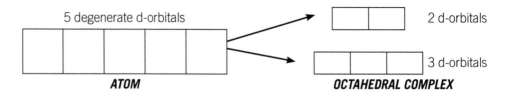

5 degenerate d-orbitals

2 d-orbitals

3 d-orbitals

ATOM

OCTAHEDRAL COMPLEX

This splitting of the d-orbitals gives two sets of orbitals close together in energy. An electron in one of the d-orbitals can move from the lower to the upper of these sets of orbitals, but to do this it needs to gain energy, which is absorbed in the form of light. It is only one frequency (colour) of light that is absorbed, which corresponds to the energy gap between the orbitals. The relationship between the energy and the frequency absorbed is given by the equation $E = hf$. The light that remains is transmitted and gives the complex its characteristic colour.

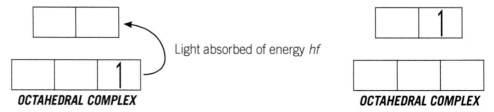

Light absorbed of energy hf

OCTAHEDRAL COMPLEX

OCTAHEDRAL COMPLEX

These d–d transitions depend on the amount of splitting between the d-orbitals, and this varies between ions of different transition metal complexes. As the splitting varies, so does the frequency (colour) of the light absorbed and this leads to different colours for different complexes. Compounds containing the complex $[Cu(H_2O)_6]^{2+}$ are typically blue as they absorb all colours apart from blue, whilst those containing $[Fe(H_2O)_6]^{3+}$ are yellow as they absorb other colours.

Different ligands give different splitting of orbitals, and so give different colours.

Why are some complexes colourless?

Copper(I) complexes have an electronic configuration with a full d sub-shell (d^{10}), which means that there are no empty orbitals to allow electrons to move between energy levels. As a consequence, Cu(I) complexes do not absorb light in the visible range, and appear colourless. Similarly Sc^{3+} ions have an empty d sub-shell so there are no electrons to move between d-orbitals.

Knowledge check

Explain why the complex $[Cr(H_2O)_6]^{3+}$ is dark green.

Knowledge check

Explain why zinc(II) compounds form colourless solutions.

Link

Catalysts and their functions are discussed in AS Topic 2.2 on pages 116–117 in the AS book.

Transition metals as catalysts

Transition metals and their compounds are used in industry as catalysts for a large range of chemical processes. Examples of where the metals themselves are used include:

Iron	The Haber process, to produce ammonia from nitrogen and hydrogen.
Nickel	The hydrogenation of vegetable oils to form margarine.

Transition metal compounds are used in the following processes:

Vanadium(V) oxide, V_2O_5	The contact process for the production of sulfuric acid.
Manganese dioxide, MnO_2	The catalytic decomposition of hydrogen peroxide.

Transition metal **catalysts** are used in many industrial processes that would be uneconomical without them, and these catalysts are therefore essential to our current economy: they are needed to make almost all plastics, artificial fibres, fertilisers, explosives, ethanoic acid and most other acids and solvents.

◀ Manganese dioxide catalyses the decomposition of hydrogen peroxide.

The ability of transition metals to act as catalysts depends on their unique properties. Catalysts act as intermediaries in chemical reactions, and provide an alternative, lower energy route for the reaction. Transition metals can do this due to their **empty orbitals** and **variable oxidation states**.

Empty orbitals can combine with other molecules. Molecules with lone pairs can form coordinate bonds to the metal atom to form complexes, and this can increase the reactivity of the species bonded to the metal, or bring two reacting molecules closer together. This makes a reaction more likely, especially in heterogeneous catalysts where a solid surface can provide an area where molecules are adsorbed and brought close together for reaction.

Variable oxidation states of the transition metal ions allow the metal ion to act as a catalyst in redox reactions. It can act as an oxidising or reducing agent, by oxidising or reducing one of the reactants. The transition metal can then be returned to its original oxidation state by reaction with another molecule. It therefore appears unchanged at the end of the reaction.

The key factors depend on whether the catalyst is homogeneous or heterogeneous. **Heterogeneous catalysts** are typically solids that provide a surface for molecules to be adsorbed and come together in an advantageous arrangement. **Homogeneous catalysts** typically form coordinate bonds with the reactants then use their variable oxidation states to oxidise/reduce them, which makes each reactant much more reactive.

Key terms

Catalysts are substances which increase the rate of a chemical reaction by providing an alternative pathway with a lower activation energy.

Homogeneous catalysts are catalysts that are in the same physical state as the reactants in the reactions that they catalyse.

Heterogeneous catalysts are catalysts that are in a different physical state from the reactants in the reactions that they catalyse.

Top tip

You will need to recall the examples of catalysts listed on the last page (Fe, Ni, V_2O_5 and MnO_2). These are all heterogeneous so adsorption is a key stage in the alternative reaction route they provide.

Reactions of transition metal ions with hydroxide ions

Transition metal ions in aqueous solution are present as the hydrated complexes, $[M(H_2O)_6]^{n+}$. Due to the high positive charge density on the complex ion, such metal ions are often acidic, readily losing H^+ ions, e.g.:

$$[Cr(H_2O)_6]^{3+} \rightleftharpoons [Cr(H_2O)_5(OH)]^{2+} + H^+$$

Addition of alkali removes the H^+ as H_2O, and the reaction can progress further to the metal hydroxide, which is insoluble:

$$[Cr(H_2O)_6]^{3+} + OH^- \rightleftharpoons [Cr(H_2O)_5(OH)]^{2+} + H_2O$$

$$[Cr(H_2O)_5(OH)]^{2+} + OH^- \rightleftharpoons [Cr(H_2O)_4(OH)_2]^+ + H_2O$$

$$[Cr(H_2O)_4(OH)_2]^+ + OH^- \rightleftharpoons [Cr(H_2O)_3(OH)_3] + H_2O$$

This behaviour is typical of transition metal ions such as Cr^{3+}, Fe^{2+}, Fe^{3+} and Cu^{2+}. All these reactions are reversible so addition of acid can reverse them to regenerate the complex ions in solution.

For chromium(III), addition of excess alkali removes the H^+ from the remaining water molecules, forming an anionic hydroxide complex:

$$[Cr(H_2O)_3(OH)_3] + 3OH^- \rightleftharpoons [Cr(OH)_6]^{3-} + 3H_2O$$

We say that chromium(III) hydroxide is amphoteric: it can react as both an acid and a base. The reaction above shows the reaction of this hydroxide as an acid as it is donating H^+ to the hydroxide ions. It reacts as a base in the reverse of the earlier reactions:

$$[Cr(H_2O)_3(OH)_3] + H_2O \rightleftharpoons [Cr(H_2O)_4(OH)_2]^+ + OH^-$$

In this reaction, the hydroxide acts as a base by accepting H^+ from the water.

Knowledge check 6

Write an equation for the formation of a precipitate on addition of sodium hydroxide solution to $[Cu(H_2O)_6]^{2+}$(aq).

▲ Addition of sodium hydroxide solution to Cr^{3+}(aq) causes a grey-green precipitate to form.

The observations seen on addition of sodium hydroxide to solutions containing each complex are given below.

Transition metal ion	Addition of some OH⁻	Addition of excess OH⁻
$[Cr(H_2O)_6]^{3+}$	Grey–green precipitate of $[Cr(H_2O)_3(OH)_3]$	Precipitate dissolves giving a deep green solution of $[Cr(OH)_6]^{3-}$.
$[Fe(H_2O)_6]^{2+}$	Dark green precipitate of $[Fe(H_2O)_4(OH)_2]$	No further reaction for the bulk. (Some red–brown colour seen at the surface due to oxidation by the air.)
$[Fe(H_2O)_6]^{3+}$	Red–brown precipitate of $[Fe(H_2O)_3(OH)_3]$	No further reaction
$[Cu(H_2O)_6]^{2+}$	Pale blue precipitate of $[Cu(H_2O)_4(OH)_2]$	No further reaction

Top tip ≫

In an exam you can write the hydroxides without the water molecules, such as $Cr(OH)_3$ or $Fe(OH)_2$.

Test yourself

1. Write the electronic configurations of the following species: V, Cr, Co, V²⁺, Mn²⁺.   [2]

2. Explain why copper is considered to be a transition element but zinc is not.   [2]

3. Explain why d-block elements can form a range of oxidation states.   [2]

4. When aqueous ammonia is added to the pale blue solution (A) formed by dissolving copper(II) sulfate, a pale blue precipitate (B) forms initially. However, upon addition of more ammonia the solid dissolves, forming a royal blue solution (C). Identify the copper-containing species present in A, B and C.   [3]

5. Explain why an aqueous solution of $CuSO_4$ is pale blue but an aqueous solution of $ZnSO_4$ is not coloured.   [3]

6. Give an example of a transition metal used as a heterogeneous catalyst. Explain why transition metals can act as heterogeneous catalysts.   [2]

7. State what is meant by a homogeneous catalyst. Explain why transition metals can act as homogeneous catalysts.   [2]

8. (a) Using the convention of representing electrons by arrows in boxes, give the outer electronic configuration of an iron(III) ion, Fe^{3+}.   [1]

   (b) Hydrated crystals of iron(III) chloride contain the purple $[Fe(H_2O)_6]^{3+}$ ion, which has the same shape as the $[Cu(H_2O)_6]^{2+}$ ion. Sketch the shape of the $[Fe(H_2O)_6]^{3+}$ ion.   [1]

   (c) Explain why aqueous ions such as $[Fe(H_2O)_6]^{3+}$ and $[Cu(H_2O)_6]^{2+}$ are coloured.   [3]

   (d) Describe what is **seen** when an aqueous solution of sodium hydroxide is added, dropwise, to separate solutions containing Fe^{3+}(aq) and Cr^{3+}(aq) ions until an **excess** of sodium hydroxide is present. You should give an equation for any reaction(s) that occur.   [4]

Chemical kinetics

Chemists have many ways of working out whether or not chemical reactions are possible, but it is also important to know about the rate of a reaction. If a reaction is possible in terms of energy, it may be too slow to be useful, or need too high a temperature to be effective. This topic includes more information on methods to study the rate of a chemical reaction and how temperature affects the rate.

Studying the way reaction rates change as concentrations change can also give information about the mechanism of a reaction. This topic includes methods of producing a rate equation and how this can be used to distinguish between different proposed mechanisms.

You should be able to demonstrate and apply knowledge and understanding of:

- Principles underlying the measurement of reaction rate, including by sampling and quenching.

- The meaning of order of reaction and how this is found from experimental results.

- Rate equations and how these are found and used.

- Rate-determining steps for reactions and how this links the kinetics to the mechanism.

- The effect of temperature and catalysts on reaction rate, and the use of the Arrhenius equation to link these.

Topic contents

Maths skills ⟫

- Plot graphs and draw tangents to find rates.
- Analyse data sets to identify patterns and relationships between rates and concentrations.
- Use and rearrange rate equations to find rate constants and their units.
- Use logarithmic rules to rearrange the Arrhenius equations to find rate constants, activation energies and temperatures.

Rates of chemical reactions

Chemical kinetics is the study of rates of chemical reactions. In Unit 2 of the course the fundamental ideas of reaction rates were introduced, as well as some methods of studying the rates of chemical reactions. In this topic, alternative approaches to studying the rates of reactions are explored, as well as using the measured rates to give key information about the mechanism of chemical reactions.

Measuring rates of reaction

To measure the rate of a reaction, we need to work out how much of a reactant has been used up, or how much of a product has been produced, in a set period of time. This is conveniently done by looking at a property of the reaction that changes during the reaction, such as mass of reactants, volume or pressure of gas, colour or other electromagnetic absorption. A specific example is the iodine clock reaction, which shows a distinct colour change after a set amount of compound has reacted.

Sampling and quenching

Many of the methods discussed in AS Topic 2.2 allow data to be collected on the progress of a reaction throughout the entire chemical reaction. This is not always possible and sometimes a more labour-intensive approach is required. This is **sampling** and **quenching**, where a small amount of the reaction mixture is removed at regular time intervals (sampling) and immediately placed into iced water (quenching). This cools and dilutes the reaction mixture, which slows the reaction down and effectively stops it.

When sampling and quenching are used, the samples collected must be analysed by an appropriate method, with titration being a common approach. Each sample must be analysed individually to obtain information on the progress of the reaction.

Advantages and disadvantages:

- Sampling and quenching can be used for a large range of reactions.

- Sampling and quenching are labour and time intensive as each sample must be analysed individually, so the time intervals used between measurements tend to be longer than in colorimetric methods, which can be automated.

- Sampling is only appropriate when a reaction mixture is homogeneous, such as reactions that are all in solution. If a reaction mixture is not homogeneous then the sample taken may not be representative of the overall mixture.

If a reaction uses a heterogeneous catalyst then sampling can be undertaken without quenching. When a solid catalyst is used in a gas or liquid mixture, the catalyst speeds up the reaction significantly. Removing a sample of the gas or liquid takes it away from the catalyst so the reaction rate is immediately reduced.

If a reaction uses a homogeneous catalyst then taking a sample also takes a sample of the catalyst with the reactants, so the reaction will continue. Quenching is needed in this case, and it can be undertaken using cooling and dilution like any other reaction. It can also be quenched by destroying the catalyst, for instance an acid catalyst can be neutralised using alkali.

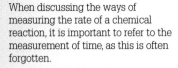

Link

Rates of reaction in AS Topic 2.2 on page 110 of the AS book.

Top tip

When discussing the ways of measuring the rate of a chemical reaction, it is important to refer to the measurement of time, as this is often forgotten.

Key terms

Sampling is taking samples from the reaction mixture at regular time intervals.

Quenching is the sudden stopping of a chemical reaction to allow for analysis to occur without the reaction proceeding further. It is usually undertaken by adding the sample to ice water as this both cools and dilutes the reactants.

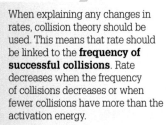

Top tip

When explaining any changes in rates, collision theory should be used. This means that rate should be linked to the **frequency of successful collisions**. Rate decreases when the frequency of collisions decreases or when fewer collisions have more than the activation energy.

Calculating rates of reaction

Once data is collected, a rate must be calculated. In most methods, the rate is calculated by using the equation:

$$\text{Rate} = \frac{\text{Change in concentration}}{\text{Time taken}}$$

The rate calculated in this manner is the average rate over the time period. The concentrations change as the reaction progresses so it is likely that the rate will also change, with the rate decreasing as the concentrations of reactants decrease.

To find the initial rate, the results should be plotted as a graph and a tangent drawn to the curve at time = 0. Once you've drawn the tangent you need to calculate its gradient or slope, as this is equal to the initial rate. To do this you need to draw a vertical line down from any point on the gradient to the time axis, forming a right-angled triangle.

Worked example

Calculate the initial rate of reaction from the graph below:

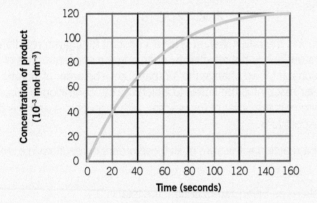

Answer

First draw a tangent at $t = 0$. This is shown as the orange line.

Next draw a line down from the tangent to meet the time axis. We choose to use a time of 40 seconds.

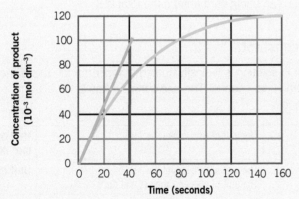

Length of horizontal (time) side of triangle = 40 s

Length of vertical (concentration) side of triangle = 92×10^{-3} mol dm^{-3}

Rate of reaction = 92×10^{-3} mol dm$^{-3} \div 40$ s = 2.3×10^{-3} mol dm^{-3} s^{-1}

Rates can be measured at different times by plotting a tangent at different places on the curve. The same method can be used when provided with a graph showing how the concentration of a reactant changes, although in this case the concentration will be decreasing.

Knowledge check 1

Explain why the rate of reaction changes during a chemical reaction.

Maths tip

To draw a tangent, a ruler is aligned with the curve at the point being studied, and a straight line is drawn which can be of any length.

Maths tip

You can draw a line down to the time axis at any point, but the maths is usually easier if you choose to do it along one of the gridlines so it touches the time axis at a round number of seconds, e.g. 20 seconds, 50 seconds or 100 seconds. It does not matter which time you choose, they should all give the same answer, but you may find it easier to read the scale for a larger triangle.

Find the concentration value that corresponds to the top point of the triangle, and the time from where the line you added meets the time axis. The gradient is the concentration divided by the time.

Top tip

When calculating rate it is important that the units of rate match the units of the measurements. For example, time can be given in alternative units such as minutes, which may give a rate unit of mol dm^{-3} min^{-1}.

Knowledge check 2

The concentration of a reactant decreases from 0.84 mol dm^{-3} to 0.50 mol dm^{-3} over the first two minutes of a reaction. Calculate the average rate over this time, giving its unit.

Key terms

Rate is the rate of change of the concentration, or of the amount, of a particular reactant or product.

Rate constant is a constant in the rate equation. It is not affected by changing the concentrations of the reactants at a particular temperature.

The **order of reaction** with respect to a particular reactant is the power to which the concentration is raised in the rate equation.

Top tip

If you are asked to give the meanings of the terms rate, rate constant or order of reaction in an exam, it is easier to write a general rate equation such as Rate = k[A]^m[B]^n and then refer to the relevant parts of this. Example: The order of the reaction is $m+n$.

Study point

An alternative way of identifying a first order reaction is to measure the half-life of the reaction. A first order reaction has a constant half-life, so if the concentration of a reactant drops from 2 mol dm^{-3} to 1 mol dm^{-3} in 85 seconds then it will drop from 1 mol dm^{-3} to 0.5 mol dm^{-3} in a further 85 seconds.

3 Knowledge check

What is the order of each of the following reactions?

(a) Rate = k [H$_2$]^1[I$_2$]^1

(b) Rate = k [CH$_3$I]^1[Br$^-$]^0

(c) Rate = k [CH$_3$COOH]^1

(d) Rate = k [CH$_3$CHO]^1

(e) Rate = k [C$_2$H$_4$]^1[Br$_2$]^2

What are the units of the rate constants in (a) to (d) above (you are not expected to calculate units for (e)).

Rate equations

The **rate** of a chemical reaction in solution depends on the concentration of the reactants. When the concentration of one reactant (shown as [A]) is doubled, scientists have found that the rate of the reaction may:

Stay the same	Rate is not proportional to concentration	rate ∝ [A]^0
Double	Rate is proportional to concentration	rate ∝ [A]^1
Increase by four times	Rate is proportional to concentration squared	rate ∝ [A]^2

This has led scientists to produce a rate equation that gives the rate of a chemical reaction at different concentrations of reactants.

For a general reaction:

$$A + B \longrightarrow products$$

The rate equation is:

$$Rate = k\,[A]^m[B]^n$$

In the rate equation, k is the **rate constant**. This is constant for a given reaction at a particular temperature and is not affected by changing the concentrations of the reactants. It is not constant if we change the temperature. The **order of reaction** with respect to a particular reactant is the power to which the concentration is raised in the rate equations, where m is the order of the reaction with respect to A and n is the order of the reaction with respect to B.

The overall order of a reaction is the sum of all these orders of reactions, i.e. $m+n$ for the reaction above.

- We describe reactions as being **zeroth order** if the total is 0.

- We describe reactions as being **first order** if the total is 1.

- We describe reactions as being **second order** if the total is 2.

- We describe reactions as being **third order** if the total is 3.

Unit of the rate constant

The rate of a reaction in solution is typically quoted in mol dm^{-3} s^{-1}: this is the change in concentration (mol dm^{-3}) per second. The rate constant has to have a unit for the units in the equation to balance.

Zeroth order reaction, e.g. Rate = k	The unit of rate is mol dm^{-3} s^{-1}.	**Unit of k is mol dm^{-3} s^{-1}**
First order reaction, e.g. Rate = k [A]	The unit of rate is mol dm^{-3} s^{-1}, The unit of concentration is mol dm^{-3}.	**Unit of k is s^{-1}**
Second order reaction, e.g. Rate = k [A]^2	The unit of rate is mol dm^{-3} s^{-1}, The unit of concentration^2 is (mol dm^{-3})^2.	**Unit of k is mol^{-1} dm^3 s^{-1}**

Obtaining rate equations

Rate equations can only be found experimentally, by studying the effects of changing the concentration of each individual reactant. **There is no way of obtaining the rate equation from the overall chemical equation for the reaction**.

How to derive a rate equation from experimental data

1. Focus on one substance at a time and look at the given information to find two experiments that differ only in the concentration of this reactant. By doing this we can identify the effect of this one substance alone.

2. If changing this reactant concentration does not affect the reaction rate, the order with respect to this reactant is zero.

3. If doubling this reactant concentration doubles the rate of reaction, the order with respect to this reactant is one.

4. If doubling this reactant concentration increases the reaction rate by a factor of four as this is two squared, the order with respect to this reactant is two.

5. Repeat this process for each reactant to find the order with respect to each one.

6. The order of the reaction is the sum of all of these orders.

Worked example 1

Derive a rate equation for the reaction of Br_2 with butadiene, C_4H_6, in solution.

Experiment number	Initial concentration of Br_2/ mol dm^{-3}	Initial concentration of butadiene/ mol dm^{-3}	Initial rate of formation of product/ mol dm^{-3} s^{-1}
1	6.0×10^{-3}	1.0×10^{-3}	3.0×10^{-3}
2	6.0×10^{-3}	2.0×10^{-3}	6.0×10^{-3}
3	6.0×10^{-3}	3.0×10^{-3}	9.0×10^{-3}
4	1.0×10^{-3}	6.0×10^{-3}	0.5×10^{-3}
5	2.0×10^{-3}	6.0×10^{-3}	2.0×10^{-3}
6	3.0×10^{-3}	6.0×10^{-3}	4.5×10^{-3}

To find the order with respect to Br_2, two sets of data where the concentration of Br_2 changes but the concentration of butadiene stays the same are needed. In this case, experiment numbers 4 and 5 are suitable. In these, the concentration of Br_2 doubles, and the rate quadruples. This shows that the reaction is second order with respect to Br_2.

To find the order with respect to C_4H_6, two sets of data where the concentration of C_4H_6 changes but the concentration of bromine stays the same are needed. In this case, experiment numbers 1 and 2 are suitable. In these, the concentration of butadiene doubles, and the rate doubles. This shows that the reaction is first order with respect to butadiene.

The rate equation is therefore: Rate = $k[Br_2]^2[C_4H_6]^1$

The overall order of reaction is: Third order

After working out the rate equation, work out a value for the rate constant, k. To do this, select any one set of data and put in the values to the rate equation. You are not expected to know the unit of the rate constant for a third order reaction, so don't worry if the unit in the calculation is unfamiliar.

$$k = \frac{Rate}{[Br_2]^2[C_4H_6]} = \frac{4.5 \times 10^{-3}}{(3.0 \times 10^{-3})^2 \times (6.0 \times 10^{-3})}$$

$$k = 8.3 \times 10^4 \ mol^{-2} \ dm^6 \ s^{-1}$$

Link

The effects of changing concentration on rate in AS Topic 2.2 on page 113 of the AS book.

Top tip

Data tables may sometimes include pH rather than the concentration of H^+ ions. It can be useful to convert these pH values to concentration of H^+, but you can sometimes just use the idea that one pH unit is a factor of 10 change in H^+ concentration.

Worked example 2

Derive a rate equation for the reaction of X with Y to produce XY_2.

$$X(aq) + 2Y(aq) \rightarrow XY_2(aq)$$

Experiment number	Initial concentration of X / mol dm^{-3}	Initial concentration of Y / mol dm^{-3}	Initial rate of formation of XY_2 / mol dm^{-3} s^{-1}
1	0.10	0.10	0.0001
2	0.10	0.20	0.0004
3	0.10	0.30	0.0009
4	0.20	0.10	0.0001
5	0.30	0.10	0.0001

(a) Find the order of the reaction with respect to X.

(b) Find the order of the reaction with respect to Y.

(c) Write an overall rate equation.

(d) Find the value and unit of the rate constant.

Answers

(a) The order is zero as changing the concentration between experiment numbers 1, 4 and 5 does not affect the rate.

(b) The order is 2 as doubling the concentration of Y from experiment 1 to experiment 2 causes the rate to increase by a factor of 4.

(c) Rate = $k[X]^0[Y]^2 = k[Y]^2$

(d) k = Rate $\div [Y]^2 = 0.0001 \div (0.1)^2 = 0.01$ mol^{-1} dm^3 s^{-1}

4 Knowledge check

Write a rate equation for the reaction:

$$H_2O_2 (aq) + 2 I^- (aq) + 2 H^+ (aq) \rightarrow I_2 (aq) + 2 H_2O (l)$$

Use the following information:

Concentration of H_2O_2 (aq) / mol dm^{-3}	Initial concentration of I$^-$ (aq) / mol dm^{-3}	Initial concentration of H$^+$ (aq) / mol dm^{-3}	Initial rate / 10^{-6} mol dm^{-3} s^{-1}
0.0010	0.10	0.10	2.8
0.0020	0.10	0.10	5.6
0.0020	0.10	0.20	5.6
0.0010	0.40	0.10	11.2

Determine the value of k in this reaction.

Mechanisms and rate-determining steps

A mechanism is a description of the series of steps that occur during a chemical reaction. Each step in a mechanism will occur at a different rate, with its own rate equation. The rate of the slowest step limits the rate of the overall reaction, and this step is called the **rate-determining step**. When studying the kinetics of a reaction, we are effectively studying the kinetics of the rate-determining step.

The rate-determining step is the slowest step in the mechanism.

What does the rate equation tell us about the rate-determining step?

Collision theory says that for a reaction to occur:

1. The reacting particles must collide.

2. The particles must have sufficient energy for reaction (the activation energy).

The rate equation tells us how many particles must collide in the rate-determining step:

- In a second order reaction, two particles must collide.

- In a third order reaction, three particles must collide.

- In a first order reaction, there is only one particle in the rate-determining step.

So if the rate equation is:	The rate-determining step has the following reactants:
Rate = $k\,[C_3H_7I][Br^-]$	$C_3H_7I + Br^- \longrightarrow$ products
Rate = $k\,[C_4H_9I]$	$C_4H_9I \longrightarrow$ products
Rate = $k\,[CH_3COOCH_3][H^+]^2$	$CH_3COOCH_3 + 2H^+ \longrightarrow$ products

It is not possible to identify conclusively the products of the rate-determining step in every case; however, it is possible to suggest products. These must balance like any other equation, such as:

$$C_3H_7I + Br^- \longrightarrow C_3H_7Br + I^-$$

All the steps in the mechanism, including the rate-determining step, must combine to form the overall equation.

If the rate-determining step has the following reactants:	The rate equation is:
$C_3H_7I + Br^- \longrightarrow$ products	Rate = $k\,[C_3H_7I][Br^-]$
$CH_3I \longrightarrow$ products	Rate = $k\,[CH_3I]$

We can use the information derived from the kinetics of a reaction to prove or disprove a reaction mechanism. From the kinetics, we can work out what the reactants are for the rate-determining step. If our proposed mechanism does not have a step with these reactants, then it cannot be the correct mechanism.

Key term

The **rate-determining step** is the slowest step in a reaction mechanism.

Knowledge check

Write equations for the rate-determining steps of the chemical reactions that have the following rate equations:

(a) Rate = $k\,[C_2H_4][Br_2]$

(b) Rate = $k\,[I_2]$

(c) Rate = $k\,[H_2O_2][I^-][H^+]$

 Key term

Activation energy is the minimum amount of energy required for a collision to be successful.

< Link >

The effects of temperature on rate are explained qualitatively in the AS course, Topic 2.2, on page 115 of the AS book. This uses the Boltzmann distribution as shown in the diagram below.

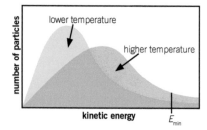

6 Knowledge check

Rearrange the Arrhenius equation to produce expressions beginning $A =$ and $T =$.

Top tip >>

Take care with the units you use in your calculations. The unit of activation energy is usually given in kJ mol⁻¹, while the value of R in the data booklet uses J mol⁻¹ K⁻¹. When calculating you must convert both to the same energy unit, either multiply the activation energy by 1000 to convert to J mol⁻¹ or divide the value of R by 1000 to give kJ mol⁻¹ K⁻¹.

The units of $e^{(-E_a/RT)}$ all cancel out so there are no units for this expression. This means that the unit of A must be the same as the unit of the rate constant.

Effect of temperature on rate

Increasing the temperature of a reaction causes the rate of a reaction to increase. This can be explained by collision theory, as particles will react when they collide with sufficient energy. Increasing the temperature means that more of the collisions will have sufficient energy to react, called the **activation energy**.

Arrhenius equation

In terms of the rate equation, temperature does not affect the concentrations of each substance, so it is the rate constant that is affected when we heat or cool a reaction mixture. We can quantify the effect of temperature on the rate constant using the Arrhenius equation:

$$k = Ae^{(-E_a/RT)}$$

k = Rate constant

A = Frequency factor, related to the frequency of collisions between particles. It can be treated as a constant over a limited range of temperatures, although it does vary if temperature changes significantly. In many cases the value of the frequency factor is calculated from a given set of data at one temperature prior to its use in calculating the rate constant at a different temperature.

e = Mathematical constant, found on all scientific calculators. It often appears as e^x above the ln button. This is because ln is the inverse of e^x.

E_a = Activation energy, in J mol⁻¹

R = Gas constant, given on the data sheet with value 8.31 J K⁻¹ mol⁻¹.

T = Temperature in Kelvin

Overall the expression $e^{(-E_a/RT)}$ is considered to show the fraction of collisions that possess an energy level above the activation energy. As the expression includes two constants (e, R) three out of the four remaining factors (k, A, E_a and T) must be known to calculate a value for the final factor. You may see the Arrhenius equation expressed in many different ways, and if you are not familiar with the rules for manipulating exponentials and logs you may not be able to see that they are equivalent. It is possible to learn two versions of the Arrhenius equation and then rearrange these to find any one of the variables. These two are:

$$k = Ae^{(-E_a/RT)} \quad \text{and} \quad E_a = -RT\ln(k/A)$$

Worked example

The reaction between iodine and hydrogen to produce HI has a rate constant of 1.37×10^{-4} mol⁻¹ dm³ s⁻¹ at a temperature of 575 K. The activation energy for this reaction is 157 kJ mol⁻¹. Calculate the value of the frequency factor giving its unit. Use this to find the rate constant at a temperature of 600 K.

Answer

STEP 1: Find the value of A by rearranging the expression.

$$k = Ae^{(-E_a/RT)} \quad so \quad A = k \div e^{(-E_a/RT)}$$

The energy terms are converted to J so $A = 1.37 \times 10^{-4} \div e^{(-157 \times 10^3/8.31 \times 575)}$

$$= 1.37 \times 10^{-4} \div e^{-32.86}$$
$$= 2.55 \times 10^{10} \text{ mol⁻¹ dm³ s⁻¹}$$

The unit of the frequency factor is the same as the unit of the rate constant.

STEP 2: Find the value of the rate constant at 600 K.

$$k = Ae^{(-E_a/RT)}$$

Fill in the values for A, E_a, R and T at 600 K.

$$k = 2.55 \times 10^{10} \times e^{(-157 \times 10^3/8.31 \times 600)}$$
$$k = 5.38 \times 10^{-4} \text{ mol⁻¹ dm³ s⁻¹}$$

Finding the activation energy graphically

The activation energy can be found by rearranging the Arrhenius equation if we have information about the frequency factor; however, it is more common to find information regarding the rate constant at different temperatures. These can be used to find both frequency factor and activation energy. The Arrhenius equation can be rearranged to give:

$$\ln k = (-E_a/RT) + \ln A$$

The structure of this version of the Arrhenius equation can be used to plot a straight line graph of $\ln k$ against $1/T$, and the intercept of this graph gives $\ln A$ with the gradient equal to $-E_a/R$.

- $\ln k$ is log to base e of the rate constant – the logarithm adjusts well for the range of values seen in the rate constant at different temperatures.

- $1/T$ uses the temperature in Kelvin in all cases.

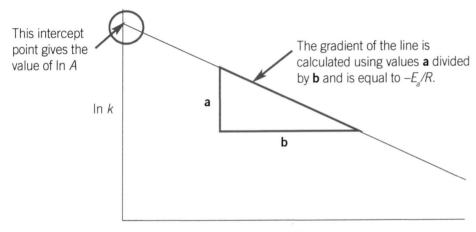

This intercept point gives the value of $\ln A$

The gradient of the line is calculated using values **a** divided by **b** and is equal to $-E_a/R$.

$\ln k$

a

b

$1/T$

The gradient is calculated at any point along the straight line. The value of the gradient is $-E_a/R$ so needs to be multiplied by -8.31 J K^{-1} mol^{-1} to find a value for the activation energy. E_a is calculated in J mol^{-1} in this case, and needs to be converted to kJ mol^{-1} for use by dividing by 1000.

The intercept on the vertical axis gives a value for $\ln A$. To find the value for A the calculation is $e^{\text{intercept value}}$. The unit of this is the same as the unit of the rate constant.

Effect on catalysts on rate

Catalysts increase the rate of chemical reactions by providing alternative routes with lower activation energies. This does not affect the concentrations in a rate equation, so it is the rate constant that is changed. Looking at the Arrhenius equation, we can see that reducing the activation energy will increase the value of $e^{(-E_a/RT)}$ and hence increase the rate constant. As the route of reaction changes it is likely that the value of the frequency factor will also change, but the combination of both changes is likely to lead to a significant increase in reaction rate.

Stretch & challenge

Chemists often use the concept that the rate of a chemical reaction doubles when the temperature increases by 10 K (which also equals 10°C). This is a simplification, as it depends on the activation energy, and the temperatures involved. For temperatures around room temperature, this rule of thumb is only true for activation energies of around 50 kJ mol^{-1}. If the activation energy is much greater than 50 kJ mol^{-1}, will the rate double, more than double or less than double if temperature increases by 10°C?

Knowledge check 7

The rate of a first order chemical reaction at 300 K is 0.0345 mol dm^{-3} s^{-1} when the initial concentration is 0.100 mol dm^{-3}. The activation energy of this reaction is 42 kJ mol^{-1}. Calculate the rate of this reaction at 320 K with an initial concentration of 0.150 mol dm^{-3}.

Test yourself

1. Suggest an appropriate method for measuring the rate of the reaction between aqueous sodium carbonate and dilute ethanoic acid. [1]

2. State what is meant by 'sampling and quenching'. [1]

3. Give the meaning of the term 'order of reaction'. [1]

4. Propanone, CH_3COCH_3, reacts with aqueous bromine, $Br_2(aq)$ to produce the colourless compound CH_3COCH_2Br.

$$CH_3COCH_3(aq) + Br_2(aq) \rightarrow HBr(aq) + CH_3COCH_2Br(aq)$$

(a) Suggest a method of studying the rate of this reaction. [1]

(b) The rate of reaction was studied using different concentrations of Br_2 and CH_3COCH_3 and the data found is shown in the table. All the experiments were performed in a solution of pH 1.

Experiment	Concentration of Br_2 / mol dm^{-3}	Concentration of CH_3COCH_3 / mol dm^{-3}	Rate / mol dm^{-3} s^{-1}
1	0.20	0.25	7.0×10^{-7}
2	0.40	0.25	7.0×10^{-7}
3	0.20	0.50	1.4×10^{-6}

Find the orders with respect to Br_2 and CH_3COCH_3. [2]

(c) Further studies show that the reaction is catalysed by H^+ ions, and the reaction is first order with respect to H^+. Find the rate of reaction you would expect for experiment 1 if it were repeated in a solution of pH 0. [2]

(d) Write the rate equation for the reaction. [1]

(e) Calculate the value of the rate constant, k, giving its unit. [2]

5. Dinitrogen pentoxide, N_2O_5, decomposes in the gas phase according to the equation shown.

$$2N_2O_5 \rightarrow 4NO_2 + O_2$$

The initial rates of this reaction for different concentrations of N_2O_5 were measured and are given in the table below.

Concentration of N_2O_5 / mol dm^{-3}	Initial rate / mol dm^{-3} s^{-1}
4.00×10^{-3}	3.00×10^{-5}
6.00×10^{-3}	4.50×10^{-5}
8.00×10^{-3}	6.00×10^{-5}

The rate equation for this reaction is: Rate = $k[N_2O_5]^1$

(a) Show that the rate equation is consistent with the data above. [1]

(b) Calculate the value of the rate constant under these conditions. Give your answer to an appropriate number of significant figures and state its unit. [2]

(c) Two possible mechanisms have been suggested for this reaction. These are shown below.

Mechanism A	Mechanism B
$N_2O_5 \rightarrow NO_2 + NO_3$	$2N_2O_5 \rightarrow 2NO_3 + N_2O_4$
$NO_3 \rightarrow NO + O_2$	$NO_3 + N_2O_4 \rightarrow NO + 2NO_2 + O_2$
$NO + N_2O_5 \rightarrow 3NO_2$	$NO + NO_3 \rightarrow 2NO_2$

Giving your reasons, state which of the mechanisms is compatible with the rate equation. [2]

6. A chemical reaction has a rate constant, k, of 3.0×10^7 mol^{-1} dm^3 s^{-1} at 350 K.

(a) Identify the order of this reaction, giving a reason for your answer. [1]

(b) The activation energy of this reaction is 23 kJ mol^{-1}.
    (i) State what is meant by the activation energy. [1]
    (ii) Find the frequency factor, A, for this reaction, giving its units. [2]
    (iii) Calculate the value of the rate constant at 400 K. [2]

(c) (i) A catalyst for the reaction halves the activation energy to 11.5 kJ mol^{-1}. Calculate the rate constant at a temperature of 350 K. [2]
    (ii) Explain why the use of catalysts is often preferred to the use of very high temperatures for reactions such as these. [2]

7. The data below shows the results of an iodine clock reaction. Calculate the rate for each experiment and hence find the rate equation of the reaction. [4]

Experiment	$[H_2O_2]$ / mol dm^{-3}	$[H^+]$ / mol dm^{-3}	$[I^-]$ / mol dm^{-3}	Time / s
1	0.30	0.10	0.10	46
2	0.60	0.10	0.10	23
3	0.60	0.20	0.10	23
4	0.30	0.20	0.20	23

Enthalpy changes for solids and solutions

When substances are formed or changed, either physically or chemically, there are energy changes. In chemistry, we are often concerned with the changes between chemical energy and heat energy, and the classification of changes as exothermic and endothermic is one that is fundamental to how chemists expect reactions to behave.

Chemists have many ways of describing specific energy changes, and the need for precise naming of these is as important as the precise naming of chemical substances. Changes such as those associated with standard enthalpies of formation and combustion should already be familiar but in this section a wider range of changes are discussed. These allow predictions about the physical and chemical properties of the substances involved, such as the solubility of ionic compounds.

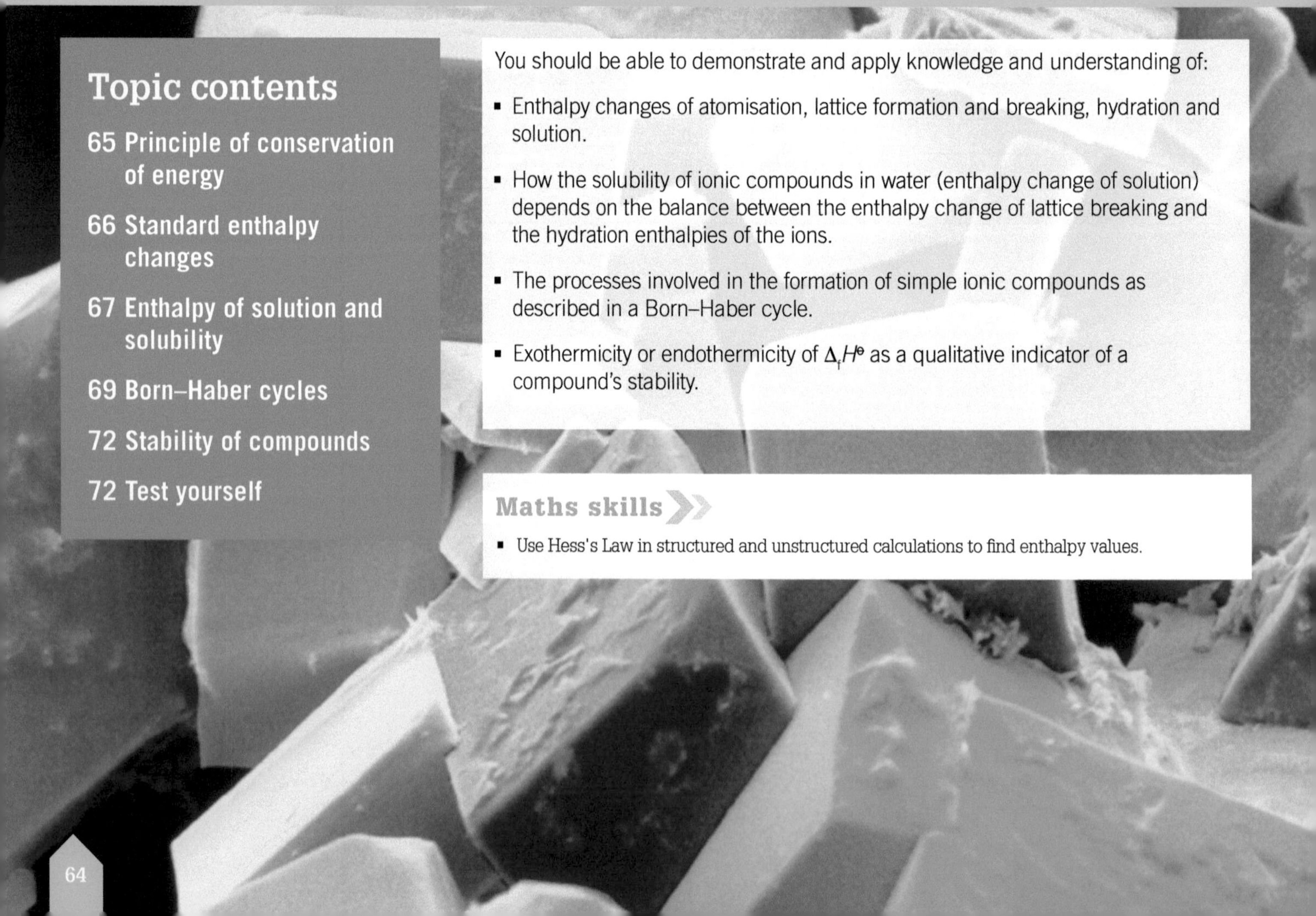

Topic contents

You should be able to demonstrate and apply knowledge and understanding of:

- Enthalpy changes of atomisation, lattice formation and breaking, hydration and solution.

- How the solubility of ionic compounds in water (enthalpy change of solution) depends on the balance between the enthalpy change of lattice breaking and the hydration enthalpies of the ions.

- The processes involved in the formation of simple ionic compounds as described in a Born–Haber cycle.

- Exothermicity or endothermicity of $\Delta_f H^\ominus$ as a qualitative indicator of a compound's stability.

Maths skills ≫

- Use Hess's Law in structured and unstructured calculations to find enthalpy values.

Enthalpy changes are a way of measuring the changes in energy during any chemical or physical change. At A-level, enthalpy and energy can be treated as being the same. Knowledge of the energy needed or released during a chemical reaction is key when planning a chemical process in a laboratory, and even more important when undertaking chemical reactions on an industrial scale.

Principle of conservation of energy

All the ideas regarding energy changes in chemistry and all other sciences are based on the principle of conservation of energy:

Energy cannot be created or destroyed, but may only be converted from one form to another.

In terms of chemical changes, this has two main consequences:

- To measure the energy change of a chemical reaction we can measure the energy given off in the reaction. The energy of the chemicals will decrease by the same amount of energy as is given out as heat or other energy types.

- Hess's law: The energy change of any chemical reaction is the same regardless of the route taken.

Energy cycles

It is often difficult to measure enthalpy values directly, but it is possible to use the principle of conservation of energy to calculate these enthalpies from values we can measure more easily. This method is based on *Hess's law*, which can be expressed as:

'If a reaction can take place by more than one route, then the total energy change for each route will be the same.'

This allows us to construct energy cycles like this one:

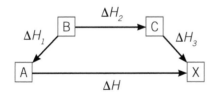

- If we have a chemical reaction that changes A directly to X, as shown above, the energy change is ΔH.

- If we change A to B to C to X, the total energy change is the same, i.e.
$\Delta H = -\Delta H_1 + \Delta H_2 + \Delta H_3$.

- If we know three out of the four terms in the energy cycle above, we can use this information to calculate the fourth energy change using the equation.

It is possible to undertake calculations involving enthalpies of formation without drawing an energy cycle. If you are given enthalpies of formation **only**, and need to work out the enthalpy change of a reaction, there is a more straightforward method of working out the enthalpy change. As a consequence of the energy cycle, the enthalpy change of any reaction is:

Enthalpy change = $\Delta_f H^\ominus$ (for all products) − $\Delta_f H^\ominus$ (for all reactants)

Link

Thermochemistry in AS Topic 2.1 on pages 96–107 of the AS book.

≪ **Maths tip**

Many enthalpy values will be negative so it will be common to use the rule 'minus minus is a plus'. To avoid mixing up signs it can be useful to put all values in brackets during your calculations

e.g. $\Delta H = -(-34) + (-56) - (+45) - (-20)$
$= -47$

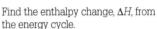

Knowledge check 1 ◀

Find the enthalpy change, ΔH, from the energy cycle.

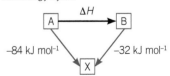

REMEMBER:

- The enthalpy change is calculated using Products – Reactants.

- If you have balancing numbers in the equation, you must include these. For example, if a reaction forms 2NaCl you need to include 2× the enthalpy of formation of NaCl.

Standard enthalpy changes

The **enthalpy change of reaction**, with the symbol ΔH, is a measure of the energy change of a reaction. It is more common to see standard enthalpy changes of reaction, $\Delta H^{\ominus}$, which represent the energy change under standard conditions.

Standard conditions are: A temperature of 298 K (25°C)

A concentration of 1 mol dm⁻³ for solutions.

A pressure of 101 kPa, or one atmosphere (1 atm), for gases.

Standard state is the physical state of a substance under standard conditions, such as oxygen gas, liquid water or sodium chloride solid.

The **standard enthalpy change of reaction, $\Delta H^{\ominus}$**, is the enthalpy change that occurs in a reaction between molar quantities of reactants in their standard states under standard conditions.

The enthalpy changes of some specific reactions are given special names, and the standard enthalpies of formation and combustion were covered in the AS work. In many of the calculations undertaken for ionic compounds, the energy cycles are constructed with the alternative route going via the gaseous state. This means that energy terms associated with converting substances to and from gaseous atoms or ions are commonly used.

Standard enthalpy change of atomisation, $\Delta_{at}H^{\ominus}$

This is the enthalpy change when one mole of atoms of an element in the gas phase are formed from the element in its standard state under standard conditions.

e.g. $Na(s) \longrightarrow Na(g)$ or $\frac{1}{2} Cl_2(g) \longrightarrow Cl(g)$ or $\frac{1}{4} P_4(s) \longrightarrow P(g)$

Standard enthalpy change of lattice formation, $\Delta_{latt}H^{\ominus}$

This is the enthalpy change when one mole of an ionic compound is formed from ions of the elements in the gas phase.

e.g. $Na^+(g) + Cl^-(g) \longrightarrow NaCl(s)$ or $Ca^{2+}(g) + 2Cl^-(g) \longrightarrow CaCl_2(s)$

You may also see the enthalpy of lattice breaking, which is the reverse of this process – the energy change when one mole of an ionic compound is broken up into ions of the elements in the gas phase.

e.g. $NaCl(s) \longrightarrow Na^+(g) + Cl^-(g)$ or $CaCl_2(s) \longrightarrow Ca^{2+}(g) + 2Cl^-(g)$

Standard enthalpy change of hydration, $\Delta_{hyd}H^{\ominus}$

This is the enthalpy change when one mole of an ionic compound in solution is formed from ions of the elements in the gas phase.

e.g. $Na^+(g) + Cl^-(g) + aq \longrightarrow NaCl(aq)$ or $Ca^{2+}(g) + 2Cl^-(g) + aq \longrightarrow CaCl_2(aq)$

First electron affinity

This is the enthalpy change when one mole of gaseous negative ions are formed from gaseous atoms of a substance by gaining an electron.

$$Cl(g) + e^- \longrightarrow Cl^-(g) \quad \text{or} \quad O(g) + e^- \longrightarrow O^-(g)$$

First ionisation energy

This is the enthalpy change when one mole of gaseous positive ions are formed from gaseous atoms of an element by losing an electron.

$$Na(g) \longrightarrow Na^+(g) + e^- \quad \text{or} \quad Cu(g) \longrightarrow Cu^+(g) + e^-$$

Enthalpy of solution and solubility

Ionic substances consist of a lattice of positive and negative ions held together by electrostatic forces. To dissolve, the forces between the ions must be broken, and this can only happen if they are replaced by other forces. Water molecules are polar – the oxygen atoms are $\delta-$ and the hydrogen atoms are $\delta+$. The oxygen atoms surround the positive ions of the substance and the hydrogen atoms surround the negative ions – we say the ions become hydrated. In the case of sodium chloride, the ions are Na^+ and Cl^- as shown below.

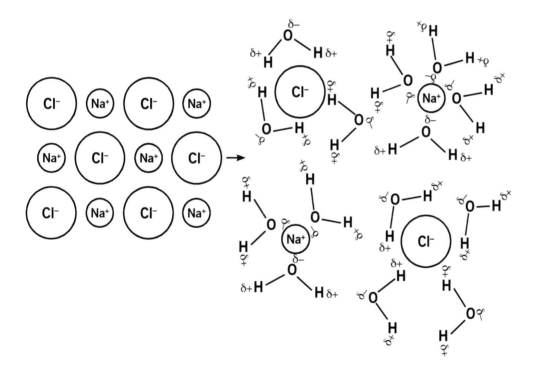

The enthalpy change during this process is called the enthalpy change of solution.

Knowledge check 3

The enthalpy change when $Ca^{2+}(g)$ is formed from $Ca(g)$ is 1735 kJ mol^{-1}. Suggest a value for the first ionisation energy of Ca.

Knowledge check 4

The standard enthalpy change of hydration of calcium ions is −1650 kJ mol^{-1} and chloride ions is −364 kJ mol^{-1}. If the enthalpy change of lattice breaking for calcium chloride, $CaCl_2$, is 2237 kJ mol^{-1}, find the standard enthalpy change of solution and explain whether you expect calcium chloride to be soluble.

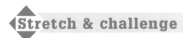
Top tip

It is not easy to predict the solubility of compounds, so you should recall the common patterns seen throughout the course. These include the fact that the following compounds are soluble:

all group 1 compounds,
all ammonium compounds,
all metal nitrates.

The following are insoluble, except when the rules above say that they would be soluble:

all metal carbonates,
all metal hydroxides,
all lead compounds.

Standard enthalpy change of solution, $\Delta_{sol}H^{\ominus}$

This is the enthalpy change that occurs when one mole of a substance dissolves completely in a solvent under standard conditions to form a solution.

The enthalpy of solution is the sum of the enthalpy of lattice breaking and enthalpy of hydration:

$$M^+X^-(s) \longrightarrow M^+(g) + X^-(g) \longrightarrow M^+(aq) + X^-(aq)$$

- The enthalpy of lattice breaking is endothermic.

- The enthalpy of hydration is exothermic.

If the enthalpy of hydration is greater than the enthalpy of lattice breaking, the salt dissolves; however, if the enthalpy of hydration is smaller than the enthalpy of lattice breaking, the salt will not usually dissolve. The more exothermic the total of these values, the more soluble a salt is likely to be. The same factors affect both the enthalpy of lattice breaking and enthalpy of solution:

- Both are increased by increasing charge on the ions.

- Both are increased by decreasing size of the ions.

This makes it very difficult to predict solubility from first principles, and so patterns in solubility are useful to predict whether particular salts are soluble or not.

Worked example

Use the data below to explain why silver chloride is insoluble:

Standard enthalpy of hydration of silver ions, Ag^+	−464 kJ mol⁻¹
Standard enthalpy of hydration of chloride ions, Cl^-	−364 kJ mol⁻¹
Standard enthalpy of lattice breaking for AgCl	905 kJ mol⁻¹

Answer

The standard enthalpy of solution is given by:

Standard enthalpy of solution = Standard enthalpy of hydration + Enthalpy of lattice breaking

$$= -464 - 364 + 905 = 77 \text{ kJ mol}^{-1}$$

The process of dissolving is highly endothermic and so the silver chloride will not dissolve.

Born–Haber cycles

Born–Haber cycles are energy cycles similar to those seen for other processes; however, there are many more steps. The energy cycle involves turning all the initial elements into gas phase atoms, then into ions and then combining these back to form a solid (using a lattice enthalpy term) or a solution (using a hydration enthalpy term). The energy changes associated with these processes can be broken down into a series of small steps. These steps are shown on the Born–Haber cycle (energy cycle) below.

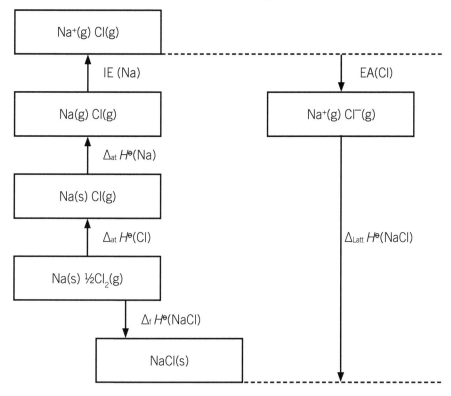

▲ Max Born and Fritz Haber, the scientists that give the Born–Haber cycle its name

The ease of formation of an ionic compound depends on the energy changes associated with each step of the process. These steps are:

Formation of gas phase atoms: the enthalpy of atomisation

This is the energy change when one mole of atoms of an element in the gas phase are formed from the element in its standard state.

$$Na(s) \longrightarrow Na(g) \quad \text{and} \quad \tfrac{1}{2} Cl_2(g) \longrightarrow Cl(g)$$

Formation of a cation: ionisation energy

Formation of a metal cation requires the atom in the gas phase to lose one or more electrons, and this requires energy to be put into the atom.

$$\text{Ionisation energy:} \qquad Na(g) \longrightarrow Na^+(g) + e^-$$

This process is always endothermic as it requires energy to remove an electron from an atom.

>> **Study point**

The enthalpy of atomisation of a diatomic gas is half the bond enthalpy.

Top tip »

When ions are formed with multiple charges (2+, 3+, 2– or 3–) then there will be several ionisation energies or several electron affinities in the calculation as each one represents the loss or gain of one electron.

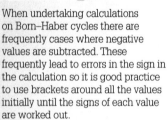

5 Knowledge check

Write chemical equations that correspond to the following standard enthalpy changes:

(a) First ionisation energy of copper.

(b) The bond energy of Cl_2.

(c) The atomisation of O_2.

(d) The lattice formation of Na_2O.

(e) The electron affinity of fluorine.

Top tip »

When undertaking calculations on Born–Haber cycles there are frequently cases where negative values are subtracted. These frequently lead to errors in the sign in the calculation so it is good practice to use brackets around all the values initially until the signs of each value are worked out.

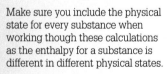

Top tip »

Make sure you include the physical state for every substance when working though these calculations as the enthalpy for a substance is different in different physical states.

Formation of anions

When a non-metal atom gains an electron to form an anion with a charge of –1, the energy change is called the first electron affinity.

Electron affinity: $Cl(g) + e^- \longrightarrow Cl^-(g)$

This process is sometimes exothermic, but second and third electron affinities (to form ions with –2 and –3 charge) are usually endothermic.

Enthalpy of lattice formation

Lattice formation enthalpies represent the energy released when the positive and negative ions in an ionic compound come together to form a solid. This process always releases energy, and it is this lattice energy that is the energy that drives the formation of ionic compounds. Since the formation of cations is always endothermic, and the formation of anions is often endothermic, there would not be any ionic compounds formed without any other energy to balance this out. This energy change is caused by the oppositely charged ions coming together to form the crystal lattice:

$$Na^+(g) + Cl^-(g) \longrightarrow NaCl(s)$$

The enthalpy of lattice formation is always exothermic, and the more exothermic it is, the more stable the ionic compound.

Calculations using given Born–Haber cycles

The Born–Haber cycle shows that the enthalpy of formation is the sum of all other energy terms. For sodium chloride this is:

$$\Delta_f H^\ominus(NaCl) = \Delta_{at} H^\ominus(Na) + \Delta_{at} H^\ominus(Cl) + IE(Na) + EA(Cl) + \Delta_{Latt} H^\ominus(NaCl)$$

The term that is usually 'unknown' is the enthalpy of lattice formation. This is the change from $Na^+(g)$ and $Cl^-(g)$ to form $NaCl(s)$, the longest arrow in the diagram on the previous page. We can calculate this value by using an alternative route around the cycle. The calculation is:

$$\Delta_{Latt} H^\ominus = -EA(Cl) - IE(Na) - \Delta_{at} H^\ominus(Cl) - \Delta_{at} H^\ominus(Na) + \Delta_f H^\ominus(NaCl)$$

In writing this equation, any time the path goes against the direction of the arrow, the value is subtracted, and when it goes with the arrow the value is added.

Producing and using simple Born–Haber cycles

If a Born–Haber cycle is not provided then one can be constructed using equations provided in a question. In this case, look for equations that contain any of the substances listed in the chemical equation and link these together to form an overall cycle.

Worked example

Calculate the enthalpy of lattice formation for calcium hydride, CaH_2, using the data provided.

$$Ca(s) + H_2(g) \longrightarrow CaH_2(s) \quad \Delta_f H^\ominus = -189 \text{ kJ mol}^{-1}$$

			Enthalpy change / kJ mol⁻¹
$Ca(s)$	$\rightarrow$	$Ca(g)$	193
$Ca(g)$	$\rightarrow$	$Ca^+(g) + e^-$	590
$Ca^+(g)$	$\rightarrow$	$Ca^{2+}(g) + e^-$	1150
$H_2(g)$	$\rightarrow$	$2H(g)$	436
$H(g) + e^-$	$\rightarrow$	$H^-(g)$	–72

Answer

In this case, start with the equation provided, and add the unknown value:

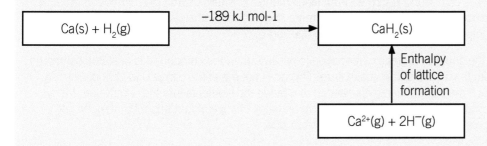

Next, identify any changes that you can link to this skeletal Born–Haber cycle. Looking at the reactants Ca(s) and H_2(g) it is possible to identify atomisation reactions for both substances from the table of data. These can be added to the cycle.

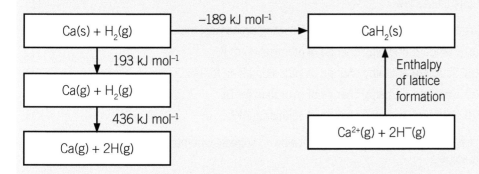

Next, we can see that the table of data includes steps to form the ions Ca^{2+} and H^- from these atoms; however, we need to make sure that the values for H are doubled as there are two of them. This allows us to complete the cycle.

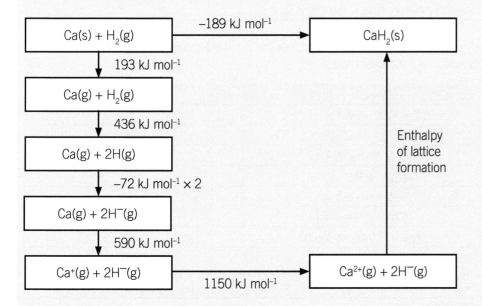

Now the calculation can be completed:

Enthalpy of lattice formation = – (1150) –(590) – (2 × –72) – (436) – (193) + (–189)

= –2414 kJ mol^{-1}

Top tip

Make sure that any diagrams you draw are clear, with the directions of arrows agreeing with the directions of the reactions given. You will need to refer to the values and the directions of the arrows when calculating the final value.

Top tip

Make sure you always state the physical state for all substances in your Born–Haber cycle.

Knowledge check 6

Use the values below to find the enthalpy of formation of copper(II) fluoride, CuF_2.

Process	Enthalpy change / kJ mol^{-1}
$\Delta_{at}H^{\ominus}$(Cu)	339
$\Delta_{at}H^{\ominus}$(F)	79
1st IE (Cu)	745
2nd IE (Cu)	1960
EA(F)	–348
$\Delta_{Latt}H^{\ominus}$(CuF$_2$)	–3037

Top tip

When transferring your information to the calculation, the most common errors are forgetting the multiplication numbers when two atoms are involved in any of the changes, and incorrectly transferring the direction of any arrows into + and – signs. Remember to include brackets as this may help you avoid errors with the signs of numbers.

Top tip ≫

When discussing the stability of compounds in terms of enthalpy of formation, it is important to emphasise that this is the stability **relative to the elements.**

Stability of compounds

If the enthalpy change of formation of a compound is negative, then energy is given out as the compound is formed from its elements. This tells us that the compound is stable compared with the elements. The more negative the enthalpy change of formation of a compound, the more stable the compound is.

If the enthalpy change of formation is positive, then the compound is unstable compared with the elements that make it up. This does not mean the compound cannot exist, but it means that energy is needed to change the elements into the compound. Many compounds that exist have a positive enthalpy change of formation, but they do not decompose because the process is too slow.

Test yourself

1. State Hess's law. [1]

2. Write chemical equations for the following processes. Include state symbols.
   (a) Standard enthalpy change of atomisation of K. [1]
   (b) Standard enthalpy change of lattice breaking for Na_2O. [1]
   (c) Standard enthalpy change of hydration for Br^-. [1]
   (d) Standard enthalpy change of solution for KI. [1]

3. Use the standard enthalpy terms below to suggest whether barium chloride is soluble. [2]

	Enthalpy change / kJ mol^{-1}
Standard enthalpy change of lattice breaking of $BaCl_2$	2018
Standard enthalpy change of hydration of Ba^{2+}	1360
Standard enthalpy change of hydration of Cl^-	335

4. Use the values in the table to calculate the standard enthalpy change of lattice formation of magnesium oxide. [3]

	Enthalpy change / kJ mol^{-1}
$Mg(s) \longrightarrow Mg(g)$	149
$O_2(g) \longrightarrow 2O(g)$	496
$Mg(g) \longrightarrow Mg^+(g) + e^-$	736
$Mg^+(g) \longrightarrow Mg^{2+}(g) + e^-$	1450
$O(g) + e^- \longrightarrow O^-(g)$	−141
$O^-(g) + e^- \longrightarrow O^{2-}(g)$	791
$Mg(s) + \frac{1}{2}O_2(g) \longrightarrow MgO(s)$	−602

5. The standard enthalpy change of hydration of Na^+ is −406 kJ mol^{-1}. Use this value and the Born–Haber cycle on the left to find the standard enthalpy change of hydration for I^-. [3]

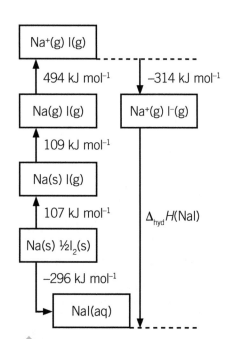

Entropy and feasibility of reactions

Chemists need to be able to predict whether a reaction is feasible without having to try every one. They use a range of theories and ideas to predict feasibility and some of the ones seen so far, such as enthalpy change of reaction, only give a guide to whether a reaction is feasible.

Entropy is one of the fundamental concepts of science. It is described in many different ways; however, the second law of thermodynamics states that the overall entropy in any system will increase during any spontaneous change. This makes an understanding of entropy essential in predicting whether reactions are feasible.

You should be able to demonstrate and apply knowledge and understanding of:

- Entropy, S, of a system as a measure of the freedom possessed by particles within it and the increase in entropy towards a maximum for all natural changes in a sealed system.

- Particles in a solid having much less freedom than those in a gas and that entropy increases in the sequence S (solid) $< S$ (liquid) $< S$ (gas).

- How to calculate an entropy change from absolute entropy values, $\Delta S = S_{final} - S_{initial}$.

- The concept of Gibbs free energy change and how it is calculated using the relationship, $\Delta G = \Delta H - T\Delta S$.

- Spontaneous reactions having a negative value for ΔG and how the effect of entropy change explains the spontaneous occurrence of endothermic processes.

Maths skills ≫

- Interconverting a range of units in Gibbs free energy calculations.
- Using and rearranging the Gibbs free energy equation to find minimum temperatures.

Entropy

Entropy is often described as the disorder in a system. The first beaker below has low entropy and the second has a much higher entropy.

However, this definition of entropy is not the main one used in chemistry. Chemists view entropy as the degree of freedom of a system – particles that can move freely in any direction have a much higher entropy than particles that are constrained.

- If a system has atoms in fixed positions, without any freedom to move, then the entropy is low. This is typical of a solid.

- If a system has atoms that are free to move in any direction and must stay close together, then the entropy is greater. This is typical of a liquid.

- If a system has atoms that are free to move in any direction to move to any position, then the entropy is high. This is typical of a gas.

The entropy is represented by the symbol S, and has units of Joules per Kelvin ($J\ K^{-1}$), with molar entropies given as $J\ K^{-1}\ mol^{-1}$. For any substance in its different physical states the entropy of the gas is greatest and the solid is least:

$$S \text{ (solid)} < S \text{ (liquid)} < S \text{ (gas)}$$

Study point

Mixtures have a greater entropy than pure substances, but the physical states have the greatest effect. A liquid mixture has a greater entropy than a pure liquid, but this will still be much lower than a gas.

Knowledge check

Place the following in order of increasing entropy:

$Cl_2(g)$, $Br_2(l)$, $I_2(s)$, $Br_2(aq)$

Entropy changes

In a chemical reaction, the entropy change can be calculated from the standard entropies of the substances involved. The standard entropy of a substance is the entropy of one mole in a given physical state under standard conditions. There are tables of data that list these for a range of substances. The calculation is:

$$\Delta S = S \text{ (all products)} - S \text{ (all reactants)}$$

Worked example

Calculate the entropy change for the combustion of methane.

$$CH_4(g) + 2O_2(g) \rightarrow CO_2(g) + 2H_2O(l)$$

Substance	Physical state	Standard entropy / $J K^{-1} mol^{-1}$
Methane, CH_4	Gas	186
Oxygen, O_2	Gas	205
Carbon dioxide, CO_2	Gas	214
Water, H_2O	Liquid	70
Water, H_2O	Gas	189

Answer

The entropy change is given by:

$$\Delta S = S(CO_2) + 2 \times S(H_2O) - S(CH_4) - 2 \times S(O_2)$$
$$= 214 + 140 - 186 - 410$$
$$= -242 \text{ J K}^{-1} \text{ mol}^{-1}$$

Note that the value for water used is the liquid phase as this matches the chemical equation.

Knowledge check 2

State, giving a reason, the effect of increasing the temperature above 100°C on the entropy change for combustion of methane.

Knowledge check 3

Calculate the entropy change for the combustion of hexane, assuming that the water is produced in the liquid phase. The standard entropy of hexane, $C_6H_{14}(l)$, is 204 J K^{-1} mol^{-1}.

Second law of thermodynamics

The second law of thermodynamics states:

Entropy will always tend to increase in any isolated system that is not in equilibrium.

In simple terms, this means that if you take any ordered system that is not in equilibrium, it will tend to increase its degree of freedom unless work is done to it. It is important that we are discussing overall entropy in this case – parts of a system may have entropy decreasing but this is because entropy has increased elsewhere.

In the combustion of methane, the entropy change calculated for the substances involved is negative – the entropy of the substances decreases. This appears to go against the second law of thermodynamics; however, the combustion releases a significant amount of heat energy to the surroundings. This increases the entropy of the surroundings, which more than compensates for the decrease in the entropy of the substances in the reaction.

Gibbs free energy, ΔG

In the previous topic, it is implied that chemical reactions tend to occur when the energy change is negative, i.e. they are exothermic. It is true that most chemical reactions are exothermic, but some endothermic reactions also occur. To explain this, both enthalpy and entropy need to be considered together, and these may be combined into the expression for Gibbs free energy, ΔG. The relationship is:

$$\Delta G = \Delta H - T\Delta S$$

- ΔG is the change in the Gibbs free energy in kJ mol⁻¹.

- ΔH is the enthalpy change in kJ mol⁻¹.

- ΔS is the entropy change in kJ mol⁻¹ K⁻¹. It is often calculated in J mol⁻¹ K⁻¹ so you would need to divide this by 1000 to convert J into kJ.

- T is the temperature in Kelvin (K).

If the free energy change is negative then a reaction will occur spontaneously, but if the free energy change is positive, the reaction will not occur spontaneously. If a reaction has a positive free energy change, we may be able to cause the reaction to occur by changing the temperature:

- If ΔS is positive (an increase in entropy) such as a gas being produced, then increasing the temperature will cause $-T\Delta S$ to become more negative until the overall change is negative.

- If ΔS is negative (a decrease in entropy) such as a precipitation, then decreasing the temperature will cause $-T\Delta S$ to become smaller and less positive until the overall change is negative.

Because the reaction will occur spontaneously if the free energy is negative, this means that a change that has a great increase in entropy will occur even though it is endothermic, as the entropy will be enough to make the Gibbs free energy change negative. Endothermic changes can be physical changes, such as boiling or dissolving, or chemical changes such as the thermal decomposition of some salts where gases are produced.

4 Knowledge check

(a) Calculate the Gibbs free energy change for a reaction where the enthalpy change is –4 kJ mol⁻¹ and the entropy change is 12 J K⁻¹ mol⁻¹ at 50°C.

(b) A reaction has an enthalpy change of 24 kJ mol⁻¹ and an entropy change of 42 J K⁻¹ mol⁻¹. Is the reaction feasible at 298 K?

◀Stretch & challenge

In some cases the Gibbs free energy of a reaction will never become negative (it will always be positive). This occurs when the reaction is endothermic and the entropy change is negative.

In other examples, where both enthalpy change and entropy change are negative, then the temperature found when $\Delta G = 0$ is actually the highest temperature at which the reaction is feasible and above this temperature the reaction is no longer feasible.

Worked example

The enthalpy change of combustion of methane is –890 kJ mol⁻¹ with an entropy change of –242 J K⁻¹ mol⁻¹. Show that the reaction is feasible at 298 K.

Answer

$$\Delta G = \Delta H - T\Delta S$$

$$= -890 - (298 \times (-242 \div 1000)) \quad \textit{The ÷ 1000 converts J into kJ}$$

$$= -818 \text{ kJ mol}^{-1}$$

As the value is negative the reaction is feasible at 298 K.

Some reactions are feasible at some temperatures but not at others. The temperature when the reaction becomes feasible is when the value of ΔG changes from positive to negative. At this point $\Delta G = 0$. We can use this to find the minimum temperature at which a reaction becomes feasible.

$$\Delta G = \Delta H - T\Delta S = 0 \qquad \text{so} \qquad \Delta H = T\Delta S$$

This rearranges to $\qquad\qquad\qquad T = \Delta H \div \Delta S$

Worked example

Find the minimum temperature at which a reaction will occur when ΔH is 46.0 kJ mol^{-1} and ΔS is 140 J K^{-1} mol^{-1}.

Answer

First we need to convert the entropy to kJ K^{-1} mol^{-1} so entropy change is $140/1000 = 0.140$ kJ K^{-1} mol^{-1}.

$$T = 46 \div 0.140 = 329 \text{ K}$$

The reaction is feasible at temperatures above 329 K. We can confirm this by calculating the value of the Gibbs free energy change above and below this temperature. This will allow us to confirm that the reaction becomes feasible at this temperature.

Knowledge check 5

When a drying agent absorbs water, the enthalpy change is −37 kJ mol^{-1} and the entropy change is −87 J K^{-1} mol^{-1}. Heating reverses these processes and regenerates the drying agent so it can be used again. Calculate the minimum temperature needed to regenerate the drying agent.

Test yourself

1. Place the following substances in order of increasing entropy under standard conditions: [1]

   Air, Magnesium, Mercury, Nitrogen

2. The standard entropy of liquid water is −286 J K^{-1} mol^{-1} and its entropy of vaporisation is +44 J K^{-1} mol^{-1}. Calculate the standard entropy of water vapour. [1]

3. Write the expression for Gibbs free energy. [1]

4. The general equation for the thermal decomposition of Group 2 carbonates is:

   $$MCO_3 \longrightarrow MO + CO_2$$

   Use the data in the table to find the minimum decomposition temperature for each carbonate and hence the pattern in the thermal stabilities of the Group 2 carbonates. [3]

	Standard enthalpy change of decomposition / kJ mol^{-1}	Standard entropy change of decomposition / J K^{-1} mol^{-1}
MgCO$_3$	117	175
CaCO$_3$	178	161
SrCO$_3$	235	171
BaCO$_3$	267	172

5. The standard enthalpies of formation and standard entropies of some substances are given in the table:

Substance	Standard enthalpy of formation / kJ mol^{-1}	Standard entropy / J K^{-1} mol^{-1}
Fe	0	27
FeO	−266	54
Mg	0	33
MgO	−602	27
CO	−111	198
CO$_2$	−394	214

(a) Give a reason why the entropy values for CO and CO$_2$ are much higher than the other substances listed. [1]

(b) Iron(II) oxide, FeO, can be reduced by carbon monoxide in the blast furnace:

$$FeO + CO \longrightarrow Fe + CO_2$$

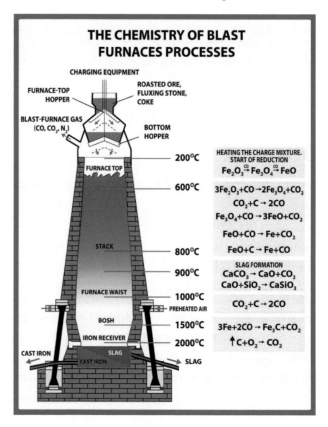

(i) Calculate the enthalpy change and entropy change for this reaction. [2]

(ii) Show that the reaction is feasible at 1100 K. [3]

(iii) A student says that the reaction is heated as it is not feasible at lower temperatures. Is the student correct? Justify your answer. [2]

(c) A scientist investigates using a similar process to extract magnesium metal from magnesium oxide. Calculate the enthalpy change and entropy change for this reaction and hence find the lowest temperature where the reaction would be feasible. [3]

Equilibrium constants

Not all chemical reactions convert 100% of reactants into products. Many reactions are reversible and in these cases the systems reach equilibrium with a mixture of reactants and products formed. In the laboratory and in industry it is important to know how much product will be formed in a mixture, how to maximise this and how to know whether a reaction is economically worthwhile. Changes in temperature affect the equilibrium mixture and using equilibrium constants allows chemists to quantify the contents of the equilibrium mixtures.

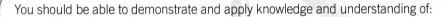

You should be able to demonstrate and apply knowledge and understanding of:

- The effect of temperature on K_p and K_c for exothermic and endothermic reactions.

- How to calculate values of K_p and K_c and quantities present at equilibrium from given data.

- The significance of the magnitude of an equilibrium constant and how this relates to the position of equilibrium.

Maths skills ⟩⟩

- Using and rearranging equilibrium constants expressions to find values and concentrations.
- Combining pressure and concentration units to find units of equilibrium constants.

General equilibria

Key term

A **dynamic equilibrium** is a reversible reaction where the rates of the forward and reverse reactions are equal.

A system in which none of the concentrations of reactants or products are changing is described as being in equilibrium. Chemical equilibria are **dynamic equilibria** as both forward and reverse reactions are occurring all the time, and at the same rate. The equilibrium mixture stays the same as long as the conditions stay the same, but may change when the conditions change. Le Chatelier's principle gives a qualitative method of identifying the effects of changing the conditions.

Le Chatelier's principle states that if a system at dynamic equilibrium experiences a change in its conditions then the equilibrium will shift to try to counteract the change.

Equilibrium constants

Practical check

The determination of an equilibrium constant for a reaction is a **specified practical task**. To do this you need to measure or calculate the concentrations of all species involved in a system at equilibrium.

To work out the exact amount of reactants and products in a mixture at dynamic equilibrium, we can use an equilibrium constant. In AS the equilibrium constant K_c was introduced, but there is also a second equilibrium constant K_p:

- K_c is the equilibrium constant in terms of concentration, and is usually used for reactions in solution, although it can be used for any reaction.

- K_p is the equilibrium constant in terms of partial gas pressures, and is used for reactions involving gases.

Link

Le Chatelier's principle, equilibria and equilibrium constants were covered initially in AS Topic 1.7.

Equilibrium constants in solution, K_c

For a reversible reaction in solution of the type:

$$a\,A(aq) \quad + \quad b\,B(aq) \quad \rightleftharpoons \quad x\,X(aq) \quad + \quad y\,Y(aq)$$

The equilibrium constant for reactions in solution is:

$$K_c = \frac{[X]^x[Y]^y}{[A]^a[B]^b} \qquad \text{where [X] represents the concentration of X}$$

- The top line has the concentrations of all the products to the powers of their balancing number, multiplied together.

- The bottom line has the concentrations of all the reactants to the powers of their balancing number, multiplied together.

Remember the formula has products over reactants.

e.g. For the reaction below:

$$Sn^{4+}(aq) \quad + \quad 2Fe^{2+}(aq) \quad \rightleftharpoons \quad 2Fe^{3+}(aq) + Sn^{2+}(aq)$$

The equilibrium constant, K_c is given by: $K_c = \dfrac{[Fe^{3+}]^2[Sn^{2+}]}{[Sn^{4+}][Fe^{2+}]^2}$

1 Knowledge check

Write expressions for the equilibrium constants K_c for these equilibria.

(a) $H_2CO_3(aq) \rightleftharpoons 2\,H^+(aq) + CO_3^{2-}(aq)$

(b) $NH_4^+(aq) + OH^-(aq) \rightleftharpoons NH_4OH(aq)$

(c) $H_2O(aq) + Cl_2(aq) \rightleftharpoons HCl(aq) + HOCl(aq)$

(d) $Fe^{3+}(aq) + NCS^-(aq) \rightleftharpoons FeNCS^{2+}(aq)$

Equilibrium constants in gases, K_p

In gases, we do not usually talk about the 'concentration of a gas', although it is possible to do this. We usually refer to the pressure of a gas, and we can do this for mixtures of gases as well. In a mixture of gases, the total pressure of the gas mixture is the sum of the pressure exerted by each of the gases in the mixture. In the production of ammonia, there is a mixture of three gases: hydrogen, nitrogen and ammonia. In this case:

| pressure of gas mixture | = | pressure exerted by nitrogen | + | pressure exerted by hydrogen | + | pressure exerted by ammonia |

We call the pressure exerted by an individual gas the partial pressure, and it is directly proportional to the concentration of the gas in the mixture. We write the partial pressure of an individual gas as p_{NH_3}. So the equation above can be written as:

$$p_{TOTAL} = p_{N_2} + p_{H_2} + p_{NH_3}$$

Because the partial pressures are related to the concentrations of each gas, we can use them in place of concentrations in an expression for the equilibrium constant. This equilibrium constant is called K_p. For a reversible reaction between gases of the type:

$$a\,A(g) \quad + \quad b\,B(g) \quad \rightleftharpoons \quad x\,X(g) \quad + \quad y\,Y(g)$$

The equilibrium constant for reactions in gases is:

$$K_p = \frac{p_X{}^x\, p_Y{}^y}{p_A{}^a\, p_B{}^b} \quad \text{where } p_X \text{ represents the partial pressure of X}$$

- The top line has the partial pressures of all the products to the powers of their balancing numbers, multiplied together.

- The bottom line has the partial pressures of all the reactants to the powers of their balancing numbers, multiplied together.

The equilibrium constant has products over reactants.

Top tip

Square brackets represent concentrations so these must not be used in any form in an expression of K_p as the values here are not concentrations.

Knowledge check

Write expressions for the equilibrium constants K_p for these equilibria.

(a) $2\,SO_2(g) + O_2(g) \rightleftharpoons 2\,SO_3(g)$

(b) $CH_4(g) + H_2O(g) \rightleftharpoons CO(g) + 3\,H_2(g)$

(c) $H_2(g) + I_2(g) \rightleftharpoons 2\,HI(g)$

(d) $COCl_2(g) \rightleftharpoons CO(g) + Cl_2(g)$

(e) $PCl_5(g) \rightleftharpoons PCl_3(g) + Cl_2(g)$

Interpreting equilibrium constants

Equilibrium constants can give us a guide to the degree that an equilibrium lies towards products or starting materials. An equilibrium that has similar amounts of starting materials and products would have $K_c = 1$.

- If K_c is a lot less than 1, then very few products are formed, and most of the mixture is starting materials. This is typically the case when ΔG for the reaction is positive, as the reaction will not occur spontaneously.

- If K_c is a lot more than 1, then most of the reactants have been converted into products. This is typically the case when ΔG for the reaction is negative, as the reaction will occur spontaneously.

Top tip

In your exams you may be asked about the effects of various changes on the value of an equilibrium constant. It is particularly important to remember that pressure does not affect K_p, although it may affect the amounts of each substance.

Stretch & challenge

Although pressure changes do not affect the value of the equilibrium constant, K_p, they may affect the position of equilibrium. If there are the same number of molecules of gas on both sides of the equilibrium then the effects of pressure cancel out and there is no effect on the position of equilibrium. If there are different numbers of molecules on both sides of the equilibrium then a change in pressure will affect the position of equilibrium even though the constant stays the same.

Equilibrium and kinetic data can both give us a lot of information about chemical reactions; however, they tell us very different things:

Equilibrium data tells us about the relative stability of the reactants and products, and the free energy changes that occur between them. It tells us nothing about how the reaction occurs.

Reaction rates give us information about the changes that occur between the reactants and transition state. This allows us to deduce what is happening during the reaction, and the order in which individual bonds are broken and made, called the **reaction mechanism**. It tells us nothing about the relative stability of reactants and products.

Kinetic, energetic and equilibrium data are all considered for any industrial process. Any reaction must aim to produce the maximum amount of product as quickly as possible with the efficient use of energy to reduce costs. Often these factors cannot all be at their ideal values at the same time, and so a compromise must be reached which gets as close to each ideal value as possible.

- Equilibrium yield of product can be changed by altering concentration, pressure or temperature. The equilibrium constant will allow us to identify which concentration or pressure values favour high yield, and the energetics can allow us to identify what temperature will favour a higher yield.
- Rates may be maximised by increasing temperature, increasing pressure or adding a catalyst. In industry, these factors would be altered to increase rate, unless that factor would decrease yield.
- Energy calculations will identify how much energy needs to be input into the system for a reaction to occur, minimising the input of excess energy. Similarly the energy generated by an exothermic reaction may be calculated and this allows a company to decide whether it is economical to harness this energy or whether it merely needs to be removed as waste.

Equilibrium constants and temperature

Both types of equilibrium constant remain the same as long as the temperature is constant:

Temperature is the only thing that changes equilibrium constants.

The way temperature affects the equilibrium constants can be deduced using Le Chatelier's principle. For any exothermic reaction, an increase in temperature will cause the equilibrium to shift in the endothermic direction, which will reduce the amount of products and increase the amount of reactants. This will decrease the value of the equilibrium constant.

For any endothermic reaction, an increase in temperature will cause the equilibrium to shift in the endothermic direction, which will increase the amount of products and decrease the amount of reactants. This will increase the value of the equilibrium constant.

- Exothermic reaction: Increasing temperature decreases K_c and K_p.
- Endothermic reaction: Increasing temperature increases K_c and K_p.

Calculating and using equilibrium constants

Calculating equilibrium constants, K_c

From information about the amounts of each substance in an equilibrium mixture, we can obtain values for the equilibrium constant.

Worked example

An equilibrium mixture of the reactants H_2 and I_2 and product HI contains 0.0035 mol dm^{-3} of each reactant and 0.0235 mol dm^{-3} HI. Calculate the value of K_c.

Answer

The reaction is:

$$H_2(g) + I_2(g) \rightleftharpoons 2HI(g)$$

So the equilibrium constant is: $K_c = \dfrac{[HI]^2}{[H_2][I_2]}$

And putting the values in gives: $K_c = (0.0235)^2 / (0.0035 \times 0.0035) = 45.1$

Because there are two concentrations on the top of the formula and two on the bottom, their units cancel out and **K_c has no unit**.

If the information given in the question does not list the concentrations of every reactant and product at equilibrium, then you need to work these out. A common situation is to give the concentrations of each reactant at the start of the reaction and then give the concentration of *one* product at equilibrium.

Worked example

If an **equimolar** solution of A and B where the concentration of each is 0.50 mol dm^{-3} is allowed to reach equilibrium then the equilibrium mixture contains 0.20 mol dm^{-3} of D. Calculate the value of K_c.

Answer

A + B → 2C + D

	[A] / mol dm^{-3}	[B] / mol dm^{-3}	[C] / mol dm^{-3}	[D] / mol dm^{-3}
At start	0.50	0.50	0	0
At equilibrium	0.30	0.30	0.40	0.20

At the start we only have A and B, both with concentration 0.50 mol dm^{-3}.

At equilibrium we have [D] = 0.20.

Since 2C are made when each D is made then [C] = 0.40 mol dm^{-3}.

To make 0.20 D we must use up 0.20 A and 0.20 B, leaving 0.30 of each behind.

We now must write an expression for K_c and put these values into it to get the value of K_c.

$$K_c = \frac{[C]^2[D]}{[A][B]} = \frac{0.40^2 \times 0.20}{0.30 \times 0.30} = \textbf{0.36 mol dm}^{-3}$$

There are three concentration terms in the numerator and two in the denominator. These cancel leaving one concentration unit for the equilibrium constant.

Top tip

When calculating the values of concentrations at equilibrium, remember that the numbers of moles of compounds are not conserved in a reaction. It is possible to get more moles of products than there are of reactants.

Top tip

When writing equilibrium constants, include all charges within the square brackets. It is common to see charges incorrectly placed outside, such as [NH$_4$]$^+$.

Knowledge check

Calculate the value of K_c for the equilibrium below if the concentrations of COCl$_2$ and Cl$_2$ are

3.2×10^{-3} mol dm^{-3} and the concentration of CO is 1.1×10^{-4} mol dm^{-3}.

$$COCl_2 \rightleftharpoons Cl_2 + CO$$

Knowledge check

Work out the value of K_c in the following system.

A mixture of Fe^{3+} and NCS$^-$ which has 0.2 mol dm^{-3} of each at the start produces a mixture containing 0.15 mol dm^{-3} FeNCS^{2+}.

$$Fe^{3+}(aq) + NCS^-(aq) \rightleftharpoons FeNCS^{2+}(aq)$$

Key term

An **equimolar mixture** is one that has equal amounts of moles (and hence equal concentrations) of each substance.

Using values of K_c

Equilibrium constants can be used to calculate the amounts of different reactants in a mixture. If given the value of an equilibrium constant, it is possible to calculate the percentage of each substance at equilibrium.

Worked example

The equilibrium constant, K_c, for the reaction below is 8 mol dm^{-3}. If there is 2 mol dm^{-3} of N_2O_4 in an equilibrium mixture, what concentration of NO_2 is present?

$$N_2O_4(g) \rightleftharpoons 2NO_2(g)$$

Answer

Step 1: Work out the expression for K_c in terms of concentration.

$$K_c = \frac{[NO_2]^2}{[N_2O_4]}$$

Step 2: Put the numbers into the equation:

$$8 = \frac{[NO_2]^2}{2} \quad \text{so } [NO_2]^2 = 16 \quad \text{giving } \mathbf{[NO_2] = 4 \ mol \ dm^{-3}}$$

Calculating and using equilibrium constants, K_p

From information about the partial pressures of each gas in an equilibrium mixture, it is possible to obtain values for the equilibrium constant K_p as well. The partial pressures represent the part of the total pressure exerted by a particular gas.

Partial pressure = % of a particular gas × pressure

Partial pressure = Pressure of a particular gas in a mixture

In a straightforward calculation, the partial pressures of all the substances present in an equilibrium mixture are provided and these simply need to be placed into the expression for the equilibrium constant.

Worked example

Ammonia is produced from hydrogen and nitrogen gases in the Haber process.

$$3H_2 + N_2 \rightleftharpoons 2NH_3$$

An equilibrium mixture is found to contain 300 Pa partial pressure of both nitrogen and hydrogen gases and 4230 Pa partial pressure of ammonia. Calculate the value of K_p under these conditions.

Answer

$$K_p = \frac{p_{NH_3}^2}{p_{N_2} \, p_{H_2}^3} \quad \text{where } p_X \text{ represents the partial pressure of X}$$

Substituting the values above gives:

$$K_p = \frac{4230^2}{300 \times 300^3} = 2.2 \times 10^{-3} \ \text{Pa}^{-2}$$

When measuring and calculating a value for K_p experimentally, it is more likely that you will know the initial pressures of the gas(es) as these can be easily measured. Measurement of one gas in the equilibrium mixture will then allow the partial pressures of all to be found and hence K_p can be calculated.

Worked example

A sample of pure PCl_5 with a partial pressure of 1.01×10^6 Pa was introduced into a vessel. The equilibrium below occurred, producing a partial pressure of 4.02×10^4 Pa of PCl_3.

$$PCl_5(g) \rightleftharpoons PCl_3(g) + Cl_2(g)$$

Calculate the value of K_p at this temperature.

Answer

First, the values for the partial pressures of PCl_5 and Cl_2 must be calculated.

Cl_2: When one mole of PCl_3 is produced, 1 mole of Cl_2 is also produced so a partial pressure of 4.02×10^4 for PCl_3 means the partial pressure of Cl_2 will be the same.

PCl_5: When 4.02×10^4 Pa of PCl_3 is produced, then 4.02×10^4 Pa of PCl_5 must have decomposed, leaving $(1.01 \times 10^6) - (4.02 \times 10^4)$ Pa $= 9.698 \times 10^5$ Pa.

Next we place these in the expression for K_p:

$$K_p = \frac{p_{Cl_2} \times p_{PCl_3}}{p_{PCl_5}} = \frac{4.02 \times 10^4 \times 4.02 \times 10^4}{9.698 \times 10^5} = 1666 \text{ Pa}$$

Study point

The units of the equilibrium constant depend on the units of pressure used. If the pressure is in Pascals then the units of K_p will be in multiples of Pa – they may commonly be Pa^{-2}, Pa^{-1}, Pa or Pa^2. The units for each partial pressure on the top of the expression can cancel with any on the bottom to leave the units of the expression. If there are the same number of molecules on both sides of the reversible reaction then there are the same units on the top and bottom, which all cancel and leave no unit.

Test yourself

1. State Le Chatelier's principle. [1]

2. Write the equilibrium constant, K_c, for the following reaction:

    $N_2(g) + 3H_2(g) \rightleftharpoons 2NH_3(g)$ [1]

3. Write the equilibrium constant, K_c, for the following reaction:

    $N_2(g) + O_2(g) \rightleftharpoons 2NO(g)$ [1]

4. Write the equilibrium constant, K_p, for the following reaction:

    $H_2(g) + I_2(g) \rightleftharpoons 2HI(g)$ [1]

5. Dinitrogen tetroxide, N_2O_4, breaks up to two molecules of NO_2 in a reversible reaction:

    $N_2O_4 \rightleftharpoons 2NO_2$

    At 100°C, the value of K_p is 8.33×10^6 Pa with the partial pressure of NO_2 being 2.80×10^5 Pa. Calculate the partial pressure of N_2O_4. [3]

6. A sealed 1 dm³ vessel is filled with a small amount of HI at a temperature of 500 K. The reversible reaction below occurs until the mixture reaches equilibrium, with $K_c = 160$:

$$2HI(g) \rightleftharpoons H_2(g) + I_2(g)$$

At equilibrium the concentration of H_2 equals the concentration of I_2 which is 0.042 mol dm⁻³. Calculate the concentration of HI. [3]

7. A chemist is studying the Haber process using a temperature of 350°C. He initially mixes 0.40 mol of N_2 and 0.80 mol of H_2 in a sealed 1 dm³ vessel. The heated mixture is allowed to reach equilibrium and then the amount of ammonia produced is measured. The vessel contains 0.24 mol of ammonia along with unreacted H_2 and N_2. Find the value of K_c under these conditions.

$$N_2(g) + 3H_2(g) \rightleftharpoons 2NH_3(g)$$ [3]

8. Phosphorus(V) chloride, PCl_5, decomposes to produce PCl_3 and Cl_2 in a reversible reaction:

$$PCl_5(g) \rightleftharpoons PCl_3(g) + Cl_2(g)$$

(a) At the start of the reaction the concentration of PCl_5 is 0.60 mol dm⁻³ and this decreases to 0.32 mol dm⁻³ in the equilibrium mixture. Calculate the value of K_c for this equilibrium. [3]

(b) This reaction is endothermic. State and explain the effect of increasing temperature on the value of K_c. [2]

9. One of the steps in the Contact Process has the following equilibrium constant, K_c.

$$K_c = \frac{[SO_3]^2}{[SO_2]^2[O_2]}$$

(a) Write the chemical equation that represents this reversible reaction. [1]

(b) At a particular temperature the value of K_c for this reaction is 4.5.

(i) Give the unit of K_c. [1]

(ii) A mixture of SO_2 and oxygen is allowed to reach equilibrium at this temperature. The initial concentration of SO_2 is 1.20×10^{-4} mol dm⁻³ with an excess of oxygen and at equilibrium the concentration of SO_2 decreases to 8.0×10^{-5} mol dm⁻³. Find the concentration of oxygen at equilibrium. [3]

(iii) At a higher temperature the value of K_c is 3.5. State, giving a reason, what information this provides about the equilibrium reaction. [2]

Acid–base equilibria

Some acids and bases are amongst the most familiar compounds; however, the range of acidic and basic compounds is very large. These include both strong and weak acids and bases and the properties of these can be very different. Weak acids such as ethanoic acid form a common part of our diet, but strong acids such as sulfuric acid can be very corrosive and harmful to living things.

Acid–base reactions can be reversible and set up dynamic equilibria. The concepts of general equilibria can be applied to acids and bases to find how these substances will behave. In this topic equilibria from weak acids, weak bases and buffers are studied.

You should be able to demonstrate and apply knowledge and understanding of:

- Lowry–Brønsted theory of acids and bases and the differences in behaviour between strong and weak acids and bases explained in terms of the acid dissociation constant, K_a.

- The significance of the ionic product of water, K_w, and how to use pH, K_w, K_a and pK_a in calculations involving strong and weak acids and pH and K_w in calculations involving strong bases.

- The shapes of the titration curves for strong acid/strong base, strong acid/weak base, weak acid/strong base and weak acid/weak base systems and how suitable indicators are selected for acid–base titrations.

- The mode of action of buffer solutions and how to use pH, K_w, K_a and pK_a in buffer calculations.

- The importance of buffer solutions in living systems and industrial processes.

- The acidity and basicity of some salt solutions and the concept of hydrolysis of salts of a strong acid/strong base, a strong acid/weak base and a weak acid/strong base.

Maths skills 》

- Using logarithmic rules to interconvert between pH and H⁺ concentration.
- Using and rearranging equilibrium constant expressions to find the concentration of H⁺ ions.

Acids and bases

There are a few definitions of acids and bases in chemistry; however, one of the most common and most useful is the Lowry–Brønsted definition.

An acid is a substance that releases or provides H+ ions, i.e. it is an H+ donor.

A base is any substance that removes or accepts H+ ions, i.e. it is an H+ acceptor.

You will frequently see acids and bases classed as strong or weak, or concentrated or dilute. Each term has a specific meaning with strong and weak referring to degree of dissociation and concentrated and dilute referring to concentration.

Strong and weak acids and bases

The differences between strong and weak acids were first noted in terms of pH, but a better approach is to work with the degree of dissociation of the acid.

A strong acid is one that almost totally dissociates (splits up) into H+ ions and negative ions in solution in water, e.g. hydrochloric acid, HCl:

$$HCl(aq) \longrightarrow H^+(aq) + Cl^-(aq)$$

A weak acid is one that only partially dissociates into H+ ions and negative ions in water. The free ions are in equilibrium with the undissociated acid molecule, e.g. ethanoic acid, CH_3COOH:

$$CH_3COOH(aq) \rightleftharpoons H^+(aq) + CH_3COO^-(aq)$$

This leads to differences in pH, so a 1 mol dm^{-3} solution of a strong acid has a pH of 0, whilst a 1 mol dm^{-3} solution of a particular weak acid has a pH of 4.0, which represents 10000 times fewer free H+ ions. Similarly, strong bases are bases that accept H+ ions in a non-reversible reaction whilst weak bases accept H+ ions in a reversible reaction. In a weak base there is an equilibrium between protonated and non-protonated forms of the base.

Reviewing pH

The strengths of acids are usually quoted on the pH scale. This ranges from 0 (strong acid) to 14 (strong alkali) through 7, which is neutral. We can measure the pH by using a pH probe, which will give a value for the pH of the solution, or more simply we can use an indicator or pH paper to obtain a guide to the pH value. The pH scale is usually shown as running from 0 to 14; however, very concentrated solutions of strong acids may have negative pH values.

0	1	2	3	4	5	6	7	8	9	10	11	12	13	14
Strong acid				**Weak acid**			**Neutral**		**Weak alkali**				**Strong alkali**	

Acids have pH values below 7.
In simple terms, the further a solution's pH value is below neutral (7), the stronger the acid.

Alkalis have pH values above 7.
In simple terms, the further a solution's pH value is above neutral (7), the stronger the alkali.

The numbers on the pH scale are obtained from the concentration of H+(aq) ions.

$$pH = -\log[H^+(aq)] \qquad \text{where log means } \log_{10}$$

Note: the value of $\log[10^x]$ is x, so if $[H^+] = 10^{-7}$, as it is for water, then $pH = -\log[10^{-7}] = 7$

You need to be very familiar with converting the concentration of H+ ions to pH and vice versa. On most calculators the steps are the same:

To convert [H+] to pH press | (-) | log | [H+] value | = |

To convert pH to [H+] press | SHIFT | log | (-) | pH value | = |

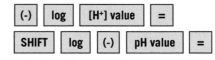

Key terms

Concentrated means that there is a large amount of an acid or base dissolved in a set volume of water.

Dilute means that there is a small amount of an acid or base dissolved in a set volume of water.

Link

Acid–base equilibria were covered initially in AS Topic 1.7.

Key terms

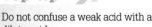

A **weak acid** is one that dissociates partially to release H+ ions in a reversible reaction.

A **strong acid** is one that dissociated fully to release H+ ions.

Top tip

Do not confuse a weak acid with a dilute acid.

Do not confuse a strong acid with a concentrated acid.

Knowledge check

Work out the pH of the following solutions:

(a) A solution that contains 0.2 mol dm^{-3} H+ ions.

(b) A solution that contains 0.03 mol dm^{-3} H+ ions.

(c) A solution that contains 10^{-9} mol dm^{-3} H+ ions.

(d) A solution that contains 3 × 10^{-11} mol dm^{-3} H+ ions.

Acid dissociation constants, K_a

When an acid dissociates it is an equilibrium process, and so it has an equilibrium constant:

$$HA(aq) \rightleftharpoons H^+(aq) + A^-(aq)$$

For the equilibrium above, $K_c = \dfrac{[H^+(aq)][A^-(aq)]}{[HA(aq)]}$ The unit of this is always mol dm^{-3}.

If we ignore the water involved we can produce the acid dissociation constant, K_a:

$$K_a = \dfrac{[H^+][A^-]}{[HA]}$$

The more dissociated the acid is, the more hydrogen ions and anions there will be, so the larger the value of K_a.

A weak acid will have a low value of K_a.

A strong acid will have a high value of K_a.

We can use the value of K_a for an acid to work out whether an acid is strong or weak. The table gives values of K_a for some acids:

Acid	pK$_a$	Formula	K_a / mol dm^{-3}
Nitric acid	−1.38	HNO$_3$	24
Sulfuric(IV) acid	1.85	H$_2$SO$_3$	1.4×10^{-2}
Methanoic acid	3.74	HCOOH	1.8×10^{-4}
Ethanoic acid	4.77	CH$_3$COOH	1.7×10^{-5}
Carbonic acid	6.35	H$_2$CO$_3$	4.5×10^{-7}

The range of values of K_a is large and so these values are often reported as pK_a values. The method of calculating this is similar to that used to calculate pH:

$$pK_a = -\log_{10}(K_a)$$

The pK_a of a strong acid is small or negative whilst the pK_a of a weak acid is larger.

Ionic product of water, K_w, and neutralisation

Water can be purified to remove all the impurities in it, to attempt to obtain 100% pure water. Even after all purification processes, water can still conduct a small amount of electricity showing that there are still ions dissolved in it. We cannot remove these ions as they are produced by the water itself. There is an equilibrium process in the water, where the water dissociates (breaks down) to form H$^+$ and OH$^-$ ions:

$$H_2O(l) \rightleftharpoons H^+(aq) + OH^-(aq)$$

The reaction lies mainly to the left-hand side, so almost all the water exists as water molecules, with a very small amount of ions. The equilibrium constant of this reaction is:

$$K_c = \dfrac{[H^+][OH^-]}{[H_2O]}$$

Knowledge check 2

Work out the concentration of H$^+$ ions in the following solutions:

(a) pH = 0.0

(b) pH = 2.7

(c) pH = 6.3

(d) pH = 10.5

(e) pH = 14.0

Top tip

When describing weak and strong acids in an exam it is important to link the strength to degree of dissociation or the value of K_a. Linking acid strength to pH is not sufficient.

Knowledge check 3

Write expressions for the acid dissociation constants, K_a, for HClO and HCN.

4 Knowledge check

The value of K_w at 25°C is 1.0×10^{-14} mol^2 dm^{-6} and at 50°C is 5.5×10^{-14} mol^2 dm^{-6}. Explain what information this provides about the enthalpy change of the reaction.

Top tip

Whenever you write values for K_a and K_w you must include the correct units. These are always the same. These equilibrium constants are different from K_c and K_p as the units for K_c and K_p are different for different equilibria.

Because the amount of water that dissociates is tiny, the concentration of water can be considered to be constant and we can combine it into the equilibrium constant and rename this K_w, the *ionic product of water*:

$$K_w = [H^+][OH^-] \qquad \text{Ionic product of water}$$

The value of K_w is a constant at a particular temperature, and at 25°C the value of K_w is approximately 10^{-14} mol^2 dm^{-6}. Since the amount of H$^+$ and OH$^-$ must be the same in pure water, we can work out the concentration of H$^+$ in the pure water.

Since $\qquad [H^+] = [OH^-]$, and $\qquad K_w = [H^+][OH^-]$

We can write: $\qquad\qquad\qquad\qquad\qquad K_w = [H^+]^2 = 10^{-14}$

Therefore $\qquad\qquad\qquad\qquad\qquad [H^+] = 10^{-7}$ mol dm^{-3}

When an acid reacts with a base, the reaction is the reverse of the equilibrium above and a neutral solution is produced in a neutralisation reaction. During the reaction, the free H$^+$ ions react with free OH$^-$ to produce water:

$$H^+(aq) + OH^-(aq) \longrightarrow H_2O\,(l)$$

This is the ionic equation for the reaction occurring in all neutralisation reactions. Ionic equations only list the ions actually involved in the reaction, and ions that remain unchanged at the end are not included. Because the neutralisation reaction is the same in all cases, the energy change is the same in every case, so all neutralisation reactions involving strong acids and bases cause the same temperature rise of the solutions.

Calculating pH

To calculate the pH of any solution, the concentration of H$^+$ ions must be found, and then the equation for pH must be applied to this. For all solutions whether they are acidic or basic, pH is defined as:

$$pH = -\log_{10}[H^+]$$

pH of strong acids

For a strong acid, all the hydrogen ions are released from the acid molecules, so the concentration of the acid gives the concentration of H$^+$ ions.

e.g. $\qquad$ In 1.0 mol dm^{-3} hydrochloric acid, $[H^+] = 1.0$ mol dm^{-3}

$\qquad\qquad$ In 0.2 mol dm^{-3} nitric acid, $[H^+] = 0.2$ mol dm^{-3}

This can then be used directly with the formula for pH.

Stretch & challenge

A dibasic acid can donate two H$^+$ ions (H$_2$SO$_4$ for example) and a tribasic acid can donate three H$^+$ ions (H$_3$PO$_4$ for example). In these cases the concentration of H$^+$ ions will be higher than the concentration of the acid. If all the H$^+$ ions in a dibasic acid were released then the concentration of H$^+$ would be two times the concentration of the acid so [H$^+$] = 2 × [dibasic acid]. This is not always the case as the degree of dissociation of each H$^+$ can become lower with each one lost. This means the K_a for each proton gets smaller. In the case of H$_2$SO$_4$ the K_a for the first H$^+$ is 1 × 10^3 mol dm^{-3} and for the second is 1 × 10^{-2} mol dm^{-3}.

Worked example

What is the pH of a 0.05 mol dm^{-3} solution of hydrochloric acid?

Answer

A solution of 0.05 mol dm^{-3} HCl will have the same concentration of H$^+$ ions, so [H$^+$] = 0.05.

$$pH = -\log_{10}[H^+] = -\log_{10}(0.05) = 1.30$$

pH of weak acids

In weak acids not all the hydrogen ions are released, but we can use the K_a values to work out the concentration of H⁺ ions. For ethanoic acid, the K_a value is 1.7×10^{-5} mol dm⁻³, so to calculate the concentration of H⁺ ions in a solution we write:

$$K_a = \frac{[H^+][CH_3COO^-]}{[CH_3COOH]}$$

Since each CH_3COOH molecule that dissociates produces one CH_3COO^- and one H⁺, then the concentrations of each must be equal. In the expression above we can state that $[H^+] = [CH_3COO^-]$ giving,

$$K_a = \frac{[H^+]^2}{[CH_3COOH]}$$

For a weak acid very little dissociation occurs, so it is possible to assume that the concentration of the acid molecules (CH_3COOH) present is the same as the concentration we put in. This allows us to rearrange the equation to give:

$$[H^+] = \sqrt[2]{K_a \times [CH_3COOH]}$$

Worked example

What is the pH of a 0.500 mol dm⁻³ solution of ethanoic acid? ($K_a = 1.7 \times 10^{-5}$ mol dm⁻³)

Answer

$$[H^+] = \sqrt[2]{K_a \times [CH_3COOH]}$$

$$[H^+] = \sqrt[2]{1.7 \times 10^{-5} \times 0.500}$$

$$[H^+] = 0.00292 \text{ mol dm}^{-3}$$

Applying the expression for pH $= -\log_{10}[H^+] = -\log_{10}(0.00292) = \underline{2.53}$

Working out K_a values for weak acids

Given the pH of a solution of known concentration, we can work out the value of K_a. This involves rearranging the expression for K_a in a similar manner to above.

Worked example

What is the K_a of an unknown monobasic acid if a 2.0 mol dm⁻³ solution has a pH of 4.5?

Answer

First convert the pH into a concentration of H⁺ ions:

$$pH = -\log[H^+] \longrightarrow [H^+] = 10^{(-pH)}$$

In this case: $[H^+] = 10^{-4.5} = 3.2 \times 10^{-5}$ mol dm⁻³

Then use the formula for K_a, remembering that each acid molecule that dissociates produces one hydrogen ion and one anion, so $[A^-] = [H^+]$.

$$K_a = \frac{[H^+][A^-]}{[HA]}$$

$$K_a = (3.2 \times 10^{-5})^2 / 2$$

$$K_a = 5.12 \times 10^{-10} \text{ mol dm}^{-3}$$

>> **Study point**

You need to be aware of some familiar strong acids such as nitric acid, sulfuric acid and hydrochloric acid and some weak acids such as ethanoic acid and other carboxylic acids. If you are provided with any other acids you will need to identify whether these are strong or weak by looking at the K_a value or by being told as part of a question.

Knowledge check 5

1. Work out the pH of the following solutions of ethanoic acid. ($K_a = 1.7 \times 10^{-5}$ mol dm⁻³)
    (a) 0.5 mol dm⁻³
    (b) 2 mol dm⁻³
    (c) 0.01 mol dm⁻³

2. Work out the pH of these solutions of chloric(I) acid, HClO. ($K_a = 2.9 \times 10^{-8}$ mol dm⁻³)
    (a) 1 mol dm⁻³
    (b) 0.5 mol dm⁻³
    (c) 5 mol dm⁻³

3. Which of these two acids is the stronger? Explain your answer in terms of K_a.

Top tip

The strength of an acid may be provided in the form of K_a or pK_a. K_a can be found from pK_a by using $K_a = 10^{-pK_a}$.

Knowledge check 6

Work out the K_a values for the following weak acids:

(a) HB where a 0.5 mol dm⁻³ solution has a pH of 2.9.

(b) HC which has a pH of 2.2 for a 1 mol dm⁻³ solution.

(c) HD where a 0.5 mol dm⁻³ solution has a pH of 3.5.

Study point

There are several different types of pH calculation here. All of them involve finding the concentration of H⁺ ions first then calculating pH. It is useful to make a list of the methods and practise them all together – list some random concentration values and work out the pH of a strong acid / weak acid / strong base of those concentrations.

 Knowledge check

7

Work out the pH of NaOH solutions of the following concentrations:

(a) 1.0 mol dm⁻³

(b) 0.2 mol dm⁻³

(c) 0.05 mol dm⁻³

(d) 0.003 mol dm⁻³

 Key term

A **buffer** resists changes in pH as small amounts of acid and alkali are added.

 Study point

Buffers only resist changes in pH when SMALL amounts of acid or alkali are added. Addition of large amounts of acid or alkali does not allow the pH to remain constant.

pH for strong bases

When a strong base dissolves in water, all the possible OH^- ions are produced, so the concentration of the base gives the concentration of OH^-.

e.g.   In 1.0 mol dm⁻³ sodium hydroxide, $[OH^-] = 1.0$ mol dm⁻³

In 0.2 mol dm⁻³ potassium hydroxide, $[OH^-] = 0.2$ mol dm⁻³

Using the ionic product of water, K_w, we can use this information to work out the concentration of H⁺ ions in the same solution.

$$K_w = [H^+] \times [OH^-] \text{ so } [H^+] = K_w \div [OH^-]$$

This provides a value for the concentration of H⁺, which can then be used in the expression for pH.

Worked example

What is the pH of a solution of NaOH of concentration 0.1 mol dm⁻³?

Answer

In this solution of NaOH, the concentration of OH^- ions is 0.1 mol dm⁻³. Using K_w we can say:

$$[H^+][OH^-] = 10^{-14}$$

so $[H^+] \times 0.1 = 10^{-14}$

so $[H^+] = 10^{-13}$ mol dm⁻³

We can work out the pH from this value, with $pH = -\log_{10}[H^+] = -\log(10^{-13}) = 13$

Buffers

Buffers are solutions whose pH stays relatively constant as a small amount of an acid or alkali is added. The buffer solution maintains a nearly constant pH by removing any added H⁺ or OH⁻.

How buffers work

Typically, a buffer solution is made from a mixture of:

- A weak acid, HA e.g. CH_3COOH.

- A salt of the same acid with a strong base, NaA e.g. CH_3COONa. This acts as a source of the anion A^-.

In the buffer solution, there is a high concentration of the anion as the sodium salt will dissociate completely:

$$CH_3COONa \longrightarrow CH_3COO^- + Na^+$$

The weak acid dissociates partially, and the reversible reaction below sets up an equilibrium.

$$CH_3COOH(aq) \rightleftharpoons H^+(aq) + CH_3COO^-(aq)$$

The ethanoic acid is almost all in the undissociated form, as all the CH_3COO^- ions released from the sodium ethanoate dissociation forces the equilibrium to the left.

When an acid is added to a buffer

The concentration of H^+ is increased, and this causes the equilibrium to shift to the left, removing the H^+ ions by reaction with CH_3COO^-.

$$CH_3COOH(aq) \rightleftharpoons H^+(aq) + CH_3COO^-(aq)$$

H⁺ ions removed

When an alkali is added to a buffer

The concentration of OH^- is increased, and this removes some of the H^+ ions present. This causes the equilibrium to shift to the right, producing H^+ ions and CH_3COO^- from CH_3COOH. The H^+ ions replace those removed by reaction with the hydroxide ions.

$$CH_3COOH(aq) \rightleftharpoons H^+(aq) + CH_3COO^-(aq)$$

H⁺ ions replaced

In both cases, the pH of the solution does not stay totally constant, but the buffer minimises any change when H^+ or OH^- are added with the change in pH being insignificant. If a larger amount of acid or alkali is added, these may react with a significant amount of the ethanoate ions or ethanoic acid molecules which then causes the pH to change by a measurable amount.

Buffers are used where pH is very important, which is common for biological systems. They are therefore used for:

- Using or storing enzymes, to ensure the pH remains at the optimum value.

- Storage of biological molecules, such as pharmaceuticals, which will be denatured at the incorrect pH.

Some industrial processes rely on biological molecules. The fermentation processes during baking and brewing can use buffers to keep the conditions appropriate for the microorganisms to survive. This buffering often uses the weakly acidic solutions formed when carbon dioxide dissolves in water. The process of dyeing also uses buffers as the pH of the system can affect the colour and absorption of a dye.

Basic buffers

A similar buffer system for maintaining alkaline pH is based on a mixture of ammonium chloride and ammonia solution. The ammonium chloride dissociates completely, releasing all the ammonium ions.

$$NH_4Cl(aq) \longrightarrow NH_4^+(aq) + Cl^-(aq)$$

The key equilibrium is:

$$NH_4^+ \rightleftharpoons NH_3 + H^+$$

- Addition of base removes H^+ ions and this causes the equilibrium to shift to the right to produce more H^+ ions.

- Addition of acid causes the equilibrium to shift to the left, and this removes the additional H^+ provided by the acid.

Knowledge check 8

Calculate the pH of the following buffer solutions:

(a) A mixture of 0.20 mol dm⁻³ CH_3COOH ($K_a = 1.7 \times 10^{-5}$ mol dm⁻³) and 0.10 mol dm⁻³ CH_3COONa.

(b) A mixture of 0.20 mol dm⁻³ CH_3COOH and 0.40 mol dm⁻³ CH_3COONa.

(c) A mixture of 1.0 mol dm⁻³ CH_3COOH and 0.2 mol dm⁻³ CH_3COONa.

(d) A mixture of 0.20 mol dm⁻³ H_2SO_3 ($K_a = 1.6 \times 10^{-2}$ mol dm⁻³) and 0.10 mol dm⁻³ $NaHSO_3$.

Study point

When a strong base is added gradually to a weak acid, such as during a titration, a salt forms. For example, adding sodium hydroxide to ethanoic acid produces sodium ethanoate. This means that a mixture of weak acid and a salt of a weak acid has been formed. This is a buffer being created as part of the reaction.

pH for buffers

To calculate the pH of a buffer, we need to know:

- The K_a value for the weak acid.
- The ratio of the concentrations of the acid and the salt.

We use the expression:

$$K_a = \frac{[H^+][A^-]}{[HA]}$$

By substituting for the values of K_a and the ratio of $[A^-]/[HA]$ we can work out the concentration of $[H^+]$ ions. This can then be converted into a value of pH.

In the buffer mixture, all the salt will dissociate and almost none of the acid will be dissociated. This means that it can be assumed that $[HA]$ is equal to the concentration of acid used, and $[A^-]$ is equal to the concentration of the salt.

Worked example

Work out the pH of a buffer that contains equal concentrations of ethanoic acid ($K_a = 1.7 \times 10^{-5}$ mol dm^{-3}) and sodium ethanoate.

Answer

Using the expression above, we substitute in the values:

$$K_a = [H^+]\frac{[CH_3COO^-]}{[CH_3COOH]} \qquad \frac{[CH_3COO^-]}{[CH_3COOH]} = 1 \text{ as the two concentrations are the same}$$

$1.7 \times 10^{-5} = [H^+] \times 1$
$[H^+] = 1.7 \times 10^{-5}$ mol dm^{-3}

Convert this to pH using   pH $= -\log[H^+]$
$= -\log(1.7 \times 10^{-5})$
pH $= 4.77$

Henderson–Hasselbalch equation

It is possible to combine all the calculation steps for buffers into one expression. This is called the Henderson–Hasselbalch equation:

$$pH_{buffer} = pK_a + \log\frac{[salt]}{[acid]}$$

In this case the value of K_a (or pK_a) and the concentrations of both the salt and acid allow the pH of the buffer to be found. When the concentration of the salt is equal to the concentration of the acid then the log term can be simplified to log (1) which equals zero.

When the concentration of the salt equals the concentration of the acid, then pH $= -\log_{10}K_a = pK_a$

When the concentrations of the salt and acid are different we can use K_a in the same method as above. It is always the ratios of these two concentrations that is key, rather than the absolute values of the concentrations. If the concentration of the sodium ethanoate salt was double that of the ethanoic acid then the calculation gives:

$$1.7 \times 10^{-5} = [H^+] \times 2$$
$$[H^+] = 0.85 \times 10^{-5} \text{ mol dm}^{-3}$$

This can then be converted to pH giving a value of 5.07.

Summary of pH calculations

All pH calculations require you to find the concentration of H^+ ions first. The method of finding the concentration of H^+ varies depending on the substance.

STRONG ACID: The concentration of H^+ ions equals the concentration of the acid for a monobasic acid.

WEAK ACID: We need to use K_a as well as the acid concentration as:

$$[H^+] = \sqrt[2]{K_a \times [\text{acid}])}$$

STRONG BASE: We need to use K_w (from the data booklet) as well as the base concentration as:

$$[H^+] = \frac{K_w}{[OH^-]} = \frac{1.00 \times 10^{-14}}{[OH^-]}$$

BUFFER: We need to use K_a and the relative concentrations of acid and salt in the formula:

$$[H^+] = K_a \times \frac{[\text{acid}]}{[\text{salt}]}$$

Following the calculation of H^+ concentration use the equation below to calculate pH:

$$pH = -\log [H^+]$$

pH of salts

When salts are dissolved in water, many form neutral solutions; however, a few salts are acidic salts, and others are basic salts.

- A salt produced from a strong acid and a strong alkali will form neutral solutions, e.g. NaCl. This is a neutral salt.

- A salt produced from a strong acid and a weak alkali will be an acidic salt, e.g. NH_4Cl.

- A salt produced by a weak acid and a strong alkali will be a basic salt, e.g. CH_3COONa.

Knowledge check 9

Ammonium sulfate is used as a fertiliser. Suggest a pH for a solution of ammonium sulfate, explaining your reasons.

Why do we see these effects?

The ions released from some salts react with water molecules to release or remove H^+ ions. This is called salt hydrolysis. When ammonium chloride, NH_4Cl, dissolves, it produces free NH_4^+ and Cl^- ions. Since ammonia is a weak base, the following equilibrium will occur in solution:

$$NH_4^+(aq) \rightleftharpoons NH_3(aq) + H^+(aq)$$

This releases free H^+ ions in solution, making the solution slightly acidic. As the concentration of H^+ ions is increased, the pH of the solution is decreased so that it is slightly lower than 7.

When sodium ethanoate, CH_3COONa, dissolves it produces free CH_3COO^- and Na^+. Since the ethanoic acid is a weak acid, the following equilibrium is always present when the ethanoate ion is in aqueous solution:

$$H^+(aq) + CH_3COO^-(aq) \rightleftharpoons CH_3COOH(aq)$$

In this case, the equilibrium removes some H^+ ions from the water, producing a slightly alkaline solution. As the concentration of H^+ ions is decreased, the pH increases slightly above 7.

▲ The acidic salt ammonium chloride is formed immediately when the weakly basic gas ammonia mixes with the strongly acidic hydrogen chloride gas.

Acid–base titration curves

When a base is added to an acid, a neutralisation reaction occurs, according to the equation:

$$H^+(aq) + OH^-(aq) \longrightarrow H_2O(l)$$

This is the reaction that occurs during all acid–base titrations. The effect of this is to reduce the concentration of H^+ ions, and therefore increase the pH. The increase in pH is not a straight line, however, and it depends on whether the acid and alkali are strong or weak.

Strong acid–strong base

The titration curve for a strong acid with a strong alkali, e.g. HCl with NaOH of the same concentration, is shown in the graph below:

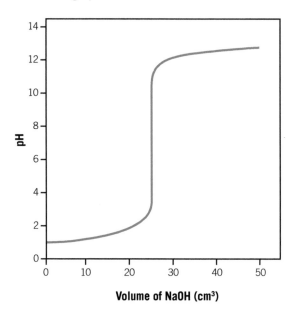

Volume of NaOH (cm³)

In this case, a 0.1 mol dm^{-3} solution of NaOH is added to 25 cm³ of a solution of 0.1 mol dm^{-3} hydrochloric acid.

- The graph starts at pH 1, the pH of a 0.1 mol dm^{-3} solution of a strong acid.

- There is a slow gradual increase in pH as the first 20 cm³ is added.

- There is a sudden increase from about pH 2 to pH 12 as a very small volume of base is added around 25 cm³, which is approximately vertical – use a ruler for this.

- There is then a slow gradual increase in pH as the last 20 cm³ of base is added.

- The graph ends at just below pH 13.

The vertical region of the curve occurs when the number of moles of alkali added equals the number of moles of acid in the original solution. This is called the equivalence point – when the concentrations of the two solutions are the same it will occur when the volume of alkali added is equal to the volume of acid.

The pattern is always similar to this for a strong acid with a strong base, but when a weak acid is used the acid quadrant (from pH 0–7) is changed and when a weak alkali is used the alkali quadrant (pH 7–14) is changed. The pattern is changed by the fact that the molecules are not totally dissociated. These give different patterns, with the sudden increase in pH being much smaller.

Weak acid–strong base

The titration curve for a weak acid with a strong alkali, e.g. ethanoic acid with NaOH of the same concentration, is shown in the graph below:

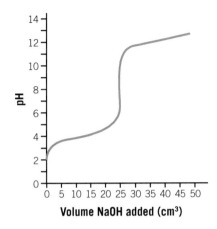

Top tip

It is often useful to label the buffer region of the weak acid–strong base titration curve so that it is clear to any examiner.

Study point

The terms equivalence point and end point are used interchangeably in many texts. Although both are very similar, they are not the same. The **equivalence point** is the point when the number of moles of alkali added is equal to the number of moles of acid. The **end point** is the point at which an indicator changes colour in a titration.

Key points

The pH increases gradually to about 4, as the base is added. When the volume of sodium hydroxide is about half the volume of acid, and hence the number of moles of sodium hydroxide is half the number of moles of acid, the pH levels off. This is because a mixture is formed containing the unreacted acid and the salt formed by neutralisation of the acid, and this acts as a buffer. The pH therefore levels out slightly over a volume range of about 5 cm³.

The pH then increases gradually towards pH 7, when volume of base added equals volume of acid before increasing vertically to about pH 12. The pH gradually increases up to about between 13 and 14.

Strong acid–weak base

The titration curve for a strong acid with a weak alkali, e.g. hydrochloric acid with ammonia of the same concentration, is shown in the graph below.

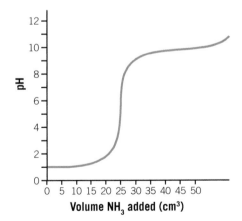

Key points

The pH up to 7 mirrors the original strong acid–strong base titration curve. After pH 7, the line becomes smoother and then levels off at a volume of about 35–45 cm³ due to the buffer effect. The pH then increases gradually up to a pH of about 12.

Top tip

When sketching any of the titration curves, it is important to start by noting any points that are known. These include:

The pH of the initial acid (at a volume of alkali of 0 cm³).

The pH at half neutralisation (for weak acid–strong base calculations). The pH at this point equals the pK_a (–log K_a).

>> **Study point**

All these diagrams are plotted for concentrations of acid and base that are the same. If a titration is undertaken with different concentrations of acid and base, the shape of the plot will be unchanged but the graph will be compressed along the volume axis (when the base has a higher concentration) or extended along the volume axis (when the acid has a higher concentration). Use the volume when the vertical region occurs to find the equivalence point and this will tell you whether the concentrations of the two are the same, as it will occur at the same volume of base as the original volume of acid.

Weak acid–weak base

The titration curve for a weak acid with a weak alkali, e.g. ethanoic acid with ammonia of the same concentration, is shown in the graph below:

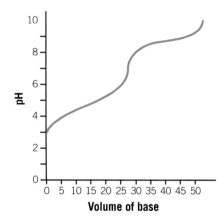

The graph for weak acids with weak bases lacks the clear vertical region that is present in the other graphs. This makes it much harder to study using an indicator in the titration. In place of an indicator, a pH probe is used, measuring the pH throughout the titration. A plot of the data is then used to identify the equivalence point, which is the point of inflection.

◀ **Stretch & challenge**

The curves already seen are drawn for monobasic acids only, such as HCl or HNO$_3$. Dibasic acids such as H$_2$SO$_4$ lose each hydrogen in turn and so there are two equivalence points, with tribasic acids showing three equivalence points. The graph below shows what this pattern would look like for a tribasic acid.

This graph allows us to check our calculations as the volume needed for the first equivalence point should be half that for the second, with the third value being three times the original.

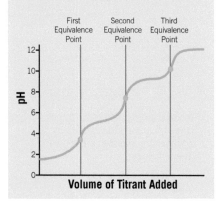

Obtaining data from titration curves

Equivalence points

Titration curves can be used to find the volume of alkali needed to neutralise the acid. For titration curves where there are vertical regions these mark the volume required for neutralisation. The value obtained can be used in the titration calculations to find the concentration of one solution where the concentration of the other solution is known.

pH of salt formed

The pH at the equivalence point of each of these titrations can be found from the midpoint of the vertical region. In the case of strong acid–strong base this will be 7; however, this will shift away from neutral when we use a weak acid or a weak base. At the equivalence point the solution only includes a salt, and the salts formed from weak acids are basic whilst those formed from weak alkalis are acidic. The pH at neutralisation is equal to the pH of the salt solution.

pK_a of a weak acid

In the discussion of buffers, the pH of a buffer was expressed as:

$$pH_{buffer} = pK_a + \log \frac{[salt]}{[acid]}$$

When half the volume of base needed for neutralisation is added to the acid, half the acid will have been converted into salt. This means that the concentration of the salt will equal the concentration of the acid, and so the log term would be log (1) which equals zero. This simplifies the equation above to:

$$pH_{buffer} = pK_a$$

So the pK_a equals the pH at the point that half the alkali has been added for neutralisation.

Indicators

Indicators are substances that change colour as the pH changes. To do this they must be weak acids, with the original molecule and dissociated ions having different colours.

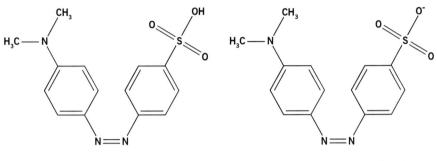

Methyl orange (red form)
In acid

Methyl orange (orange form)
In alkali

The solutions of these ions change colour as the pH of the solution changes, due to the relative amounts of the two forms changing.

$$HInd \rightleftharpoons Ind^- + H^+$$

Colour A　　　Colour B

- In acid, most of the indicator exists as the neutral form (HInd).

- In alkali, most of the indicator exists as the anion (Ind⁻).

The indicator does not suddenly change from one form to another at a specific pH – it changes over a range of pH values. This range is different for each indicator. The reason we see a sharp colour change in a titration is that the pH of the solution changes sharply at the end point. To make sure that the indicator is suitable, we have to check that the range where the indicator changes colour lies within this sharp increase in pH value.

Strong acid–strong base

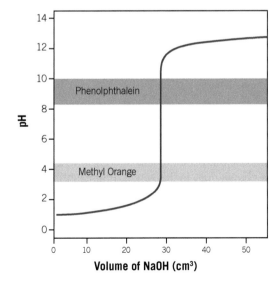

For a strong acid–strong base the pH increases sharply from 2 to 12. All the indicators change fully in this range, so any of them could be used, although the clear colour change of phenolphthalein means it is frequently the one chosen.

Key term

An **indicator** is a substance that has different colours in low and high pH solutions. This can be used to differentiate between an acid and an alkali but cannot distinguish between acids of different pH or alkalis of different pH.

Link

Acid–base titrations in AS Topic 1.7

Indicator	Approximate colour change range
Phenolphthalein	8.3–10.0
Bromothymol blue	6.0–7.5
Litmus	4.0–6.5
Methyl orange	3.2–4.4

Stretch & challenge

Dibasic acids and bases show two different equivalence points, both having small vertical regions. Indicators can be selected that match each separate vertical region, which allows a titration to be performed with one indicator to find the first equivalence point, and then a second indicator can be added to find the second equivalence point. A similar approach can be used to analyse a mixture of a strong and weak acid (or a strong and weak alkali).

Top tip »

The two most common indicators in use are phenolphthalein and methyl orange. If you have to suggest an indicator, then phenolphthalein is the indicator of choice for any titration using a strong base, as the colour change is sharp and clear. When the base is weak, phenolphthalein cannot be used and so methyl orange is chosen.

10 Knowledge check

Different acid–base indicators change colour over different pH ranges. A few are listed in the table:

Indicator	pH range
Pentamethoxy red	1.2–2.3
Bromophenol blue	3.0–4.6
Phenol red	6.4–8.0
Thymol blue	8.0–9.6

Use the pH curves to identify which of these indicators could be used for a strong acid–strong base titration, which for a weak acid–strong base titration and which for a strong acid–weak base titration. Give a reason for your choices.

Weak acid–strong base

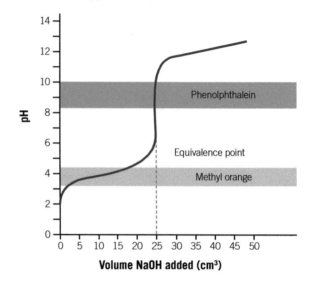

For a weak acid–strong base the pH increases sharply from 5 to 12 so phenolphthalein or bromothymol blue can be used as they change colour completely in the right pH range. Methyl orange would not change colour at all in this vertical range – it would have changed colour before reaching this point. Phenolphthalein is the most commonly used indicator for this type of titration.

Strong acid–weak base

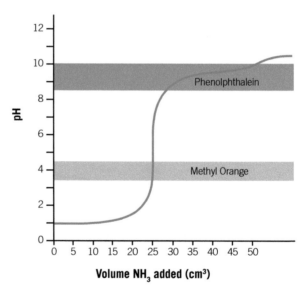

For a strong acid–weak base the pH increases sharply from 2 to 8. Methyl orange, litmus and bromothymol blue change fully in this range. Phenolphthalein starts to change at the end of this range, but it will not change fully, so we can't use it for a titration using a weak alkali. Methyl orange is the most commonly used indicator for this type of titration.

Weak acid–weak base

There is no significant vertical area showing a significant and sudden change in pH. No indicator would therefore work and alternative methods must be used for titration. This is usually carried out by using a pH probe. The results are plotted on a graph which then allows us to identify the equivalence point.

Practical check »

The titration of a weak acid–weak base using a pH probe is a **specified practical task**. This section explains why a pH probe is used rather than an indicator as is the case for all other types of titration. A weak acid–weak base titration is only used when no other method is available as the technique is more challenging than titration using an indicator.

Test yourself

1. State what is meant by the term weak acid. [1]

2. Define pH. [1]

3. Write an expression for the ionic product of water, K_w. [1]

4. Calculate the pH of a 0.20 mol dm^{-3} solution of the acid, H_2SO_4. You may assume the acid is fully dissociated. [2]

5. Find the concentration of H$^+$ ions in a solution of pH 2.45. [1]

6. The K_a value for the weak acid propanoic acid is 1.34×10^{-5} mol dm^{-3}.

   (a) Write an expression for K_a for propanoic acid, CH_3CH_2COOH. [1]

   (b) Calculate the pH of a 0.100 mol dm^{-3} solution of propanoic acid. [2]

   (c) A mixture of propanoic acid and sodium propanoate is classed as a buffer.

      (i) State what is meant by the term buffer. [1]

      (ii) Explain how this mixture functions as a buffer. [3]

      (iii) A student makes a buffer solution that has 0.100 mol dm^{-3} propanoic acid and 0.200 mol dm^{-3} sodium propanoate. Calculate the pH of this solution. [3]

7. Calculate the pH of a 0.25 mol dm^{-3} solution of the strong base NaOH. [2]

8. Suggest a pH value for a solution of sodium ethanoate, giving a reason for your answer. [2]

9. A 25.0 cm^3 solution of weak acid HX is titrated against a sodium hydroxide solution of concentration 0.100 mol dm^{-3}. The titration curve obtained is shown below.

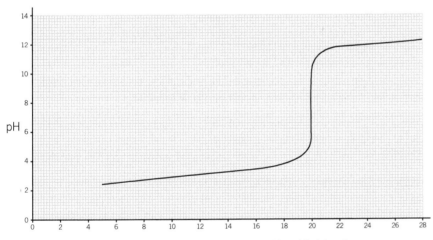

Volume of aqueous sodium hydroxide added / cm^3

   (a) State the volume of sodium hydroxide solution needed to neutralise the acid and hence calculate the concentration of the acid. [2]

   (b) State which, if any, of the following indicators would be suitable for this titration. Give a reason for your answer. [2]

Indicator	pH range
Methyl orange	3.2–4.4
Cresol red	7.2–8.8
Phenolphthalein	8.2–10.0

   (c) Use the graph to find the K_a value of acid HX. Show clearly how you obtained your value. [2]

Unit 3

When approaching the questions in Unit 3, it is important to read each question carefully and pick out the key elements required. The command word (in **bold** below) will guide you to what is needed, but there may be more than one command in a question or bullet points indicating the areas to be discussed, and there will be marks for each. Some examples:

1 (a) **State**, giving a reason, the difference(s) in the observations when excess sodium hydroxide is added to separate solutions of magnesium nitrate and lead nitrate. [2]

   (b) The chemistry of lead is very different from that of carbon. **Discuss** the differences in the chemistry of these elements and their compounds. You should include:
   - The bonding present in both and the effect on their physical properties.
   - The stabilities of the oxidation states.
   - The acid–base properties of the oxides of the elements. [6 QER]

In part (a) you need to state differences in the observations and you should always refer to both species. You must also give a reason for the differences, and in some questions you need the reason to gain any marks. Part (b) is a longer 'Quality of extended response' question and when bullet points are given, you need to make sure you address each one.

Many questions in Unit 3 require numerical responses, or conclusions based on the results of calculations. In these questions it is important to identify the numerical values you are given **including their units** and the value to be calculated. In many cases you will need to change between different units (such as Pa/atm or °C/K) or between different SI prefixes (such as J/kJ or $cm^3/dm^3/m^3$). It is often helpful to list all values and change the units as required as a first step. It is good practice to show your workings in calculations. However if a question states that you **must** show your workings, then you can lose marks if you do not, even if the answer is correct. Here is an example:

   (c) Many metal carbonates decompose when heated. Use the data given to find the minimum temperature for the decomposition of lead carbonate, giving your answer to an appropriate number of significant figures.

   $$PbCO_3 \rightarrow PbO + CO_2 \qquad \Delta H^\ominus = 97.0 \text{ kJ mol}^{-1} \qquad \Delta S^\ominus = 151 \text{ J K}^{-1} \text{ mol}^{-1}$$ [2]

In this case, the entropy must be divided by 1000 to convert it to kJ K^{-1} mol^{-1} and the answer given to three significant figures as this is the fewest significant figures in any of the data.

In some questions, particularly AO3 questions, all the observations and data may be given at the start of the question and you may need to select appropriate data or observations for each part question. It can be useful to annotate each part of the table with what it tells you. An example of a question like this is given below:

2 A mineral was studied in a series of tests. The results are given below.

Test	Result	
Addition of nitric acid to a solid sample	The solid effervesces giving off a gas that turns lime water milky.	*Test for carbonates – result positive*
Heating the mineral to constant mass	The mass of a 0.0200 mol sample of the solid loses 2.16 g upon heating to constant mass.	*Use this to find xH₂O in hydrated mineral*
Flame test	A brick-red flame is seen.	*Calcium flame colour*
Addition of concentrated sulfuric acid to the solid	Coloured fumes are formed with the smell of rotten eggs.	*Smell of rotten eggs = iodide?*

   (a) **State** which cation is present in the mineral. [1]

   (b) The compound contains two anions. **State** which anions are present, giving reasons for your answers. [4]

   (c) **Calculate** the number of moles of water present in each formula unit. [2]

   (d) Use your answers to parts (a)–(c) to **suggest** a chemical formula for the mineral. [2]

1 The reversible reaction shown below was one of the first to be studied in detail in the gas and liquid phases and in solution:

$$N_2O_4 \rightleftharpoons 2NO_2$$

(a) One study of the liquid mixture showed that it contained 0.714% NO_2 by mass. Calculate how many moles of NO_2 would be present in 25.0 g of this liquid. Give your answer to an appropriate number of significant figures. [2]

(b) Both N_2O_4 and NO_2 are soluble in a range of solvents and the equilibrium constant, K_c, can be measured in these.

(i) Give the expression for the equilibrium constant, K_c, for this equilibrium, giving its unit if any. [2]

(ii) The equilibrium constants for the N_2O_4 / NO_2 equilibrium measured at room temperature in some different solvents are listed below:

Solvent	Equilibrium constant, K_c (unit not shown)
CS_2	17.8
CCl_4	8.05
$CHCl_3$	5.53
C_2H_5Br	4.79
C_6H_6	2.03
$C_6H_5CH_3$	1.69

I. Samples of 0.4000 mol of N_2O_4 are dissolved separately in 1 dm^3 of each of these solvents at room temperature and the reactions allowed to reach equilibrium. In one solution the concentration of N_2O_4 present at equilibrium is 5.81×10^{-2} mol dm^{-3}. Find the value of the equilibrium constant in this solution and hence identify the solvent. [3]

II. The Gibbs free energy change, ΔG, of this reaction is different in different solvents. Explain how the data shows this and state, with a reason, which solvent would have the most negative ΔG value for this reaction. [2]

(c) Nitrogen dioxide is used in the production of nitric acid. It is produced in two stages:

Stage 1: $4NH_3(g) + 5O_2(g) \rightleftharpoons 4NO(g) + 6H_2O(g)$ $\Delta H^\theta = -900$ kJ mol^{-1}

Stage 2: $2NO(g) + O_2(g) \rightleftharpoons 2NO_2(g)$ $\Delta H^\theta = -115$ kJ mol^{-1}

(i) Explain why the use of a catalyst is essential in industrial processes involving exothermic equilibria such as stage 1. [2]

(ii) A student attempted to perform stage 2 using a temperature of 450 K and a pressure of 5 atm. He obtained only a small yield. Suggest how the yield could be improved, explaining your reasons. [4]

(Total 15 marks)

[*WJEC Unit 3 2017 Question 9*]

2 (a) A student planned to distinguish between the following eight compounds:

magnesium hydroxide   magnesium carbonate
iron(II) hydroxide   iron(II) carbonate
chromium(III) hydroxide   chromium(III) carbonate
lead(II) hydroxide   lead(II) carbonate

He used the following method:

Step 1:   Add dilute acid until all the solid has disappeared. Record any effervescence.

Step 2   Add 1 cm³ of sodium hydroxide solution to each solution formed in step 1. Record any precipitate observed.

(i)   State which compound(s) would give effervescence and why they do so.   [2]

(ii) The student plans to use dilute hydrochloric acid in step 1. His teacher tells him that this is not the correct acid to use. Explain why hydrochloric acid should not be used and suggest an appropriate acid to use in its place.   [2]

(iii) Give the colours of the precipitates formed when sodium hydroxide solution is added to solutions containing the following ions.   [2]

$Mg^{2+}(aq)$

$Fe^{2+}(aq)$

$Cr^{3+}(aq)$

$Pb^{2+}(aq)$

(iv) The method given in part (a) is incomplete.

Suggest an additional step that would allow the remaining solutions formed in step 2 to be identified. Give the expected observations. State the property of the metals that allows them to be distinguished in this way.   [3]

(b) Many rocks that contain carbonate anions also contain a mixture of cations. Atomic absorption spectroscopy can be used to find the ratio of the amounts of different cations present.

Analysis of an acid solution formed from the carbonate mineral huntite shows that it contains 91 ng cm⁻³ of magnesium ions and 50 ng cm⁻³ of calcium ions. These are the only two metal cations present.

(i)   Find the formula of the mineral huntite.   [3]

(ii) The initial solution was prepared using 220 µg of huntite. Calculate the volume of aqueous acid used to form the solution for analysis.   [3]

(Total 15 marks)

[WJEC Unit 3 2019 Question 12]

3   Hydrogen peroxide, H_2O_2, decomposes slowly at 300 K to form water and oxygen gas:

$$2H_2O_2(l) \rightleftharpoons 2H_2O(l) + O_2(g)$$

The reaction profile for this reaction is shown opposite:

(a) The rate of decomposition of hydrogen peroxide can be influenced by a range of catalysts. The activation energy when using MnO_2 as a catalyst is 58 kJ mol⁻¹.

Draw the reaction profile for the catalysed reaction **on the grid**.   [1]

(b) Another catalyst was used and this gave a value for the rate constant, k, of 1.68 mol⁻¹ dm³ s⁻¹ and the frequency factor, A, of 1.41×10^4 mol⁻¹ dm³ s⁻¹ at a temperature of 300 K.

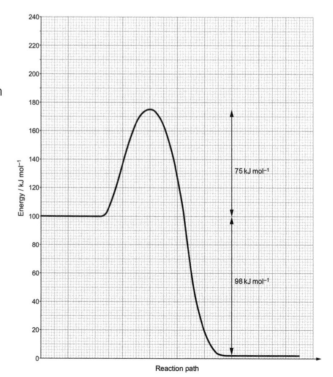

75 kJ mol⁻¹

98 kJ mol⁻¹

Energy / kJ mol⁻¹

Reaction path

(i)   State the Arrhenius equation.   [1]

(ii) Calculate the activation energy in kJ mol^{-1} using this catalyst and hence state whether it is a more effective catalyst than MnO_2.   [3]

(c) The standard enthalpy change of formation, $\Delta_f H^\oplus$, of water is -286 kJ mol^{-1}.

Use this information and the graph to calculate the standard enthalpy change of formation of hydrogen peroxide in kJ mol^{-1}.   [2]

(d) State whether you would expect the entropy change for the decomposition of hydrogen peroxide to be positive or negative. Give a reason for your answer.   [1]

(e) One way of assessing whether a reaction is feasible is to use standard electrode potentials.

	Standard electrode potential / V
$H_2O_2 + 2H^+ + 2e^- \rightleftharpoons 2H_2O$	+1.77
$Cr_2O_7^{2-} + 14H^+ + 6e^- \rightleftharpoons 2Cr^{3+} + 7H_2O$	+1.33

(i)   The apparatus below can be used to measure the standard electrode potential for the $Cr_2O_7^{2-}/Cr^{3+}$ half-cell.

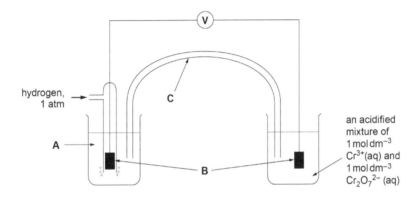

I.   State what is represented by **A** and **B** on the diagram.   [1]

II.   **On the diagram** show the direction of flow of electrons in the external circuit.   [1]

III. State what is represented by **C** on the diagram and state its function.   [1]

IV.   The concentrations of both $Cr_2O_7^{2-}$ and Cr^{3+} ions are 1 mol dm^{-3}. State and explain how the value shown on the high resistance voltmeter would change if the concentration of the Cr^{3+} ions were increased whilst the concentration of the $Cr_2O_7^{2-}$ was left unchanged.   [2]

(ii) It is suggested that hydrogen peroxide could be used to oxidise Cr^{3+} ions in acidic solution to form dichromate ions.

I.   Write an equation for this proposed reaction.   [1]

II.   Use the standard electrode potential values given to predict whether this reaction is feasible.   [2]

III. Another method of finding whether a reaction is feasible is to use the Gibbs free energy calculated from standard enthalpy of formation and standard entropy values.

State, giving a reason, whether Gibbs free energies or electrochemical methods are more appropriate for finding whether this reaction is feasible.   [2]

(Total 18 marks)

[WJEC Unit 3 2018 Question 11]

4 Cobalt forms a range of complex ions. Two of these are $[Co(H_2O)_6]^{2+}$ and $[CoCl_4]^{2-}$.

(a) Draw the structures of the $[Co(H_2O)_6]^{2+}$ and $[CoCl_4]^{2-}$ ions. [2]

(b) Explain why the $[Co(H_2O)_6]^{2+}$ ion is coloured. [3]

(c) A student was given a pink-coloured solution containing $[Co(H_2O)_6]^{2+}$ ions.

Upon addition of hydrochloric acid the solution turned blue as $[CoCl_4]^{2-}$ ions formed according to the equilibrium shown below:

$$[Co(H_2O)_6]^{2+} + 4Cl^- \rightleftharpoons [CoCl_4]^{2-} + 6H_2O$$

Aqueous silver nitrate was added to the solution containing $[CoCl_4]^{2-}$.

State and explain the observation(s) expected. [4]

(d) These two cobalt-containing ions absorb different frequencies of light. The spectra below show the absorbance at different wavelengths of visible light by the two ions:

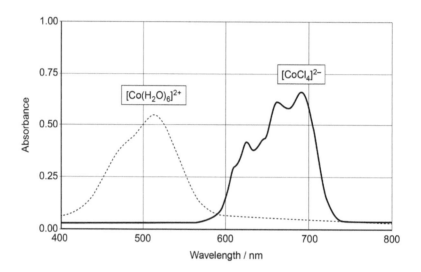

(i) The maximum absorbance for the $[Co(H_2O)_6]^{2+}$ ion occurs at a wavelength of 518 nm. Calculate the energy of this absorbance in kJ mol^{-1}. [3]

(ii) A colorimeter was used to study the equilibrium between the two ions.

Suggest, giving a reason, an appropriate wavelength to use for the experiment. [1]

(iii) Write the expression for the equilibrium constant, K_c, for this reaction **giving its unit**. [2]

$$[Co(H_2O)_6]^{2+} + 4Cl^- \rightleftharpoons [CoCl_4]^{2-} + 6H_2O$$

(iv) A student carries out an experiment starting with separate non-aqueous solutions of $[Co(H_2O)_6]^{2+}$ and Cl^-.

When the two solutions are mixed the initial concentrations of both $[Co(H_2O)_6]^{2+}$ and Cl^- in the mixture are 0.720 mol dm^{-3}. When the mixture reaches equilibrium the concentration of the water is 0.744 mol dm^{-3}.

Calculate the value of the equilibrium constant, K_c. [3]

(Total 18 marks)

[WJEC Unit 3 2018 Question 9]

5 Brass is an alloy of copper and zinc only. The copper content of the alloy can be found by volumetric or gravimetric analysis. The brass is dissolved by adding highly acidic mixtures to the alloy which forms $Cu^{2+}(aq)$ and amphoteric $Zn^{2+}(aq)$.

(a) Redox titration is one method to find the mass of copper in a known mass of alloy.

A 2.877 g sample of alloy is dissolved in concentrated nitric acid. The mixture is neutralised and then made up to a volume of 250.0 cm³.

Samples of the solution with a volume of 25.00 cm³ are removed and excess potassium iodide solution added, before titration with 0.105 mol dm⁻³ sodium thiosulfate solution. The mean volume of sodium thiosulfate needed to completely reduce the iodine in solution is 26.75 cm³.

Calculate the percentage by mass of copper in this alloy. You must show your working.   [4]

(b) An alternative method is gravimetric analysis.

Another sample of alloy is dissolved in concentrated nitric acid. The solution is neutralised and aqueous sodium hydroxide is added until all the copper(II) and zinc(II) ions form metal hydroxide precipitates. This sample is then filtered, dried and weighed (weighing 1).

The solid sample is then treated with excess aqueous sodium hydroxide and the remaining solid is removed by filtration, dried and weighed (weighing 2).

The results are given below:

Mass of empty vessel = 23.34 g

Mass of vessel and precipitate (weighing 1) = 25.12 g

Mass of vessel and precipitate (weighing 2) = 24.45 g

Calculate the percentage by mass of copper in this alloy. You **must** show your working.   [4]

(c) A student suggests that the alloys in parts (a) and (b) are the same. State and explain whether the evidence supports this statement and suggest what further evidence should be collected to confirm your conclusion.   [2]

(d) (i) Concentrated nitric acid is used to dissolve the alloy in the experiments above. The pH of this strong acid is typically −1.2.

Calculate the concentration of this nitric acid in mol dm⁻³.   [2]

(ii) The acidic solution is neutralised using aqueous sodium hydroxide of concentration 2.00 mol dm⁻³.

Calculate the pH of this sodium hydroxide solution.   [2]

[ionic product of water, $K_w = 1.00 \times 10^{-14}$ mol² dm⁻⁶]

(Total 14 marks)

[Eduqas Component 1 2018 Question 13]

Stereoisomerism

Molecules of organic compounds often contain a large number of atoms. For example, a molecule of the explosive TNT, $C_6H_2(CH_3)(NO_2)_3$, has 21 atoms from four different elements, and a molecule of sucrose, $C_{12}H_{22}O_{11}$, has 45 atoms. Even in a simpler molecule having a molecular formula C_2H_6O, the atoms can be arranged to give ethanol, C_2H_5OH or bonded together with a central oxygen atom to give methoxymethane, CH_3OCH_3. Compounds that have the same molecular formula but are bonded differently are called structural isomers. There is another type of isomerism where differences arise because of the position the atoms take up in space. We call this stereoisomerism. In the first year of this course, we have met *E-Z* isomerism and we now look at another form of stereoisomerism, optical isomerism.

Topic contents

You should be able to demonstrate and apply knowledge and understanding of:

- How stereoisomerism is distinct from structural isomerism and that stereoisomerism encompasses *E-Z* isomerism and optical isomerism.

- The terms chiral centre, enantiomer, optical activity and racemic mixture.

- Optical isomerism in terms of an asymmetric carbon atom.

- The effect of an enantiomer on plane polarised light.

Maths skills ▶▶

- Visualise and represent 2D and 3D forms including 2D representations of 3D objects

Structural isomerism and stereoisomerism

Structural isomers are compounds that have the same molecular formula but differ in their arrangement of atoms. One of the simplest types is chain isomerism, where the chains are straight or branched.

$$CH_3CH_2CH_2CH_2CH_3$$

pentane

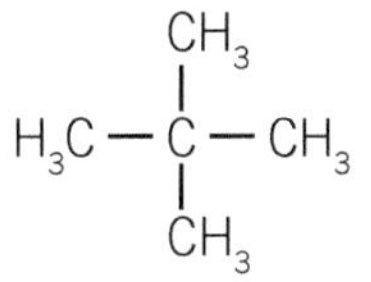

$$H_3C - \underset{\underset{CH_3}{|}}{\overset{\overset{CH_3}{|}}{C}} - CH_3$$

dimethylpropane

If the compounds contain a functional group then this may be at a different position on the chain.

$$CH_3CH_2CH_2OH \qquad CH_3CH(OH)CH_3$$

propan-1-ol   propan-2-ol

The compounds may have the same molecular formula but a different functional group is present. For example, both cyclohexanol (a secondary alcohol) and hexanal (an aldehyde) have the molecular formula $C_6H_{12}O$.

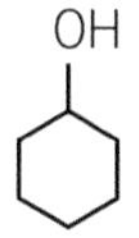

cyclohexanol

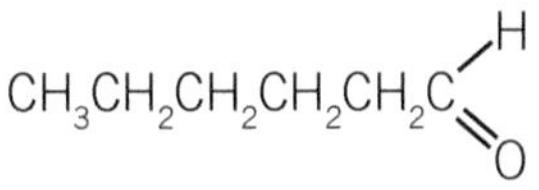

$$CH_3CH_2CH_2CH_2CH_2C\overset{\diagup H}{\underset{\diagdown O}{}}$$

hexanal

One form of stereoisomerism is *E-Z* isomerism. Stereoisomerism occurs where the isomers have the same shortened structural formula but have a different arrangement of the atoms in space. *E-Z* isomerism is studied during the first year of this course. This type of isomerism is shown in alkenes. Dichloroethene can have both chlorine atoms bonded to the same carbon atom or each carbon atom can be bonded to one chlorine atom.

$$H_2C = CCl_2 \qquad\qquad ClHC = CHCl$$

1,1-dichloroethene   1,2-dichloroethene

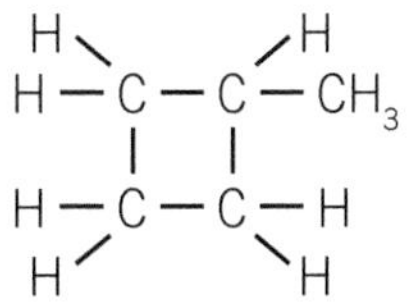

These two compounds are structural isomers. However, 1,2-dichloroethene can be bonded so that each chlorine is opposite each other or both atoms are on the same side of the double bond.

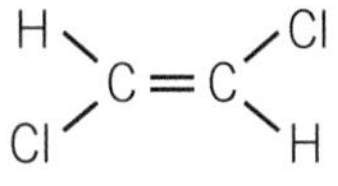

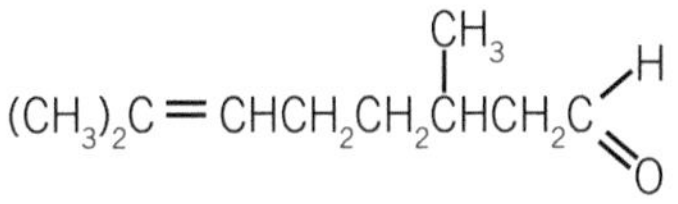

(*E*)-1,2-dichloroethene 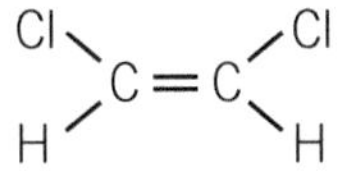   (*Z*)-1,2-dichloroethene

The π bond restricts rotation about the double bond and, as a result, 1,2-dichloroethene exists in two different forms, called *E-Z* isomers.

The other type of stereoisomerism is called optical isomerism and occurs where the two isomers have different effects on plane polarised light.

Optical isomerism terms

>> **Key terms**

A **chiral centre** is an atom in a molecule that is bonded to four different atoms or groups.

Enantiomers are non-superimposable mirror image forms of each other that rotate the plane of plane polarised light in opposite directions.

Optical activity occurs in molecules that possess chiral centre(s). These molecules rotate the plane of plane polarised light.

A **racemic mixture** is an equimolar mixture of both enantiomers that produces no overall rotation of the plane of plane polarised light.

Chiral centre

A **chiral centre** is an atom that is bonded to four different groups or atoms. This is often a carbon atom and we call this a chiral carbon atom. In older books these chiral carbon atoms were called asymmetric carbon atoms. The four different atoms or groups bonded to this chiral centre can be bonded in two different ways, which are mirror images of each other.

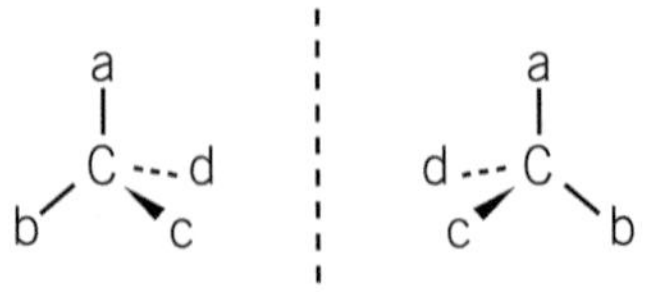

The dotted lines go backward into the page and the wedge shaped lines come out of the page towards you. A common example of a compound with a chiral centre is butan-2-ol, $CH_3CH_2CH(OH)CH_3$.

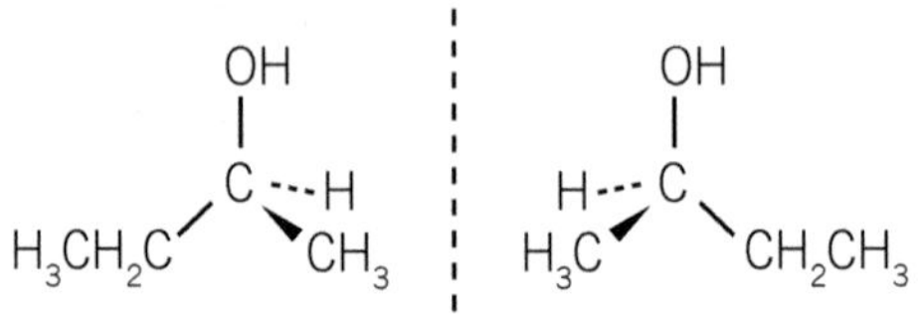

Enantiomer

Enantiomers are stereoisomers that are non-superimposable mirror image forms of each other. These can also be called optical isomers. If a compound contains two or more chiral centres, it is possible to have stereoisomers that are not mirror image forms of each other. These are called diastereoisomers, but are not within the scope of this specification. However, you should be able to identify chiral centres within this type of molecule. The diagram shows the formula of butane-1,2-diol with the chiral centres indicated by asterisks.

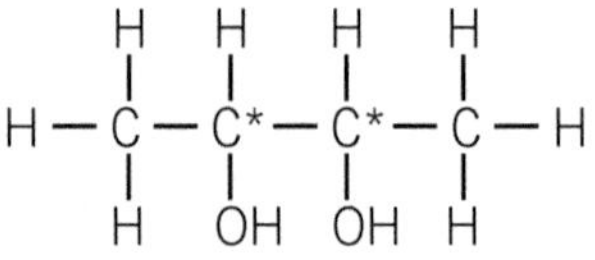

>> **Study point**

Carbon atoms that have multiple bonds cannot act as chiral centres as they cannot be bonded to four other atoms or groups. As a result these compounds do not have a chiral centre.

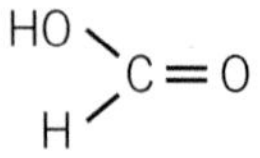

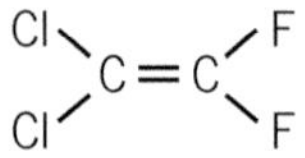

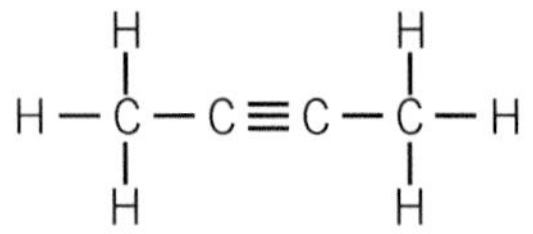

◀**Stretch & challenge**

The formula of 1,3-dichloropropan-2-ol is

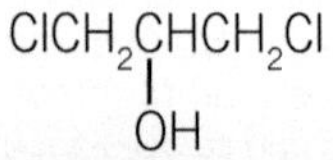

$CICH_2CHCH_2Cl$
|
OH

Give the displayed formula of another isomer of formula $C_3H_6Cl_2O$ that has a chiral centre.

Optical activity

Optical activity occurs in molecules with chiral centres. Light consists of waves that are vibrating in all planes. If light is passed through a polarising filter (e.g. a piece of Polaroid), the light that emerges is vibrating in one plane only. This light is called plane polarised light. If a solution of an enantiomer is placed in a beam of plane polarised light, the beam is rotated. The instrument used to measure the amount of rotation is called a polarimeter.

Details of how a polarimeter is used and the factors affecting the angle of rotation do not form part of the specification.

An enantiomer may rotate the plane of polarised light to the right (+) or to the left (–). Equal amounts of each enantiomer in a solution produce no overall rotation since the rotation effects cancel each other out. This mixture results from external compensation, as the effect is caused by two different compounds. This equimolar mixture is called a **racemic mixture.**

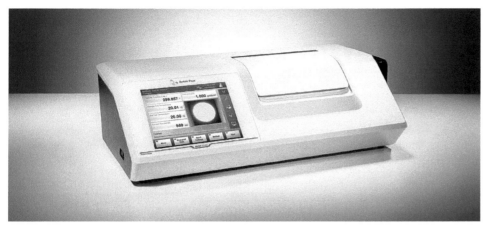

▲ Automatic digital polarimeter

The painkiller ibuprofen was first made in 1961.

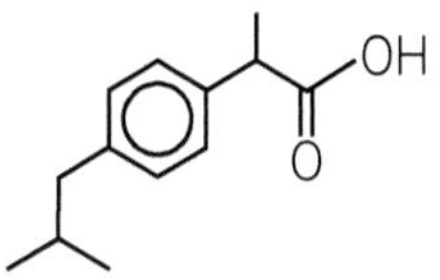

Ibuprofen contains a chiral centre and is sold as a racemic mixture. Since biological systems usually respond to one enantiomer in the required way rather than the other, there is a danger that the 'unwanted' form could have serious and unwanted side effects. Separation of the two enantiomers (called resolution) would be difficult and expensive. When ibuprofen is taken, one of the enantiomers is much more biologically active than the other but fortunately the less active form is converted by an enzyme in the body into the other, more active enantiomer.

Knowledge check 3

Explain why a racemic mixture can be described as an equimolar mixture or as a mixture containing equal amounts of both enantiomers.

Knowledge check 4

Malic acid occurs in unripe apples. Its systematic name is 2-hydroxybutanedioic acid. Write the displayed formula for malic acid and identify any chiral centre present in its formula.

Knowledge check 5

Explain why the unsaturated acid that has the formula $CH_3(CH_2)_7CH=CH(CH_2)_7COOH$ exists as E-Z isomers.

Test yourself

1. Use an asterisk(s) to indicate any chiral centres in the following formulae. [6]

   (a) 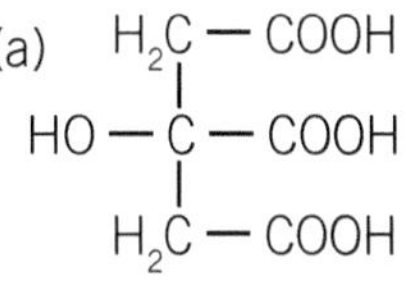
   $H_2C - COOH$
   $HO - C - COOH$
   $H_2C - COOH$

   (b) 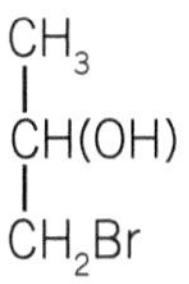
   CH_3
   $CH(OH)$
   CH_2Br

   (c)

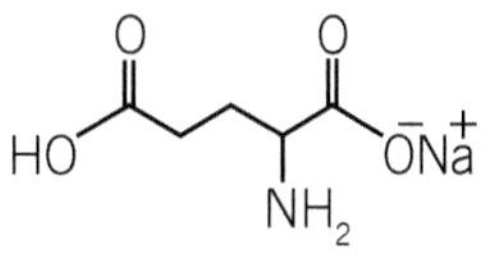

   (d)

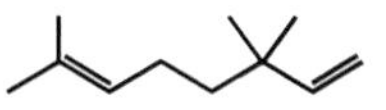

   (e) 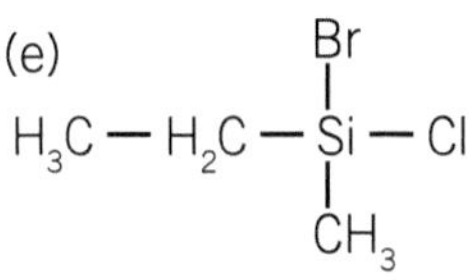
   $H_3C - H_2C - Si - Cl$ with Br above and CH_3 below

   (f) $(H_3C)_3SiCl$

2. Draw the two mirror image forms of pentan-2-ol. [1]

3. Calculate how many grams of each enantiomer there are in a solution containing a total of 0.100 mol of phenylalanine, $C_6H_5CH_2CH(NH_2)COOH$. [3]

4. Give the labelled structures of the E- and Z- forms of 2-chlorohex-2-ene. [2]

5. (a) 1-Hydroxypropane-1,2,3-tricarboxylic acid contains a chiral centre. Give the structure of this acid, showing the position of the chiral centre. [2]

   (b) This acid can be dehydrated to give an unsaturated tricarboxylic acid. Give the structure of this unsaturated acid. [1]

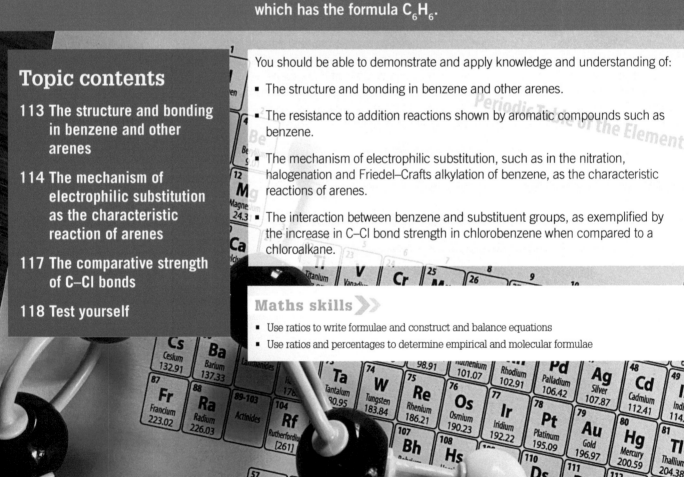

4.2 Aromaticity

The first year of this course includes work on aliphatic and alicyclic compounds. Aliphatic compounds are traditionally classed as compounds where the carbon atoms form open chains, rather than rings. Alicyclic compounds do have a ring structure but they react as if they are aliphatic compounds. In the second year of this course, aromatic compounds are introduced. The word 'aromatic' comes from the Latin word 'aroma' meaning fragrance. Not all aromatic compounds have a fragrant smell but they do all contain an aromatic ring system. The simplest aromatic compound is benzene, which has the formula C_6H_6.

Topic contents

You should be able to demonstrate and apply knowledge and understanding of:

- The structure and bonding in benzene and other arenes.

- The resistance to addition reactions shown by aromatic compounds such as benzene.

- The mechanism of electrophilic substitution, such as in the nitration, halogenation and Friedel–Crafts alkylation of benzene, as the characteristic reactions of arenes.

- The interaction between benzene and substituent groups, as exemplified by the increase in C–Cl bond strength in chlorobenzene when compared to a chloroalkane.

Maths skills >>

- Use ratios to write formulae and construct and balance equations
- Use ratios and percentages to determine empirical and molecular formulae

The structure and bonding in benzene and other arenes

In 1825 Michael Faraday isolated a colourless flammable liquid from whale oil, whilst obtaining a suitable gas for street lighting.

He found that this flammable liquid had the empirical formula CH. Nine years later the liquid, now called benzene, was found to have a relative molecular mass of 78 and the molecular formula C_6H_6. Later workers found that benzene was present in coal tar, from which it was obtained for many years. In 1865 Kekulé suggested that the structure of benzene was a six-membered ring, containing alternating single and double carbon-to-carbon bonds.

Unfortunately, the Kekulé model for the structure of benzene does not explain some of benzene's reactions. For example, it should react as an alkene and easily undergo addition reactions, such as the reaction with aqueous bromine, where the bromine is decolourised. However, it does not react in this way. To explain this discrepancy, Kekulé proposed that benzene had two forms and suggested that one form changed to the other so quickly that an approaching molecule would have no time to react with it by addition.

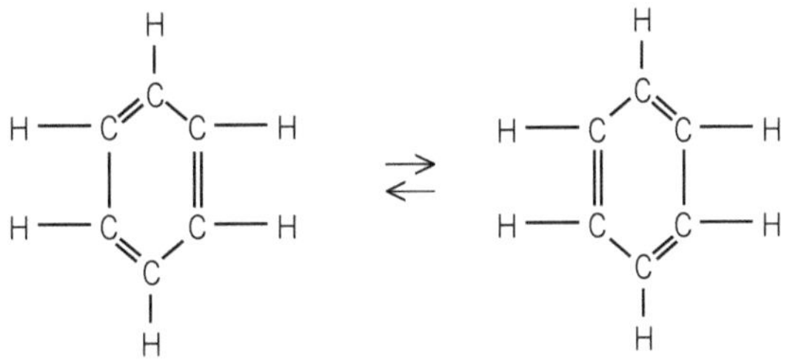

If benzene existed as the Kekulé forms above, then there would be two different bond lengths between carbon atoms in the molecule. In the early 20th century, x-ray crystallography showed that each carbon-to-carbon bond was the same length at 0.140 nm. This distance is between a carbon-to-carbon double bond at 0.135 nm and a carbon-to-carbon single bond at 0.147 nm.

When cyclohexene is hydrogenated, the energy released is 120 kJ mol⁻¹.

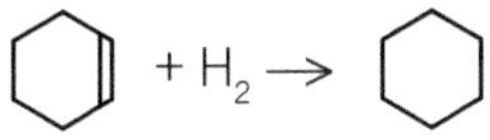

If benzene existed as the Kekulé structure and all three double bonds were hydrogenated, then the enthalpy change would be –360 kJ mol⁻¹. However, when benzene is fully hydrogenated the enthalpy change is –208 kJ mol⁻¹.

This is 152 kJ mol⁻¹ less than expected, suggesting that benzene is more stable than the Kekulé structure, and it suggests that the Kekulé structure is incorrect. This difference in energy values is called the resonance energy.

The delocalised bonding model of benzene

Modern studies of benzene have shown that it is a planar molecule and that the angles between three adjacent carbon atoms are 120°. In an ethene molecule, C_2H_4, each carbon atom is bonded to two hydrogen atoms and a carbon atom by sigma (σ) bonds. The remaining outer p electrons of each carbon atom overlap above and below the plane of the molecule, giving a localised pi (π) orbital.

Knowledge check

Look at the Kekulé structure of benzene below.

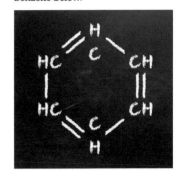

Why might it lose a mark in the examination?

Knowledge check

Choose the correct statement about the structure of benzene.

(a) Each carbon atom is bonded to two hydrogen atoms and one carbon atom.

(b) The C–C–C bond angle is 120°.

(c) The C=C bond length is longer than the C–C bond length.

(d) The smaller the resonance energy, the more stable the molecule.

Stretch & challenge

Ethene reacts with cold dilute aqueous potassium manganate(VII) to produce ethane-1,2-diol, CH_2OHCH_2OH. Suggest why benzene cannot react in a similar way to give the diol shown below.

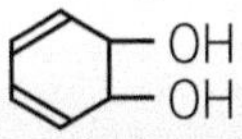

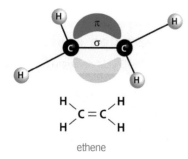

ethene

In benzene each carbon atom is bonded to two other carbon atoms and one hydrogen atom by sigma bonds. The fourth outer shell electron of a carbon atom is in a 2p orbital, above and below the plane of the carbon atoms. These p-orbitals overlap to give a delocalised electron structure, above and below the plane of carbon atoms. This delocalised electron π structure is often represented by a circle inside the ring hexagon.

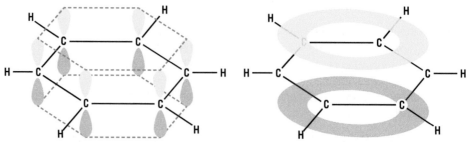

Compounds containing a single benzene ring represent the simplest type of aromatic system but there are a number of other aromatic ring systems, some with fused sides. These include naphthalene and anthracene. There are also aromatic compounds whose rings contain atoms other than carbon, for example pyridine, C_5H_5N.

naphthalene   anthracene   pyridine

If benzene underwent an addition reaction, the process would disrupt the stable delocalised electron system and the resulting product would be less stable. Benzene can be made to react by addition but more forcing conditions are needed. We have already seen that benzene can be hydrogenated to give cyclohexane but the reaction needs high temperatures and a nickel or platinum catalyst. Benzene will also react with chlorine in an addition reaction to give hexachlorocyclohexane, $C_6H_6Cl_6$. This is a radical reaction that needs bright sunlight to take place. One isomer of formula $C_6H_6Cl_6$, γ-hexachlorocyclohexane, was produced under the trade name of Lindane© and used as an insecticide. The use of Lindane© is now restricted as this toxic material is very persistent in the environment. Benzene generally reacts by substitution reactions where the delocalised system of electrons is retained.

The mechanism of electrophilic substitution as the characteristic reaction of arenes

Benzene has a delocalised ring of electrons above and below the plane of the carbon atoms. This area of high electron density makes it susceptible to attack by an **electrophile**. A hydrogen atom (as H⁺) is replaced by the incoming electrophile. This type of reaction mechanism is called electrophilic substitution. To replace a hydrogen atom by an electrophile X, the stability of the ring needs to be disturbed, giving an unstable intermediate (called a Wheland intermediate), which then loses a hydrogen ion, H⁺, in a rapid step.

intermediate

Nitration

In a nitration reaction, a hydrogen atom is replaced by a nitro group, NO_2. The electrophile is the nitryl cation (or nitronium ion), NO_2^+, and this is produced by the reaction of concentrated nitric acid and concentrated sulfuric(VI) acid (sometimes called a nitrating mixture).

$$HNO_3 + 2H_2SO_4 \rightleftharpoons NO_2^+ + H_3O^+ + 2HSO_4^-$$

The nitryl cation reacts with benzene giving nitrobenzene, $C_6H_5NO_2$, as the organic product.

One mechanism for the nitration is

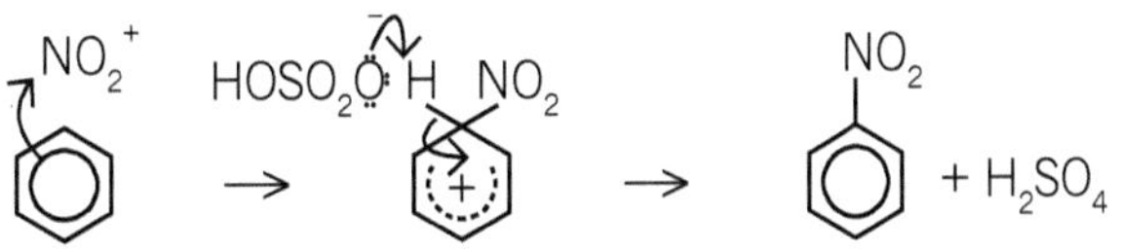

The overall reaction is

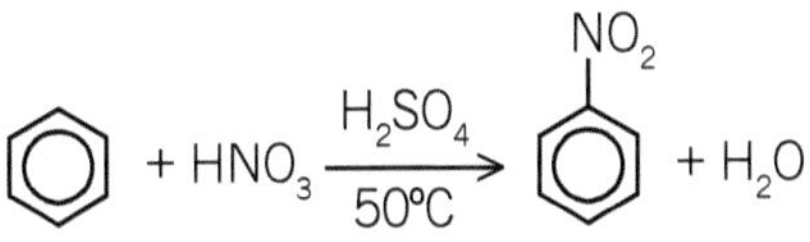

Nitrobenzene is a yellow liquid that is reduced to phenylamine, $C_6H_5NH_2$, by using tin metal and hydrochloric acid.

If the temperature of nitration exceeds 50°C then some 1,3-dinitrobenzene is also produced.

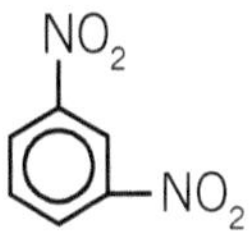

Halogenation

Benzene and bromine do not react together unless a catalyst such as iron filings, iron(III) bromide or aluminium bromide is present.

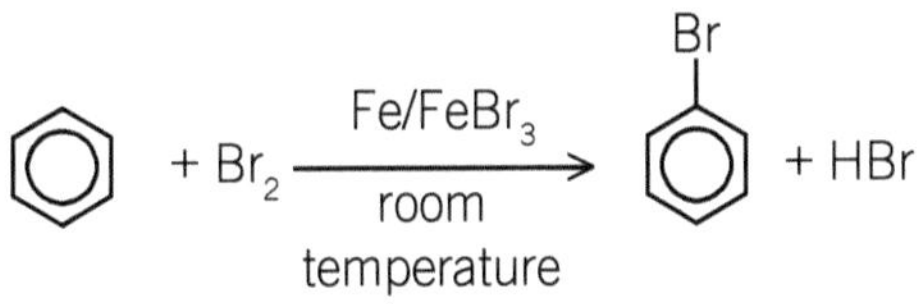

The π-electron system is not sufficiently nucleophilic to polarise the bromine molecule to any extent to give $Br^{\delta+}- Br^{\delta-}$. In the presence of iron(III) bromide the polarisation of the bromine molecule is more pronounced.

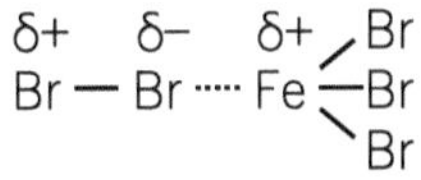

If iron filings are used then these react with bromine to give iron(III) bromide. One representation of this electrophilic substitution of bromine into the ring is

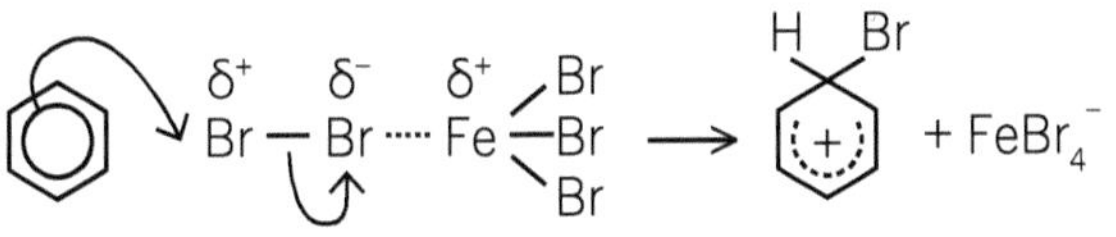

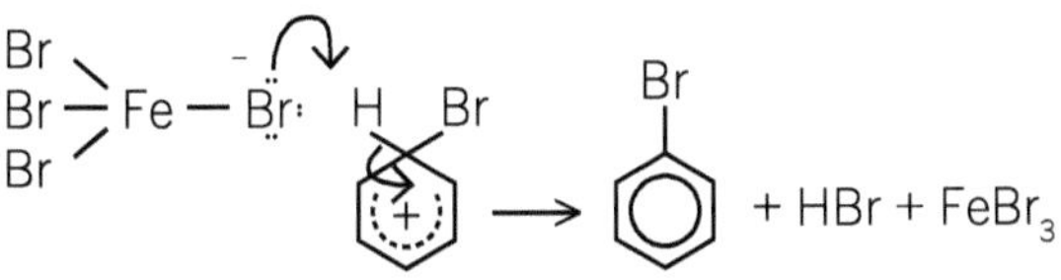

The role of the iron(III) bromide is catalytic – encouraging greater polarisation of the Br–Br bond so that attack by the ring electrons can occur but being regenerated at the end of the reaction.

Stretch & challenge

Benzene reacts with sulfuric(VI) acid by electrophilic substitution to give benzenesulfonic acid, $C_6H_6O_3S$.

$$C_6H_6 + H_2SO_4 \rightarrow C_6H_5SO_3H + H_2O$$

Top tip

When you draw electrophilic aromatic substitution reactions, the dotted line in the formula of the positively charged intermediate should not extend all the way around the ring. It should not reach the carbon atom at which substitution is occurring.

Chlorination of benzene can be carried out in a similar way, often using anhydrous aluminium chloride or iron(III) chloride as the catalyst. If an excess of chlorine is used, a mixture of 1,2-dichlorobenzene and 1,4-dichlorobenzene is produced.

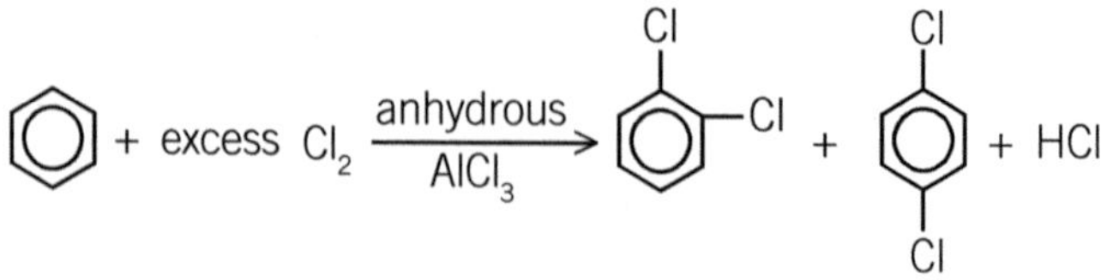

In industry, the chlorination of benzene is carried out as a continuous process, so that any polychlorination is kept to a minimum.

Chlorobenzene is an important industrial chemical. It is nitrated and the nitro- derivatives converted to 2-nitrophenol and 2-nitrophenylamine. The insecticide DDT can be manufactured by the reaction of chlorobenzene with trichloroethanal.

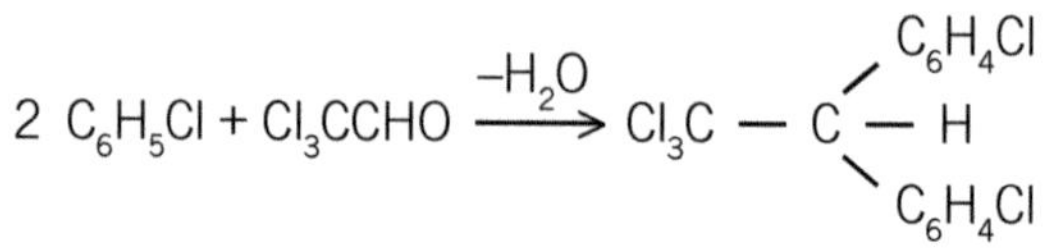

The use of DDT is now heavily restricted because of problems connected with its toxicity, its persistence in the environment and its presence in the food chain. This preparation of DDT is not a requirement for the examination.

Friedel–Crafts alkylation

The Friedel–Crafts alkylation method provides a way of producing a new carbon-to-carbon bond, giving alkyl derivatives of benzene such as methylbenzene, $C_6H_5CH_3$ and ethylbenzene, $C_6H_5CH_2CH_3$. The reaction is similar to the halogenation of benzene but uses a halogenoalkane in place of the halogen. Anhydrous aluminium chloride is often used as the catalyst.

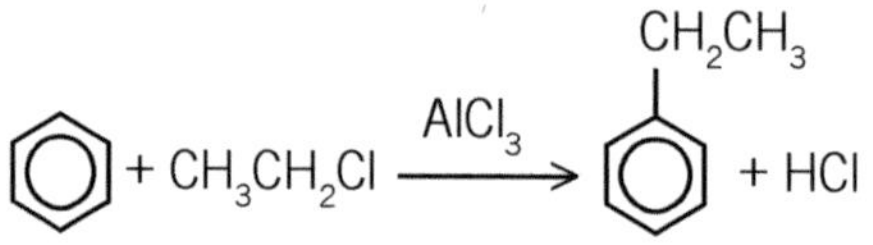

The catalyst reacts with the chloroethane to form the ethyl carbocation which acts as the electrophile for reaction with the benzene ring.

$$CH_3CH_2Cl + AlCl_3 \longrightarrow CH_3CH_2^+AlCl_4^-$$

One problem with this reaction is that the introduction of an alkyl group onto the ring activates the ring towards further alkylation. As a result, the product may also contain 1,2- and 1,4-diethylbenzenes. To reduce the chance of any polyalkylation, the halogenoalkane is added slowly to the benzene and the catalyst.

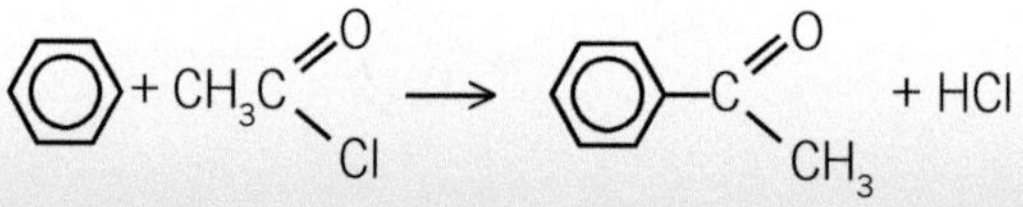

$$\text{e.g. } CH_3CH_2\overset{+}{C}H_2 \longrightarrow CH_3\overset{+}{C}HCH_3$$

Stretch & challenge

Acid chlorides, such as ethanoyl chloride, also react with benzene in a similar Friedel–Crafts reaction, giving a ketone.

Study point

When giving a complete equation for the halogenation of benzene, you should also show the other product of the reaction (usually HBr or HCl).

3 Knowledge check

A student was asked to give two reasons how he knew that iron(III) bromide was acting as a catalyst in the bromination of benzene. What do you think he gave as his correct answers?

Stretch & challenge

Give the systematic name of the choroalkane that will react with benzene to give

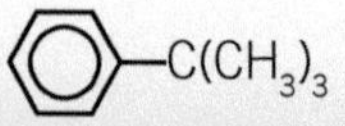

The comparative strength of C–Cl bonds

Chloroalkanes such as 1-chlorohexane, react with aqueous sodium hydroxide on heating under reflux, giving hexan-1-ol as the main organic product.

$$CH_3(CH_2)_4 CH_2Cl + NaOH \longrightarrow CH_3(CH_2)_4CH_2OH + NaCl$$

The mechanism for this nucleophilic substitution reaction is:

$$CH_3(CH_2)_4 - \overset{\overset{\displaystyle H}{|}}{\underset{\underset{\displaystyle H}{|}}{C}} - \overset{\delta+}{\underset{\overset{\displaystyle \ddot{O}^- - H}{}}{\overset{\delta-}{Cl}}} \longrightarrow CH_3(CH_2)_4 - \overset{\overset{\displaystyle H}{|}}{\underset{\underset{\displaystyle H}{|}}{C}} - \ddot{O} - H + \ddot{C}l^-$$

Aromatic compounds with a halogen atom on a side chain react in a similar way.

$$C_6H_5CH_2Cl + NaOH \longrightarrow C_6H_5CH_2OH + NaCl$$

The rate of hydrolysis of the carbon-halogen bond in this type of reaction with sodium hydroxide is C–I > C–Br > C–Cl. Benzene tends to react by electrophilic substitution. It has little tendency to react with nucleophiles, as these would be repelled by the stable π-system of electrons.

The bond energies and bond lengths of aliphatic and aromatic carbon-to-chlorine bonds are shown in the table.

Bond	Bond length / nm	Bond energy / kJmol^{-1}
aliphatic C — Cl	0.177	346
aromatic C — Cl	0.169	399

The stronger (and shorter) bond between carbon and chlorine in chlorobenzene results from a non-bonding p electron pair on chlorine overlapping with the ring π-system of electrons. The resulting bond needs much more energy to be broken and is stronger that an aliphatic C–Cl bond. Forcing conditions are therefore needed to produce phenol from chlorobenzene.

$$\underset{}{\text{Cl}} \quad \xrightarrow[\text{300°C/increased pressure}]{\text{NaOH(aq)}} \quad \underset{}{\text{OH}} \quad + \text{NaCl}$$

This reaction is not an environmentally viable method of producing phenol commercially. In industry, phenol is generally made from (1-methylethyl)benzene (cumene). This method has the advantage that propanone is a useful co-product of the reaction.

◀ **Stretch & challenge**

Compound M contains both aliphatic and aromatic C–Cl bonds. On heating with aqueous sodium hydroxide, it gives a new compound of formula $C_7H_3Cl_5O$. Deduce the structural formula of compound M, giving reasons for your answer.

Knowledge check 4

A sample of 1-bromo-4-(bromomethyl)benzene was heated under reflux with aqueous sodium hydroxide. Give the displayed formula of the most likely organic product.

Test yourself

1. Explain why benzene will not react with hydroxide ions, OH⁻, to give phenol as one of the products. [1]

2. Complete this mechanism using curly arrows and show the formulae of the two products. [2]

3. Explain why iron(III) bromide is used as a catalyst in the bromination of benzene. [1]

4. Methylbenzene can be nitrated to give a mixture of 1-methyl-2-nitrobenzene and 1-methyl-4-nitrobenzene.

   (a) State the name of the reagent A. [1]

   (b) Give the formula of the electrophile produced from reagent A [1]

   (c) State the name of a practical method that can be used for separating the products. [1]

5. The compound below is heated with aqueous sodium hydroxide. Give the formula of the organic product.

6. State which of the following will react with benzene by substitution. [1]

7. Explain why benzene is resistant to addition reactions. [1]

8. Naphthalene can be nitrated to give 1-nitronaphthalene as one of the products.

   (a) Give the molecular formula of naphthalene. [1]

   (b) In an experiment 25.6g of naphthalene was nitrated to give 31.1 g of 1-nitronaphthalene. Show that the percentage yield of 1-nitronaphthalene was 90%. [3]

Alcohols and phenols

An introduction to the chemistry of the alcohols is part of the AS course and is studied in Topic 2.7. Alcohols contain the –OH group bonded to a carbon atom and these are studied to a greater depth in this section. The A2 course provides an introduction to aromatic chemistry and this section also considers phenols where an –OH group is bonded directly to a benzene ring.

You should be able to demonstrate and apply knowledge and understanding of:

- The methods of forming primary and secondary alcohols from halogenoalkanes and carbonyl compounds.

- The reactions of primary and secondary alcohols with hydrogen halides, ethanoyl chloride and carboxylic acids.

- The acidity of phenol and its reactions with bromine and ethanoyl chloride.

- The test for phenols using aqueous iron(III) chloride.

Maths skills »

- Calculate percentage yields
- Construct and/or balance equations using ratios
- Carry out structured and unstructured mole calculations

Topic contents

Forming primary and secondary alcohols

The difference between primary and secondary alcohols is that primary alcohols have no more than one carbon atom directly bonded to the carbon of the C–OH group. Secondary alcohols have two carbon atoms bonded to the carbon atom of the C–OH group. All primary alcohols contain the –CH_2OH group whereas all secondary alcohols contain the –$CH(OH)$ group. There are also tertiary alcohols, e.g. 2-methylpropan-2-ol, $(CH_3)_3COH$.

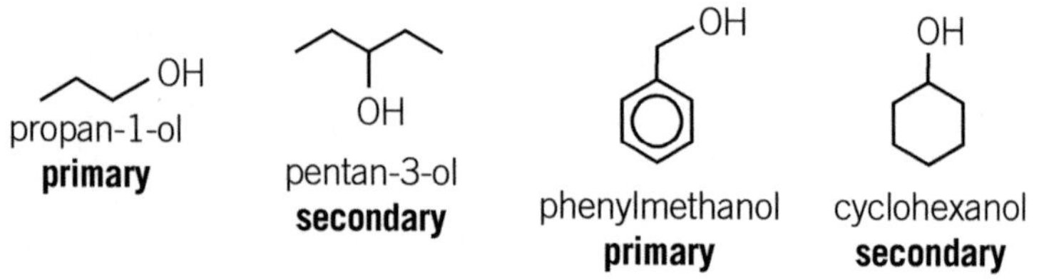

propan-1-ol
primary

pentan-3-ol
secondary

phenylmethanol
primary

cyclohexanol
secondary

There are two common methods of forming primary and secondary alcohols:

- From halogenoalkanes by a substitution reaction.

- By reducing aldehydes, ketones or carboxylic acids.

From a halogenoalkane

This reaction is carried out by **refluxing** together the halogenoalkane and an **aqueous** solution of an alkali (usually sodium or potassium hydroxide).

For example, butan-1-ol can be produced from 1-bromobutane.

$$CH_3CH_2CH_2CH_2Br + NaOH \longrightarrow CH_3CH_2CH_2CH_2OH + NaBr$$

The products of the reaction have to be separated. The organic reactant and the product, butan-1-ol, are both liquids but have different boiling temperatures (100 °C and 118 °C respectively). **Fractional distillation** can be used to separate butan-1-ol from unreacted 1-bromobutane.

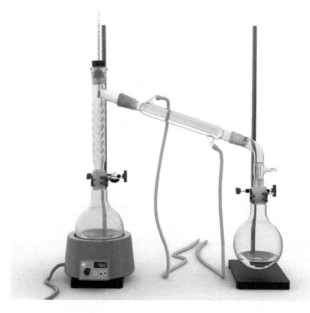
▲ Fractional distillation

1 Knowledge check

Laevulinic acid, $CH_3COCH_2CH_2COOH$, is easily obtained by heating fructose or glucose with concentrated hydrochloric acid. Write the formula of the organic compound produced when laevulinic acid reacts with an aqueous solution of $NaBH_4$.

2 Knowledge check

Name the alcohol produced when

HOOC—CH_2—C—CH_2—COOH (with O double bonded to C)

is reduced by lithium tetrahydridoaluminate(III).

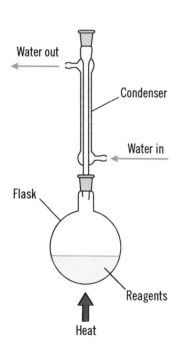

Water out

Condenser

Water in

Flask

Reagents

Heat

▲ A typical method for refluxing

The mechanism of this reaction is nucleophilic substitution where the hydroxide ion (⁻OH) acts as the **nucleophile** and attacks the relatively positive carbon atom (δ+) of the C–Br bond.

$$CH_3 - CH_2 - CH_2 - \overset{\delta+}{\underset{H}{\overset{H}{C}}} - \overset{\delta-}{Br} \longrightarrow CH_3 - CH_2 - CH_2 - \overset{H}{\underset{H}{C}} - \ddot{O} - H$$

$$+ \ddot{B}r^-$$

In general the rate of hydrolysis is C–I > C–Br > C–Cl. It is important to remember that if the concentration of the alkali is too high, or an ethanolic solution of the alkali is used, then the yield of alcohol is reduced as larger amounts of alkenes are formed.

By the reduction of a carbonyl compound

Both aldehydes and ketones can be reduced to primary and secondary alcohols respectively by using an aqueous solution of sodium tetrahydridoborate(III) (sodium borohydride), $NaBH_4$. In an equation, the reducing agent is often represented by using [H].

Aldehyde:

phenylmethanal
(benzaldehyde)   phenylmethanol

Ketone:

$$CH_3 - \overset{\overset{O}{\|}}{C} - CH_2 - CH_3 \ + \ 2[H] \longrightarrow CH_3 - \overset{\overset{H}{|}}{\underset{OH}{C}} - CH_2 - CH_3$$

butan-2-one   butan-2-ol

Sodium tetrahydridoborate(III) is not powerful enough to reduce carboxylic acids so the stronger reducing agent lithium tetrahydridoaluminate(III) (lithium aluminium hydride or 'lithal'), $LiAlH_4$, dissolved in ethoxyethane is used instead.

$$\overset{O}{\underset{HO}{\overset{\|}{C}}} - (CH_2)_4 - \overset{O}{\overset{\|}{C}}\underset{OH}{} + 8[H] \longrightarrow HOCH_2 - (CH_2)_4 - CH_2OH + 2H_2O$$

hexane-1,6-dioic acid   hexane-1-6-diol

In general the reduction using sodium tetrahydridoborate(III) is a much safer reaction as an excess of lithium tetrahydridoaluminate(III) is difficult to dispose of in an easy way and the solvent ethoxyethane is very flammable. To obtain a good yield of the alcohol an excess of the reducing agent is used. The excess reducing agent can be removed by adding a dilute acid. The organic product needs to be separated from the aqueous mixture. This is carried out using a separating funnel if the alcohol produced is immiscible with water, or separated by adding an organic solvent such as ethoxyethane, by the technique of solvent extraction.

▶▶ Key term

Nucleophiles are ions or compounds possessing a lone pair of electrons that can seek out a relatively positive site (often a δ+ carbon atom). Common nucleophiles include ⁻OH, ⁻CN and NH_3. Nucleophiles can be negatively charged, or neutral molecules.

▶▶ Study point

The term 'carbonyl compound' in this section refers to aldehydes, ketones and carboxylic acids, and not to esters or acid chlorides. Only the reduction of aldehydes and ketones is required for the A2 course.

▶▶ Study point

When giving an equation for the reduction of a carboxylic acid with lithium tetrahydridoaluminate(III) you must not forget that water is also a product of the reaction.

Knowledge check 3

A ketone is reduced with sodium tetrahydridoborate(III). The number of hydrogen atoms in the molecular formula for the alcohol produced is 6 more than in the starting ketone. State and explain the number of carbonyl groups present in the ketone.

▲ Separating funnel

Reactions of primary and secondary alcohols

Reaction with hydrogen halides

Halogenoalkanes are produced by reacting primary or secondary alcohols with hydrogen halides. Unfortunately, it is not as simple as this sentence implies. These reactions are generally slow and reversible, and often give poor yields. The most appropriate method depends on which halogen is to be substituted.

Chlorination

One method is to pass hydrogen chloride gas into the alcohol in the presence of anhydrous zinc chloride, which acts as a catalyst.

$$CH_3CH_2CH_2CH_2OH \xrightarrow[\substack{ZnCl_2 \\ heat}]{HCl} CH_3CH_2CH_2CH_2Cl + H_2O$$

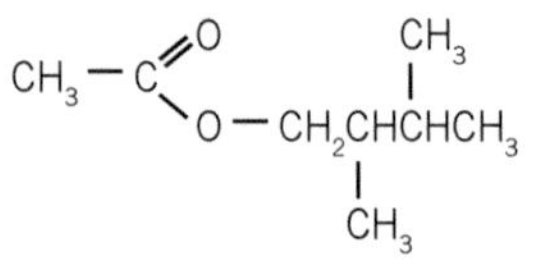

In chlorination reactions, the separation of the products can be a problem. Alternative reagents that can be used in chlorination include phosphorus(V) chloride and sulfur(VI) oxide dichloride, $SOCl_2$. The method using $SOCl_2$ is often preferred because the co-products are gaseous and separation is easier.

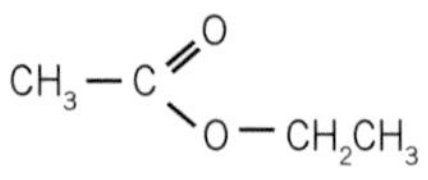

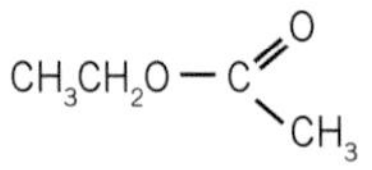

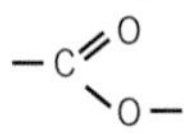

Bromination

The most convenient way of producing a bromoalkane from a primary or secondary alcohol is to carry out an 'in situ' reaction. In one method, a mixture of the alcohol, potassium bromide and 50% sulfuric(VI) acid is heated. The overall equation for the preparation of 1-bromobutane in this way is:

$$CH_3CH_2CH_2CH_2OH + KBr + H_2SO_4 \longrightarrow CH_3CH_2CH_2CH_2Br + KHSO_4 + H_2O$$

Iodination

One method is to reflux the alcohol with an excess of hydriodic acid.

$$CH_3CH_2CH_2OH + HI \longrightarrow CH_3CH_2CH_2I + H_2O$$

Reaction with ethanoyl chloride

An alcohol reacts rapidly with ethanoyl chloride giving an **ester**. During this reaction misty fumes of hydrogen chloride are seen. This method gives a better yield of an ester than by using a carboxylic acid, as the reaction is not reversible. However, the cost of acid chlorides means that this is not a cost-efficient process in industry.

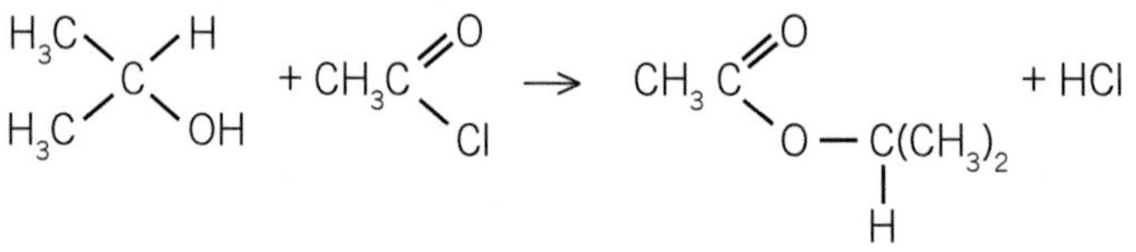

$$\underset{H_3C}{\overset{H_3C}{\diagdown}}\!C\!\underset{OH}{\overset{H}{\diagup}} + CH_3C\overset{O}{\underset{Cl}{\lessgtr}} \longrightarrow CH_3C\overset{O}{\underset{O - C(CH_3)_2}{\underset{|}{\lessgtr}}} + HCl$$

1-methylethyl ethanoate

Reaction with carboxylic acids

Primary and secondary alcohols react with carboxylic acids to give esters.

$$\text{alcohol} + \text{carboxylic acid} \rightleftharpoons \text{ester} + H_2O$$

The reaction is reversible and eventually the mixture will reach a position of equilibrium. To increase the equilibrium yield of the ester, a little concentrated sulfuric(VI) acid is added to the mixture of the alcohol and the carboxylic acid and the mixture is heated under reflux. The products can then be distilled and the ester collected at its boiling temperature. The distillate generally consists of the ester and water, together with a little unreacted alcohol and carboxylic acid. Many esters are immiscible with water and the distillate often consists of two layers. A separating funnel is used to extract the ester, which is then shaken with sodium hydrogencarbonate solution to remove any remaining carboxylic acid. The ester is then dried with anhydrous calcium chloride, which reacts with any remaining alcohol and water. The ester can then be redistilled to give a pure product.

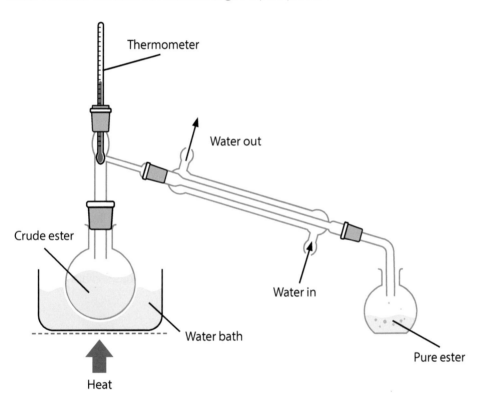

1-Butyl ethanoate can be made in this way from butan-1-ol and ethanoic acid.

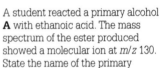

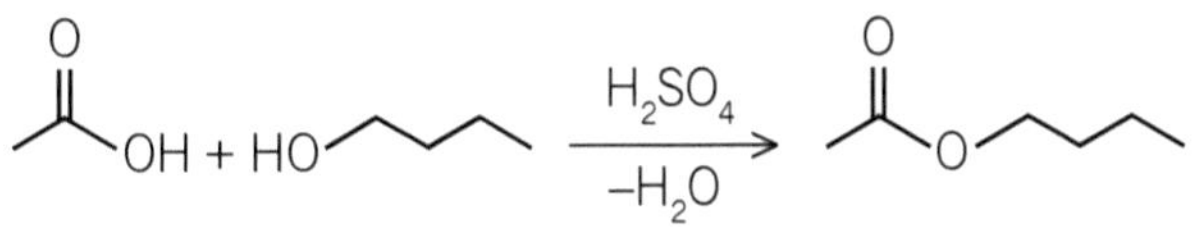

Many esters occur naturally and are extensively used in the cosmetic industry as components of perfumes. Esters also have important uses as solvents, for example in nail varnish.

Link

Link to the esterification of carboxylic acids, page 138 and AS Section 2.7.

Stretch & challenge

Butane-1,4-dioic acid reacts with an excess of methanol. Give the **empirical** formula of the resulting ester.

Knowledge check 5

A student reacted a primary alcohol **A** with ethanoic acid. The mass spectrum of the ester produced showed a molecular ion at m/z 130. State the name of the primary alcohol **A**.

Knowledge check 6

An ester, containing one ester group, contains 36.4% by mass of oxygen. Show that the ester could be ethyl ethanoate.

Knowledge check 7

Write the name and structural formula of the ester that is isomeric with ethanoic acid.

▲ Esters are used as the solvent for nail varnish

The acidity and reactions of phenol

The acidity of phenol

Link

Link to acid-base equilibria on page 87

Phenols are aromatic compounds where –OH groups are bonded directly to a benzene ring:

OH

phenol

OH

CH(CH$_3$)$_2$

H$_3$C

thymol
5-methyl-2-(propan-2-yl)phenol

OH

H$_3$C

Cl

CH$_3$

'PCMX'
the active ingredient in Dettol™
(4-chloro-3,5-dimethylphenol)

The reactivity of phenols is quite different from that of alcohols. This is partly because one of the lone pairs of the oxygen atom can overlap with the delocalised π system to form a more extended delocalised system. As a result the C–O bond in phenols is shorter and stronger than in an alcohol. This makes C–O bond fission in a phenol harder than C–O bond fission in an alcohol. The extended delocalisation creates a higher electron density in the ring and makes the ring structure more susceptible to attack by electrophiles. Phenols are much more acidic than alcohols. This means that phenol is a stronger acid than ethanol but a considerably weaker acid than ethanoic acid. The ionisation of phenol gives the phenoxide ion, C$_6$H$_5$O$^-$ but the position of equilibrium lies to the left, as shown by the bold arrow to the left:

$$\text{OH} + H_2O \rightleftharpoons \text{O}^- + H_3O^+$$

Delocalisation of the negative charge on this ion gives added stability to the phenoxide anion. Phenol itself is not very soluble in water, owing to the –OH group being a small part of a largely hydrophobic molecule. It does dissolve readily in aqueous sodium hydroxide giving a solution of sodium phenoxide:

$$\text{OH} + NaOH \rightarrow \text{O}^-Na^+ + H_2O$$

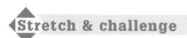

Stretch & challenge

2,4,6-Tribromophenol has only two signals in its low-resolution ^{1}H spectrum. This is because of the single O–H proton and that the two hydrogen protons in the benzene ring are in the same environment as each other.

This is an example of phenol reacting as an acid, losing a proton to the aqueous hydroxide ion present from the sodium hydroxide. The presence of substituents on the benzene ring affects the acidity of phenol. The pK_a of the active component in Dettol is 10.2, which is similar to that of phenol. 2,4-Dinitrophenol has a pK_a value of 4.30, showing a greater acifity then Dettol, due to the presence of two electron withdrawing nitro groups.

Although phenol is a weak acid it is not strong enough to react with sodium carbonate or sodium hydrogencarbonate to produce carbon dioxide. A simple test to distinguish between a simple aliphatic acid, such as ethanoic acid, and phenol is to add sodium carbonate solution. The ethanoic acid will react to give bubbles of carbon dioxide but phenol will not.

Stretch & challenge

Phenol acts as an acid when it dissolves in aqueous sodium hydroxide to give a colourless solution of sodium phenoxide, C$_6$H$_5$O$^-$Na$^+$.

The reaction of phenol with bromine

The presence of an –OH group bonded directly to a benzene ring activates the ring to attack by electrophiles. Each position is activated towards attack but the 2-, 4- and 6- positions are activated to a greater extent. As a result an incoming substituent is likely to replace a hydrogen atom at one, or more, of these positions. For example, when phenol is treated with dilute nitric acid, both 2- and 4-nitrophenol are formed.

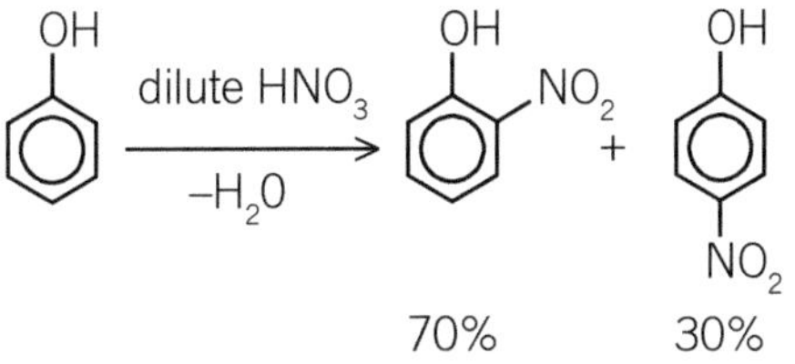

When phenol reacts with bromine, the increased electron density in the ring polarises the bromine molecules giving $Br^{\delta+}$– $Br^{\delta-}$. Aqueous bromine (bromine water) reacts with phenol to produce a white precipitate of 2,4,6-tribromophenol.

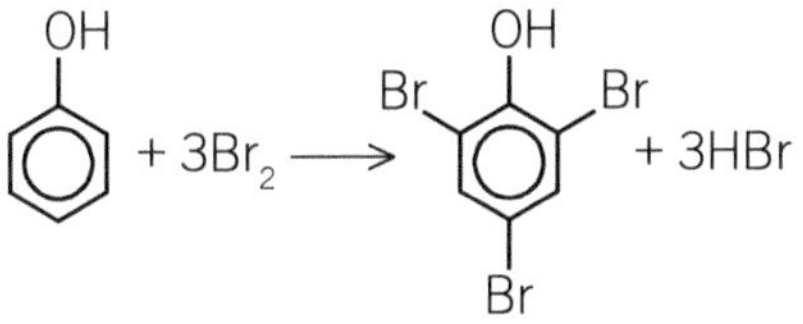

Since aqueous bromine is an orange solution and the products of the reaction are a colourless solution and a white precipitate, the orange colour disappears – 'bromine is decolourised'. This reaction can be used as a test for phenol. This result is different from the reaction of bromine with an alkene, as the reaction with phenol results in the decolourisation of bromine with the additional formation of a white precipitate. The mechanism for this reaction with phenol is electrophilic substitution where the electrophile is $Br^{\delta+}$.

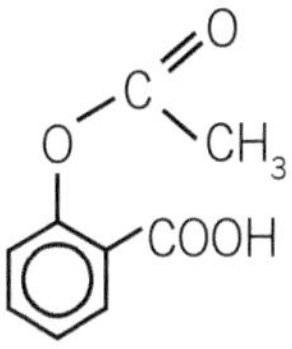

If the 2-,4- or 6- positions of a phenol are already blocked by a substituent, the bromine generally substitutes in the remaining 2-,4- or 6- positions. For example, 2-methylphenol gives 4,6-dibromo-2-methylphenol:

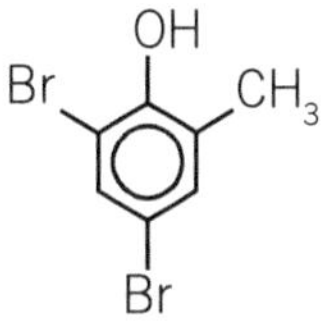

whereas 3-methylphenol gives 2,4,6-tribromo-3-methylphenol:

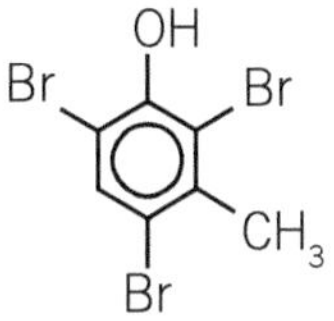

In 2,4-dinitrophenol only the 6-position is free for substitution in this way and 6-bromo-2,4-dinitrophenol is the product:

Knowledge check 8

Aspirin (2-ethanoyloxybenzenecarboxylic acid)

can be produced by reacting 2-hydroxybenzenecarboxylic acid (salicylic acid) with ethanoic anhydride. Describe, using a chemical test, how you would show that no remaining salicylic acid was present as an impurity in the sample of aspirin.

Study point

When testing for phenol using aqueous bromine, the usual method is to add aqueous bromine dropwise to the phenol solution. The bromine is at first decolourised and then when more bromine is added, a white precipitate of 2,4,6-tribromophenol is seen. If phenol was added to aqueous bromine, it would be more difficult to see the immediate decolourisation of the bromine.

Stretch & challenge

Phenol reacts with dilute nitric acid to give 2-nitrophenol and 4-nitrophenol in yields of 70% and 30% respectively. Suggest two reasons that may explain this ~2:1 proportion of products.

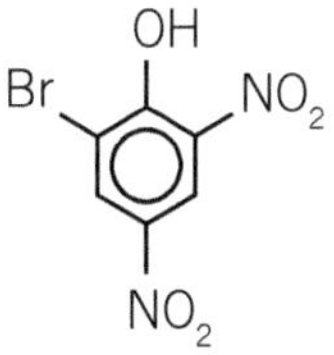

Study point

The reaction of phenol with ethanoyl chloride or with ethanoic anhydride is an esterification reaction that is similar to the corresponding reactions of alcohols. When using an acid chloride, a reagent is added that will react with the hydrogen chloride produced. This needs to be a base which cannot react readily with the acid chloride and this is the reason why pyridine is often preferred. If sodium hydroxide was used then this would immediately react with the acid chloride leading to little, if any, ester production.

When ethanoic anhydride is used, a little sulfuric acid is used to move the position of equilibrium to the right and increase the yield of the ester.

The reaction of phenol with ethanoyl chloride

Alcohols and phenols can react as nucleophiles by the use of their oxygen lone pairs. However, the delocalisation of an electron pair from the oxygen atom in a phenol means that it is more difficult for a phenol to react as a nucleophile, e.g. in a reaction with a carboxylic acid to give an ester. As a result carboxylic acids are not suitable reagents to make an ester with phenol, even the reaction of phenol with ethanoyl chloride is quite slow at room temperature.

A base, such as pyridine, C_5H_5N, can be added to speed up this reaction. The pyridine reacts with the hydrogen chloride co-product to give pyridinium chloride, $C_5H_5NH^+Cl^-$. Ethanoyl chloride is an expensive reagent and ethanoic anhydride is often used in preference.

If phenol is reacted with a less reactive acyl chloride, such as benzoyl chloride, C_6H_5COCl, then aqueous conditions can be used as the acyl chloride is only slowly hydrolysed by water. In this reaction phenol is added to aqueous sodium hydroxide and the mixture shaken.

Testing for phenols with aqueous iron(III) chloride

Phenol will react with iron(III) chloride to produce a purple colour in aqueous solution. The colour is produced by a complex being formed between the two reagents. Any compound that contains an –OH group bonded directly to a benzene ring will give a brightly coloured complex when reacted with iron(III) chloride – generally these are purple, blue or green in colour. This test can be used to differentiate between an aliphatic alcohol and a phenol.

Study point

Iron(III) chloride will also react with carboxylic acids but a red / brown colour or precipitate is formed rather than the brightly coloured solution given by a phenol.

Test yourself

1. (a) An alcohol has the formula $(C_6H_5)_2CHCH_2OH$. This formula shows that it is a .... alcohol. [1]

   (b) Give the structure of the compound formed by the reduction of HOOC–COOH by $LiAlH_4$. [1]

   (c) When a carbonyl compound reacts with $NaBH_4$ the alcohol $CH_3CH(OH)CH_3$ is formed. What is the name of the carbonyl compound? [1]

2. The skeletal formulae of some alcohols are shown below. For each one give its systematic name and state whether it is a primary or secondary alcohol. [3]

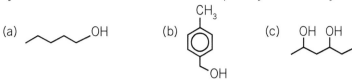

(a) ⌄⌄⌄OH (b) (c) OH OH

3. State the name of a reagent used to produce butan-2-ol from 2-chlorobutane. [1]

4. State the formulae of two reducing agents that convert pentan-3-one into pentan-3-ol. [2]

5. Complete the equation below, where the reducing agent is represented by [H]. [2]

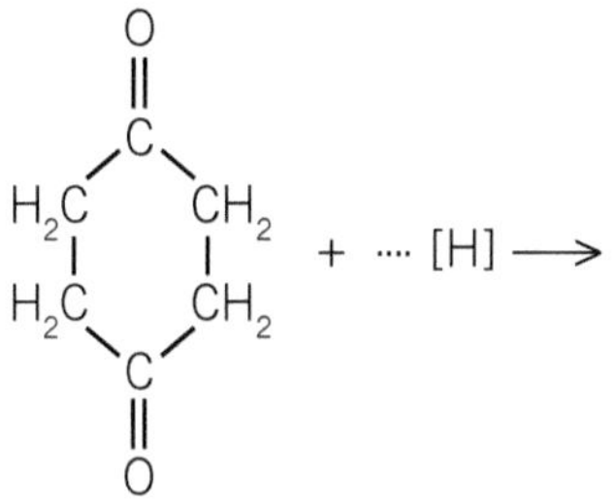

$$H_2C \underset{H_2C}{\overset{O}{\parallel}} \overset{C}{\underset{C}{}} \underset{CH_2}{\overset{CH_2}{}} + \cdots [H] \longrightarrow$$

6. Explain why sulfur(VI) dichloride oxide (thionyl chloride) is the reagent of choice when chlorinating an alcohol. Illustrate your answer by giving the equation for the reaction of 2-methylbutan-1-ol with sulfur(VI) dichloride oxide. [2]

7. An equation showing the reaction of ethanol with propanoyl chloride, CH_3CH_2COCl, is shown below. Explain what is wrong with the equation. [1]

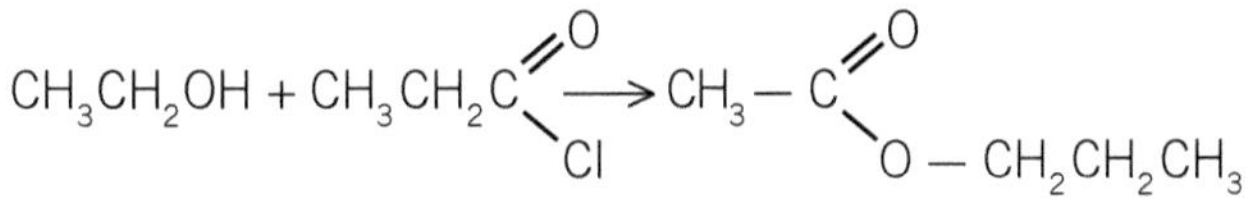

$$CH_3CH_2OH + CH_3CH_2C\overset{\displaystyle O}{\underset{\displaystyle Cl}{\diagup}} \longrightarrow CH_3 - C\overset{\displaystyle O}{\underset{\displaystyle O - CH_2CH_2CH_3}{\diagup}}$$

8. The molecular formula of an ester is $C_6H_{12}O_2$. This ester is produced from a carboxylic acid whose formula is $C_3H_6O_2$.

   Deduce the formula for the alcohol and name your choice. [2]

9. Three unknown compounds **B**, **C** and **D** are provided. One is an alcohol, another is a carboxylic acid and the third is a phenol. Their reactions with pH paper and with aqueous sodium hydrogencarbonate are shown in the table.

Compound	Colour with pH paper	Reaction with NaHCO₃ (aq)
B	Red	No reaction
C	Green	No reaction
D	Red	Colourless gas bubbles

   Use the results to explain which compound is which. [3]

10. An excess of aqueous bromine was added to a solution containing 0.050 mol of phenol (M_r 94). Calculate the mass of 2,4,6-tribromophenol that would be produced if the yield was 100%. [2]

11. Two aromatic compounds are isomers of formula C_7H_8O. Neither of these compounds reacts with iron(III) chloride to give a coloured complex. Suggest displayed formulae for these compounds. [2]

Aldehydes and ketones

Aldehydes and ketones are carbonyl compounds (containing a C=O bond). Other organic compounds, such as carboxylic acids and esters, also contain a carbonyl group but in a ketone the carbonyl group is bonded directly to two carbon atoms, rather than oxygen atoms. In an aldehyde the carbonyl group is bonded to a hydrogen atom, and to another hydrogen atom or a carbon atom. Aldehydes and ketones are common in nature and have important domestic and industrial uses. It is the polar nature of this carbonyl bond that makes aldehydes and ketones react in a different way to the alkenes, which contain a C=C bond.

Topic contents

You should be able to demonstrate and apply knowledge and understanding of:

- The formation of aldehydes and ketones by the oxidation of primary and secondary alcohols respectively.

- How aldehydes and ketones may be distinguished by their relative ease of oxidation using Tollens' reagent and Fehling's reagent.

- The reduction of aldehydes and ketones using sodium tetrahydridoborate(III).

- The mechanism of nucleophilic addition, such as in the addition of hydrogen cyanide to ethanal and propanone.

- The reaction of aldehydes and ketones with 2,4-dinitrophenylhydrazine and its use in testing for a carbonyl group and the identification of specific aldehydes and ketones.

- The triiodomethane (iodoform) test and its use in identifying $CH_3C=O$ groups or their precursors.

Maths skills ⟩⟩

- Calculate percentage yields
- Construct and/or balance equations using ratios
- Carry out structured and unstructured mole calculations

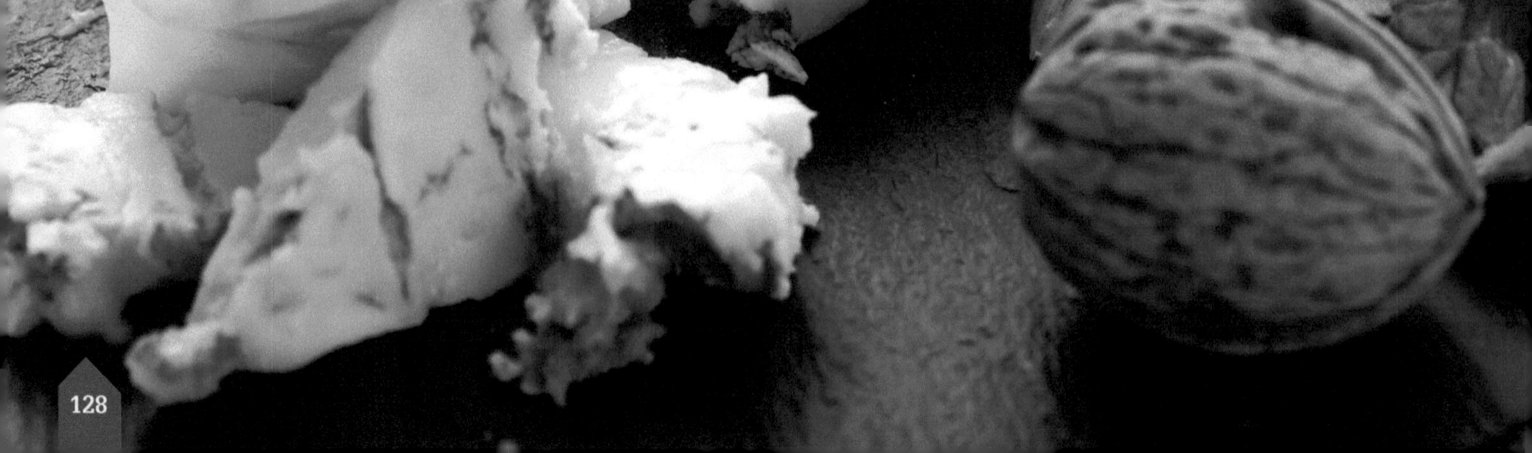

Structure and naming of aldehydes and ketones

Both aldehydes and ketones contain the polar $C^{\delta+}=O^{\delta-}$ group. This occurs because oxygen in this bond has a greater electronegativity, leaving the carbon atom slightly electron deficient. The double bond comprises a σ-bond, and a π-bond above and below the plane of the sigma bond, caused by p-p orbital overlap, as in an alkene. In an aldehyde, the carbon atom of the carbonyl group bonds to at least one hydrogen atom. An aldehyde has the suffix -al at the end of the name and a ketone ends in -one.

methanal   propanal   benzaldehyde (benzenecarbaldehyde)

A ketone has the carbonyl carbon atom directly bonded to two other carbon atoms.

propanone   phenylethanone   1,5-dichloropentan-3-one

Aldehydes and ketones are very common in both flora and fauna. For example the Pacific sea slug *Navanax intermis* excretes an alarm pheromone that contains Navenone B.

Navenone B

Formation by the oxidation of alcohols

Some alcohol chemistry has been studied during the first year of this course and this included their oxidation. The usual oxidising agent is acidified potassium (or sodium) dichromate, shown as [O] in an equation. Primary alcohols are oxidised to an aldehyde and then, on further oxidation, to a carboxylic acid. The orange colour of aqueous dichromate(VI) ions becomes green as $Cr^{3+}(aq)$ ions are produced.

Under normal conditions a ketone cannot be further oxidised by this method. Acidified potassium manganate(VII) can also be used as the oxidising agent – the purple colour of the manganate(VII) ion is lost and a colourless solution containing aqueous manganese(II) ions remains. If an aldehyde is required then this must be removed from the reaction mixture before further oxidation to the carboxylic acid can occur. One common method when producing the aldehyde is to add the oxidising agent slowly to the alcohol.

Knowledge check 1

Look at the formula of the sea slug pheromone, Navenone B. If the C=C bonds in this compound are fully hydrogenated, state the molecular formula of the product.

Knowledge check 2

Cyclohexanol is oxidised by acidified dichromate to cyclohexanone. Complete the equation for this reaction.

Knowledge check 3

(a) Describe a test that could be carried out on compound E to show that it is an aldehyde.

Compound E

(b) Give the structural formula of another isomer of formula $C_4H_8O_2$ that will not undergo the reaction that you have described for the isomer in (a).

Top tip >>

When drawing diagrams that show distillation, as shown here, it is important to ensure that the drawing does not show a closed system. If it is a closed system then it will blow apart when heated! However, there is often a real chance of fire as many distillates are very flammable. A useful tip is to draw the receiving flask with cotton wool in the neck of the flask. This reduces the risk of flammable vapour escaping but allows for the expansion of gases when the system is heated. The use of an ice bath also reduces evaporation of volatile products such as ethanal.

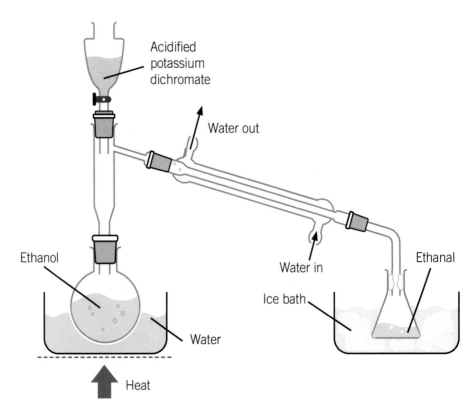

The temperature and rate of addition are controlled so that ethanal vapour reaches the top of the column, and then condenses to be collected in a cooled flask.

A secondary alcohol is similarly oxidised to a ketone.

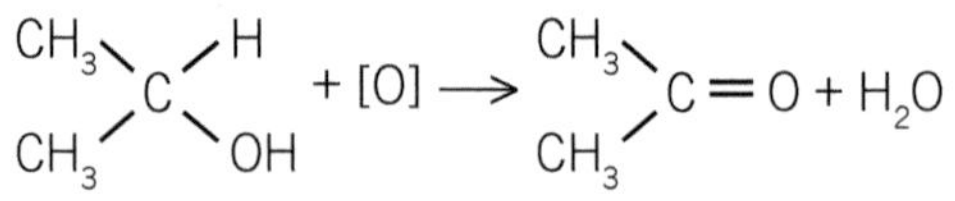

$$CH_3\!\!\diagdown\!\!\underset{CH_3\diagup}{C}\!\!\diagup^H_{OH} + [O] \longrightarrow CH_3\!\!\diagdown\!\!\underset{CH_3\diagup}{C}\!\!=\!O + H_2O$$

 4 Knowledge check

An alcohol is oxidised by acidified potassium dichromate to give a ketone of formula C_4H_8O. Give the structure of the alcohol.

 5 Knowledge check

27.6 g of ethanol (M_r 46) was oxidised by acidified sodium dichromate to produce 9.8 g of ethanal (M_r 44). Calculate the percentage yield.

Distinguishing between aldehydes and ketones

Aldehydes can be further oxidised to carboxylic acids, but the further oxidation of ketones is more difficult. If a mild oxidising agent is added to an aldehyde, then the aldehyde will be oxidised, and the oxidising agent itself is reduced. Several simple tests show whether a compound is an aldehyde or a ketone. One of these tests uses Tollens' reagent (sometimes called ammoniacal silver nitrate). It is made by adding aqueous sodium hydroxide to silver nitrate solution until a brown precipitate of silver(I) oxide is formed. Aqueous ammonia is then added until the silver(I) oxide just redissolves. The suspected aldehyde is then added to this reagent and the tube is gently warmed in a beaker of water. If the compound is an aldehyde, a silver mirror coats the inside of the tube as the Ag^+ ions are reduced to silver. Less soluble aldehydes, like benzaldehyde, are reluctant to react in this way. A ketone will not reduce Tollens' reagent.

Another mild oxidising agent that will react with an aldehyde (but not a ketone) is Fehling's reagent. This test also shows the presence of an aldehyde group in reducing sugars such as glucose. The reagent is freshly prepared by mixing together two solutions, Fehling's A and Fehling's B. On mixing the two solutions a deep blue solution containing a complex copper(II) ion is produced. If an aldehyde is added to this deep blue solution and the mixture warmed in a water bath, then the aldehyde reduces the complex copper(II) ions and the deep blue solution is replaced by an orange-red precipitate of copper(I) oxide.

Benedict's reagent is a similar reagent containing a complex copper(II) ion and is similarly reduced to copper(I) oxide. The picture (right) shows the use of Benedict's reagent in identifying an aldehyde. The aldehyde has reduced aqueous blue copper(II) ions to an orange-brown precipitate of copper(I) oxide. This solution has the advantage that it is stable and is safer to use, as it is less alkaline. The test using these solutions is effective in identifying aliphatic aldehydes but it does not react with aromatic aldehydes such as benzaldehyde (C_6H_5CHO).

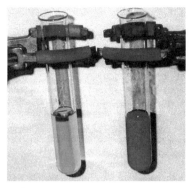

▲ Benedict's solution identifying an aldehyde

Reduction of aldehydes and ketones

Aldehydes and ketones can be reduced to primary and secondary alcohols respectively. The reducing agents used are sodium tetrahydridoborate(III), $NaBH_4$, or lithium tetrahydridoaluminate(III), $LiAlH_4$. Of these two reagents, $NaBH_4$ is preferred as it is safer and can be used in aqueous conditions. In writing equations to show this reduction it is acceptable to use [H] to represent the formula of the reducing agent.

$$\underset{O}{\overset{H}{\diagdown}}C=CCH_2CH_2CH_2C\underset{O}{\overset{H}{\diagup}} + 4[H] \longrightarrow HOCH_2CH_2CH_2CH_2CH_2OH$$

More details of this reaction are seen in Topic 4.3 on alcohols.

Nucleophilic addition reactions

Aldehydes and ketones contain a polar carbonyl bond, $C^{\delta+}=O^{\delta-}$. The relatively electron-deficient carbon atom can be attacked by nucleophiles, such as a cyanide ion, ^-CN. The reaction of propanone with hydrogen cyanide is an example of this reaction.

$$\underset{H_3C}{\overset{H_3C}{\diagdown}}C=O + H-C\equiv N \longrightarrow \underset{H_3C}{\overset{H_3C}{\diagdown}}\underset{C\equiv N}{\overset{OH}{\diagup}}C$$

2-hydroxy-2-methylpropanenitrile

In practice, the reaction is not as straightforward as the equation suggests. Hydrogen cyanide is a very weak acid and the concentration of cyanide ions present is relatively low.

$$H-C\equiv N(aq) \rightleftharpoons H^+(aq) + {}^-C\equiv N{:}(aq)$$

The reaction depends on the initial attack of cyanide ions on propanone. The reaction mixture contains propanone together with aqueous sodium or potassium cyanide, which provide a source of cyanide ions. Aqueous sulfuric acid is then slowly added to the stirred mixture. The mechanism for this reaction is nucleophilic addition. It involves attack by the nucleophile ^-CN and the overall addition of HCN across the C=O bond.

Link

Preparation of alcohols by reduction page 121.

▶▶ **Study point**

Aldehydes, ketones and carboxylic acids can all be reduced to alcohols. Sodium tetrahydridoborate(III) will reduce aldehydes and ketones but not carboxylic acids. The stronger reducing agent, lithium tetrahydridoaluminate(III), will reduce carboxylic acids as well as aldehydes and ketones.

Knowledge check 6

State the systematic name of the diol produced when the compound below is reduced by $NaBH_4$.

$$H_3C-\underset{O}{\overset{O}{\underset{\|}{C}}}-CH_2-C\underset{O}{\overset{H}{\diagup}}$$

7 Knowledge check

Deduce the displayed formula of a carbonyl compound that:

(a) Does not give a silver mirror with Tollens' reagent.

(b) Gives a yellow precipitate when treated with alkaline iodine.

(c) Reacts with hydrogen cyanide to give a product that is hydrolysed to an acid that contains 6 carbon atoms in each molecule.

▲ A positive test with Brady's reagent

8 Knowledge check

The formula of a 2,4-dinitrophenylhydrazone of a ketone is

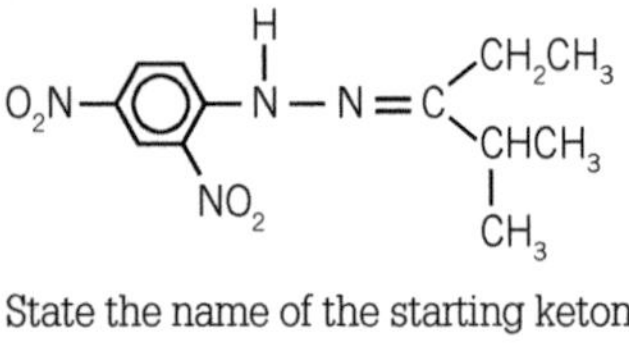

$$O_2N-\bigcirc-\overset{H}{\underset{}{N}}-N=C\overset{CH_2CH_3}{\underset{CHCH_3}{\underset{|}{CH_3}}}$$

State the name of the starting ketone.

9 Knowledge check

Which of these compounds will undergo the triiodomethane reaction?

(a) Hexan-2-one

(b) Cyclohexanone

(c) 1-Phenylethanol

(d) 2-Methylpropan-2-ol

This reaction is important as it forms a method for extending the length of the carbon chain. Hydrolysis of the hydroxynitrile produced by this addition, by refluxing with a dilute acid, gives a hydroxyacid.

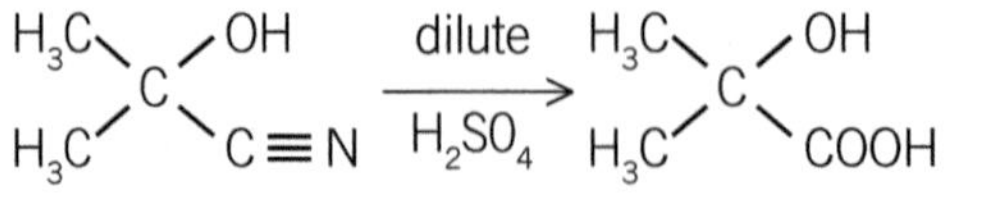

$$\underset{H_3C}{\overset{H_3C}{\diagdown}}\underset{C\equiv N}{\overset{OH}{\diagup}} C \xrightarrow[H_2SO_4]{\text{dilute}} \underset{H_3C}{\overset{H_3C}{\diagdown}}\underset{COOH}{\overset{OH}{\diagup}}C$$

2-hydroxy-2-methylpropanoic acid

The identification of aldehydes and ketones

Many common aldehydes and ketones are liquids at room temperature and pressure. The identification of a suspected aldehyde or ketone can be carried out by finding its boiling temperature but some of the boiling temperatures are too similar to each other or are too low or too high, and this makes identification difficult. If an aldehyde or ketone reacts with a suitable reagent to give a solid with a definite melting temperature within an acceptable temperature range, then it makes identification of the aldehyde or ketone easier. One reagent that is used to produce a suitable solid is 2,4-dinitrophenylhydrazine, $C_6H_3(NO_2)_2NHNH_2$. This material is dissolved in an acid to give a solution called Brady's reagent or simply 2,4-DNPH. Mixing with a suspected aldehyde or ketone gives an orange-red solid, which is filtered off and purified. The formation of this precipitate indicates that the starting material is an aldehyde or ketone. The melting temperature of the purified product (a 2,4-dinitrophenylhydrazone) is taken and compared with a table of known values, to identify the aldehyde or ketone present. This process is called 'making a derivative' or 'derivatisation'. The equation for this reaction with propanone is shown below but will not be tested in the examination.

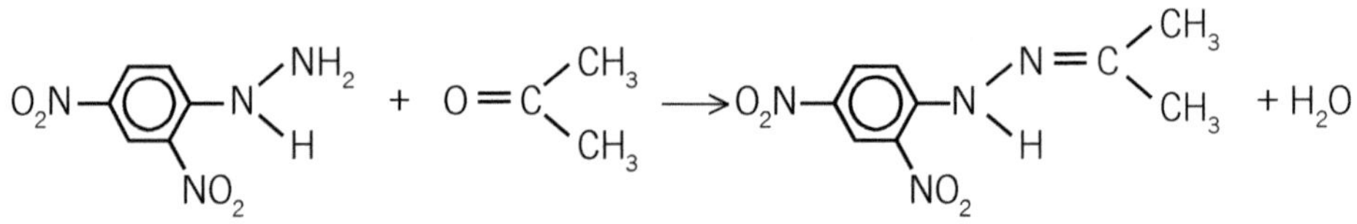

$$O_2N-\bigcirc-\underset{NO_2}{\overset{N}{\underset{}{N}}}\underset{H}{\overset{NH_2}{\diagup}} + O=C\overset{CH_3}{\underset{CH_3}{\diagup}} \longrightarrow O_2N-\bigcirc-\underset{NO_2}{\overset{N}{\underset{}{N}}}\underset{H}{\overset{N=C}{\diagup}}\overset{CH_3}{\underset{CH_3}{\diagup}} + H_2O$$

The melting temperatures for some 2,4-dinitrophenylhydrazones are shown in the table below:

Compound	Formula	Boiling temperature /°C	Melting temperature of 2,4-dinitrophenylhydrazone /°C
propanone	$\underset{CH_3}{\overset{CH_3}{\diagdown}}C=O$	56	187
butanone	$\underset{CH_3}{\overset{CH_3CH_2}{\diagdown}}C=O$	80	111
pentan-2-one	$\underset{CH_3}{\overset{CH_3CH_2CH_2}{\diagdown}}C=O$	102	143
pentan-3-one	$\underset{CH_3CH_2}{\overset{CH_3CH_2}{\diagdown}}C=O$	102	156

The identification of aldehydes and ketones in this way is traditionally called a 'wet' method. The use of GC/MS (gas chromatography/mass spectrometry) and NMR has made the identification of these compounds an easier and quicker process.

The triiodomethane (iodoform) reaction

Triiodomethane (traditional name 'iodoform'), CHI_3, is a yellow solid that has some uses as an antiseptic. The formation of this yellow solid is used to identify the presence of a $CH_3C=O$ group (sometimes called a methyl carbonyl group), or a $CH_3CH(OH)$ group in a molecule. The organic compound is warmed with a colourless solution of iodine in aqueous sodium hydroxide ('alkaline iodine') or alternatively with an aqueous mixture of potassium iodide and sodium chlorate(I),NaOCl. These reagents are sometimes represented by I_2/NaOH and I^-/OCl$^-$. This reaction is not completely restricted to carbonyl compounds that contain this grouping. For example, the reaction is also given by ethanol, this is because the reagent used is an oxidising mixture and will oxidise the $CH_3CH(OH)$ grouping in ethanol to ethanal, which contains a $CH_3C=O$ group.

An equation for the reaction of propanone with alkaline iodine is:

$$CH_3COCH_3 + 3I_2 + 4NaOH \longrightarrow CHI_3 + CH_3COO^-Na^+ + 3NaI + 3H_2O$$

⫸ Study point

Common compounds that undergo the triiodomethane reaction include ethanol, ethanal, propan–2-ol, propanone, butanone, butan-2-ol and phenylethanone.

Among compounds that can be distinguished because they do **not** undergo the triiodomethane reaction are methanol, propan-1-ol and propanal.

⫷ Top tip

Equations showing the formation of CHI_3 are not required for the examination.

Test yourself

1. Explain which of these four compounds is/are ketone(s) and why the remainder are not. [4]

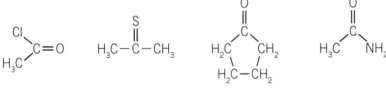

Compound A   Compound B   Compound C   Compound D

2. Using [O] as the formula of the oxidising agent, give the equation for the oxidation of butane-1,3-diol by acidified potassium dichromate. [1]

3. Write the skeletal formula of the ketone that produces pentan-3-ol on reduction with sodium tetrahydridoborate(III). [1]

4. In the reaction between pentan-3-one and hydrogen cyanide, give the formula of the nucleophile taking part in the reaction, showing any outer lone pairs of electrons present. [1]

5. Pentan-3-one reacts with hydrogen cyanide to give a hydroxynitrile, which is then hydrolysed using aqueous sulfuric acid. Give the structures of the hydroxynitrile, the resulting hydroxyacid, and the empirical formula of the hydroxyacid produced. [3]

6. Name the compounds represented by letters in this reaction sequence. [3]

Compound E $\xrightarrow{\text{H}^+/\text{Cr}_2\text{O}_7{}^{2-}\text{(aq)}}$ [benzene ring]$-CH_2-\overset{\displaystyle O}{\overset{\|}{C}}-CH_3$

$\downarrow$ Reagent F

yellow solid G

7. Name the compound that can be reduced to butane-1,4-diol by $NaBH_4$. [1]

8. Explain why the yield of ethanal will be reduced if ethanol is dropped slowly onto an excess of hot acidified potassium dichromate solution, rather than the other way round. [1]

4.5

Carboxylic acids and their derivatives

Carboxylic acids contain a –COOH group, where the carbon atom of a C=O group is also bonded to an O–H group. Many carboxylic acids are familiar substances; for example, ethanoic acid is the active ingredient in vinegar, citric acid in lemons and malic acid in apples. The –OH group present in a carboxylic acid can be replaced by chlorine, giving an acid chloride, which, in turn, can be converted to an amide where the –OH group has been replaced by an –NH$_2$ group. Derivatives of carboxylic acids also include esters where the –OH group has been replaced by an –OR group.

Topic contents

You should be able to demonstrate and apply knowledge and understanding of:

- The relative acidity of carboxylic acids, phenols, alcohols and water.

- The formation of carboxylic acids by the oxidation of alcohols and aldehydes.

- The reduction of carboxylic acids using LiAlH$_4$.

- The formation of aromatic carboxylic acids by the oxidation of methyl side-chains.

- The decarboxylation of carboxylic acids.

- The conversion of carboxylic acids to esters and acid chlorides and the hydrolysis of these compounds.

- The conversion of carboxylic acids to amides and nitriles.

- The formation of nitriles from halogenoalkanes and hydroxynitriles from aldehydes.

- The hydrolysis of nitriles and amides.

- The reduction of nitriles using LiAlH$_4$.

Maths skills ▶▶

- Calculate percentage yields
- Construct and/or balance equations using ratios
- Carry out structured and unstructured mole calculations

The structure and naming of carboxylic acids

All carboxylic acids contain at least one –COOH group. The carbon atom or atoms in this group(s) are counted in the carbon chain when they are named. Thus propanoic acid is CH_3CH_2COOH rather than $CH_3CH_2CH_2COOH$ (which has a three carbon **alkyl** chain but is named butanoic acid). If other functional groups are present in the acid then they are named as derivatives of the parent acid.

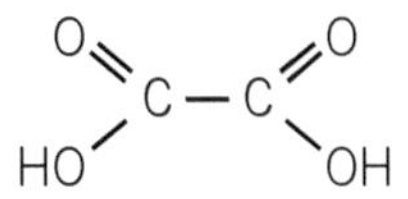

Knowledge check 1

Give the skeletal formulae of:
(a) 2-methylbutanoic acid
(b) 2-hydroxypropanoic acid
(c) heptane-1,7-dioic acid.

ethanedioic acid

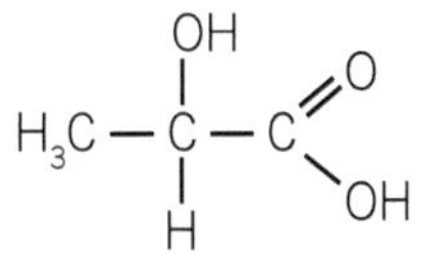

2-hydroxypropanoic acid

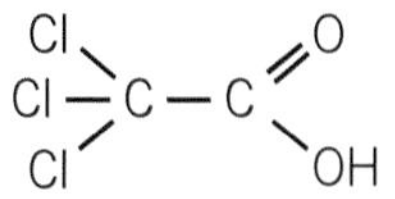

trichloroethanoic acid

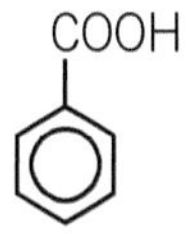

benzoic acid
(benzenecarboxylic acid)

benzene-1,2-dicarboxylic acid

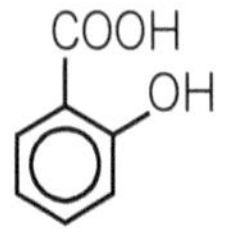

2-hydroxybenzenecarboxylic acid

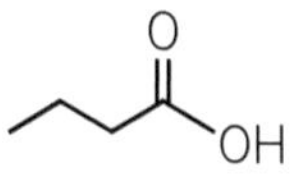

butanoic acid

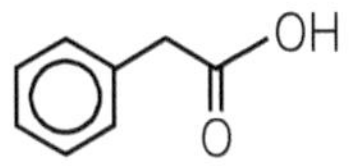

phenylethanoic acid

Acidity of carboxylic acids

Carboxylic acids are weak acids and the extent of their ionisation in aqueous solution is very small.

$$CH_3C\substack{\diagup O \\ \diagdown OH}(aq) + H_2O(l) \rightleftharpoons CH_3C\substack{\diagup O \\ \diagdown O^-}(aq) + H_3O^+(aq)$$

< **Link** >

Section 4.3 page 124.

At 25°C in an aqueous solution of ethanoic acid of concentration 0.1 mol dm^{-3} only about 0.4% of the acid is dissociated into ions. Carboxylic acids are, however, stronger acids than most phenols. In general the relative order of acidity is

carboxylic acids > phenols > water / alcohols

This relative difference in acidity can be shown by their reaction with sodium hydrogencarbonate solution. Only carboxylic acids are strong enough acids to produce colourless bubbles of carbon dioxide gas. The presence of other substituent groups may markedly alter the acidity of the carboxylic acid or phenol. Substituting one of the hydrogen atoms in ethanoic acid, giving chloroethanoic acid, $ClCH_2COOH$, increases the dissociation of the acid into ions one hundred times. Trichloroethanoic acid, Cl_3CCOOH, is a relatively strong acid. Similarly, chlorophenols and nitrophenols are stronger acids than phenol itself.

Knowledge check 2

Name the carboxylic acid that is reduced to give an unbranched alcohol of formula $C_5H_{12}O$.

Knowledge check

Give the balanced equation for the oxidation of phenylmethanol, $C_6H_5CH_2OH$, to benzoic acid, C_6H_5COOH, using [O] to represent the formula of the oxidising agent.

 Study point

Aldehydes can be oxidised by milder oxidising agents. These include Tollens' reagent and Fehling's reagent. The organic product of these reactions is the carboxylic acid anion, as the reaction is carried out in basic conditions. These reactions are discussed further in the aldehydes and ketones section.

< Link >

Section 4.4 page 129.

◄ Stretch & challenge

State the reagent(s) used to carry out the oxidation reactions below.

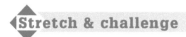

Practical check >>

When benzenecarboxylic acid (benzoic acid) is produced by oxidising methylbenzene with alkaline potassium manganate(VII) solution, followed by subsequent acidification, a practical problem is the separation of the solid carboxylic acid from the brown-black sludge of manganese(IV) oxide. One method is to heat the mixture, when the carboxylic acid dissolves. The mixture is filtered hot and the filtrate is then allowed to cool, when white crystals of benzenecarboxylic acid (benzoic acid) are produced.

Formation of carboxylic acids by the oxidation of alcohols and aldehydes

Primary alcohols can be oxidised firstly to aldehydes and then, on further oxidation, to carboxylic acids.

$$R-CH_2OH \xrightarrow{[O]} R-C\overset{\displaystyle O}{\underset{\displaystyle H}{}} \xrightarrow{[O]} R-C\overset{\displaystyle O}{\underset{\displaystyle OH}{}}$$

Acidified potassium dichromate can be used as the oxidising agent and this turns from orange $Cr_2O_7^{2-}$ ions to green Cr^{3+} ions as the oxidation of the alcohol proceeds. Another oxidising agent that can be used is alkaline potassium manganate(VII) solution, and this turns from a purple solution containing MnO_4^- ions, to a brown-black sludge of manganese(IV) oxide as it is reduced by the alcohol. Aldehydes are more volatile than the corresponding carboxylic acid and they should not be allowed to escape from the refluxing reaction mixture, to ensure that further oxidation proceeds to the carboxylic acid. Secondary alcohols are oxidised to ketones and any further oxidation to a carboxylic acid does not occur in this way.

Reduction of carboxylic acids

Carboxylic acids are relatively stable compounds and their reduction to a primary alcohol requires the use of the powerful reducing agent lithium tetrahydridoaluminate(III), $LiAlH_4$. This reagent reacts violently with water so the reduction is carried out using the solvent ethoxyethane. Sodium tetrahydridoborate(III), $NaBH_4$, is a milder reducing agent and, although it will reduce both aldehydes and ketones, it will not reduce a carboxylic acid.

The equation for this reduction can be shown using [H] as representing the reducing agent lithium tetrahydridoaluminate(III).

$$CH_3C\overset{\displaystyle O}{\underset{\displaystyle OH}{}} + 4[H] \longrightarrow CH_3CH_2OH + H_2O$$

propane-1,3-diol

3-hydroxypropanoic acid

136

Making aromatic carboxylic acids

An aromatic carboxylic acid must have the acid group directly bonded to the benzene ring.

benzenecarboxylic acid

4-methylbenzenecarboxylic acid

(4-hydroxymethyl)
benzenecarboxylic acid

Knowledge check 4

Compound L is oxidised by alkaline potassium manganate(VII) solution.

Show that the organic product contains 57.8% of carbon by mass.

The oxidation of a primary alcohol or aldehyde using acidified potassium dichromate or acidified potassium manganate(VII) solutions produces the carboxylic acid.

$$2[O] \longrightarrow + H_2O$$

Another method is to oxidise a methyl side chain. One way of doing this is to heat the compound with alkaline potassium manganate(VII) solution.

This represents the overall equation. The initial product is the potassium salt of benzenecarboxylic acid. The reaction mixture is then acidified, producing the acid itself. Another product of the reduction is a brown-black sludge of manganese(IV) oxide. Some benzaldehyde is also produced. Benzenecarboxylic acid is the end product of this oxidation, irrespective of the length of the carbon side chain. For example, ethylbenzene also produces benzenecarboxylic acid.

Stretch & challenge

Benzene-1,2-dicarboxylic acid is obtained by oxidising a hydrocarbon with alkaline potassium manganate(VII) solution. The hydrocarbon contains 89.5% by mass of carbon. Use this information to suggest a displayed formula for the hydrocarbon.

Decarboxylation of carboxylic acids

The removal of carbon dioxide from a compound is called decarboxylation. As carbon dioxide is an acidic oxide, it will react with a base. In decarboxylation reactions, the compound generally used is soda lime. This is a mixture that is largely calcium hydroxide, together with a little sodium/potassium hydroxide. The material absorbs both carbon dioxide and water but remains as a solid.

In a decarboxylation reaction, the organic acid or its sodium salt is heated with soda lime. A hydrocarbon is produced that contains one less carbon atom in the chain than the starting material. This is one of the methods for reducing the length of the carbon chain (descending the homologous series). The equation for decarboxylation using soda lime can be represented using calcium hydroxide, calcium oxide or sodium hydroxide.

Knowledge check 5

State the name of the carboxylic acid that produces propane when it is strongly heated with soda lime.

Knowledge check 6

Give the name of the nitrile that has the formula

$$H_3C - \underset{\underset{CH_3}{|}}{\overset{\overset{CH_3}{|}}{C}} - C \equiv N$$

$$CH_3COOH + Ca(OH)_2 \longrightarrow CH_4 + CaCO_3 + H_2O$$

$$\text{⬡} - COO^-Na^+ + NaOH \longrightarrow \text{⬡} + Na_2CO_3$$

$$\text{⬡} - CH_2COOH + CaO \longrightarrow \text{⬡} - CH_3 + CaCO_3$$

In these reactions, the organic product is an alkane or an arene.

In an extension of this topic, decarboxylation also occurs if the calcium salt of the acid is heated in the absence of soda lime. A ketone is the product. However, in this reaction the chain length is increased from a derivative of the two carbon ethanoic acid to a ketone containing three carbon atoms.

$$(CH_3COO)_2Ca \xrightarrow{\text{heat}} CH_3-\overset{\overset{\displaystyle O}{\|}}{C}-CH_3 + CaCO_3$$

Calcium benzenecarboxylate (benzoate) is similarly decarboxylated on heating, producing diphenylmethanone.

$$(C_6H_5COO)_2Ca \longrightarrow C_6H_5COC_6H_5 + CaCO_3$$

Decarboxylation reactions also occur if the salts of dicarboxylic acids are heated with soda lime. For example, disodium pentanedioate produces propane.

$$Na^+{}^-OOC-CH_2-CH_2-CH_2-COO^-Na^+ + 2NaOH \longrightarrow CH_3-CH_2-CH_3 + 2Na_2CO_3$$

Substituted aromatic carboxylic acids can also be decarboxylated in a similar way by heating them with soda lime.

$$H_2N-\text{⬡}-COOH + 2NaOH \longrightarrow H_2N-\text{⬡} + Na_2CO_3 + H_2O$$

Esters and acid chlorides and the hydrolysis of these compounds

Making esters from carboxylic acids and their hydrolysis

Carboxylic acids can be converted to esters by heating the carboxylic acid with an alcohol in the presence of a little sulfuric acid.

$$CH_3C\overset{\displaystyle O}{\underset{\displaystyle OH}{<}} + CH_3CH_2CH_2OH \rightleftharpoons CH_3C\overset{\displaystyle O}{\underset{\displaystyle OCH_2CH_2CH_3}{<}} + H_2O$$

Experiments using methanol, with some oxygen atoms being the ^{18}O isotope, have shown that the ester contains all the ^{18}O oxygen atoms that were originally in the methanol. It is the acid's –OH group that is lost, rather than the –OH group from the alcohol.

$$C_6H_5C\overset{\displaystyle O}{\underset{\displaystyle \boxed{OH}}{<}}\!\!HOCH_3 \longrightarrow C_6H_5-C\overset{\displaystyle O}{\underset{\displaystyle OCH_3}{<}} + H_2O$$

Knowledge check

7 A student suggested that either potassium cyanide or sodium cyanide can be used to react with 1-bromobutane to produce pentanenitrile. Explain why this suggestion is correct.

Practical check »

In decarboxylation reactions where soda lime is used, either the acid or its salt (usually the sodium salt) can be used. Although this seems a useful method for reducing the overall carbon chain length, the reaction is complicated and yields are often poor and a number of side products are also obtained.

Top tip »

The formulae of esters are usually drawn with the 'acid part' first, which is the opposite of how it is named. However, you may come across it written the other way round. For example, the formula of ethyl propanoate can be written as

$$CH_3CH_2C\overset{\displaystyle O}{\underset{\displaystyle OCH_2CH_3}{<}} \quad \text{or} \quad CH_3CH_2O-C\overset{\displaystyle O}{\underset{\displaystyle CH_2CH_3}{<}}$$

Knowledge check

8 Give the empirical formula of the diethyl ester of ethanedioic acid.

‹ Link ›

Reaction with carboxylic acids pages 123.

The mechanism of esterification will not be tested in the examination. Experimental details of ester preparation are discussed in the alcohol section of this book.

Esterification is a reversible reaction and the hydrolysis of esters can be carried out under basic or acidic conditions. In hydrolysis using basic conditions the ester is heated to reflux with aqueous sodium hydroxide and the mixture is then acidified to produce the carboxylic acid.

$$C_6H_5-C{\overset{O}{\underset{OCH_3}{}}} + NaOH \longrightarrow C_6H_5C{\overset{O}{\underset{O^-Na^+}{}}} + CH_3OH$$

$$\underset{HCl}{dilute} \Big\downarrow$$

$$\longrightarrow C_6H_5C{\overset{O}{\underset{OH}{}}} + NaCl$$

The rate of hydrolysis depends on the particular ester and the concentration of alkali used. Soap making is a process involving the hydrolysis of esters of the trihydric alcohol glycerol (propane-1,2,3-triol). A simplified equation for the hydrolysis of glyceryl esters is:

$$\begin{array}{l} CH_3(CH_2)_{15}C{\overset{O}{\underset{O-CH_2}{}}} \\ CH_3(CH_2)_{15}C{\overset{O}{\underset{O-CH}{}}} \\ CH_3(CH_2)_{15}C{\overset{O}{\underset{O-CH_2}{}}} \end{array} + 3NaOH \longrightarrow 3CH_3(CH_2)_{15}C{\overset{O}{\underset{O^-Na^+}{}}} + \begin{array}{l} CH_2OH \\ CHOH \\ CH_2OH \end{array}$$

'soap'
sodium heptadecanoate

▲ Soap making in the 19th century

In practice, the three carboxylic acid groups of the glyceryl ester are often different from each other.

Making acid chlorides from carboxylic acids and the hydrolysis of these compounds

To make an acid chloride from a carboxylic acid, the –OH group of the acid needs to substituted by a –Cl atom. Acid chlorides are reactive compounds that are easily hydrolysed if water is present, so a method of production is needed that excludes water. The reagents that can be used for this reaction include phosphorus trichloride, phosphorus pentachloride and sulfur dichloride oxide, $SOCl_2$.

▲ Traditional soap

$$3CH_3COOH + PCl_3 \longrightarrow 3CH_3COCl + H_3PO_3$$

$$CH_3COOH + PCl_5 \longrightarrow CH_3COCl + POCl_3 + HCl$$

$$CH_3COOH + SOCl_2 \longrightarrow CH_3COCl + SO_2 + HCl$$

The method that uses phosphorus trichloride produces small quantities of organic compounds of phosphorus. If phosphorus pentachloride is used then phosphorus trichloride oxide, $POCl_3$, is also produced and needs to be separated from the acid chloride. The third method, using sulfur dichloride oxide, is often preferred as both the co-products, sulfur dioxide and hydrogen chloride, are gases.

In an acid chloride, the carbonyl carbon atom is electron deficient (δ+) as it is bonded to the more electronegative chlorine and oxygen atoms. As a result, acid chlorides are very susceptible to attack by nucleophiles, such as the oxygen atoms in water and alcohols and the nitrogen atoms in ammonia and amines.

Stretch & challenge

The formula of the insecticide DEET is

H3C
O
C
N(CH2CH3)2

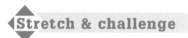

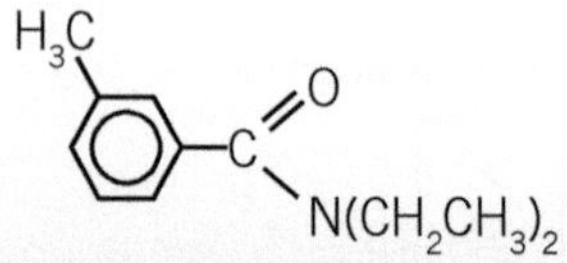

Give the displayed formula of the acid chloride and the amine that can be used to make DEET.

9 Knowledge check

Write the empirical formula of the organic compound produced when hexanoyl chloride reacts with water.

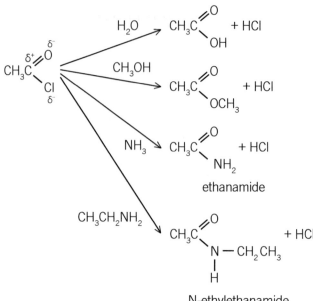

CH3C
δ+
O
δ−
Cl
δ−
H2O
CH3C
O
OH
+ HCl
CH3OH
CH3C
O
OCH3
+ HCl
NH3
CH3C
O
NH2
+ HCl
ethanamide
CH3CH2NH2
CH3C
O
N − CH2CH3
H
+ HCl

N-ethylethanamide

The reaction between smaller molecule acid chlorides such as ethanoyl chloride and water is very vigorous, forming the carboxylic acid and hydrogen chloride.

Benzoyl chloride, C_6H_5COCl, reacts with water much more slowly and its reactions can be carried out in aqueous conditions, whereas the use of ethanoyl chloride needs a fume cupboard and anhydrous conditions.

Amides and nitriles from carboxylic acids

Carboxylic acids, such as ethanoic acid, react with ammonia to give the ammonium salt of the acid. When this salt is heated, water is lost and the amide is produced.

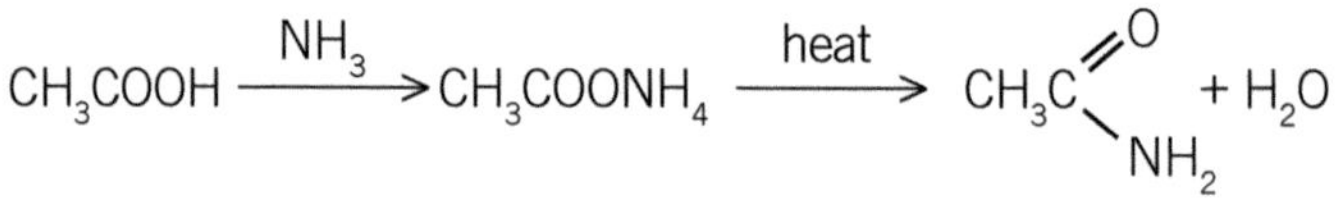

CH3COOH
NH3
CH3COONH4
heat
CH3C
O
NH2
+ H2O

ethanamide

10 Knowledge check

Write the displayed formula of the amide that will produce phenylethanenitrile when it is heated with phosphorus(V) oxide.

Amides can be dehydrated by heating with phosphorus(V) oxide, P_4O_{10}, giving the corresponding nitrile.

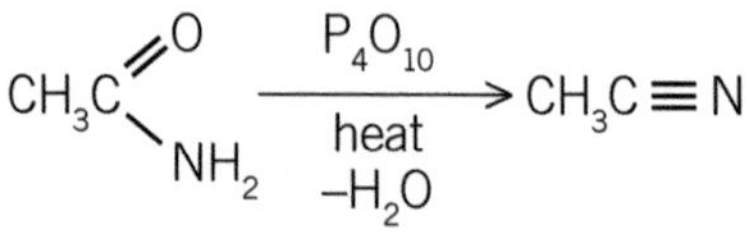

CH3C
O
NH2
P4O10
heat
−H2O
CH3C ≡ N

11 Knowledge check

A straight chain alkyl nitrile, R–CN, contains 14.4% by mass of nitrogen. Write the displayed formula of the nitrile.

When naming nitriles it is important to remember that the carbon atom of the nitrile group is counted as part of the carbon chain. For example, CH_3CH_2CN is propanenitrile but CH_3CH_2Br is bromoethane. Benzonitrile has the formula C_6H_5CN.

Formation of nitriles and hydroxynitriles

Formation of nitriles from halogenoalkanes

Nitriles can be made from halogenoalkanes by their reaction with potassium cyanide using an alcohol–water mixture as solvent.

$$CH_3(CH_2)_3CH_2Br + KCN \longrightarrow CH_3(CH_2)_3CH_2CN + KBr$$

1-bromopentane   hexanenitrile

This is a nucleophilic substitution reaction where the cyanide ion acts as the nucleophile.

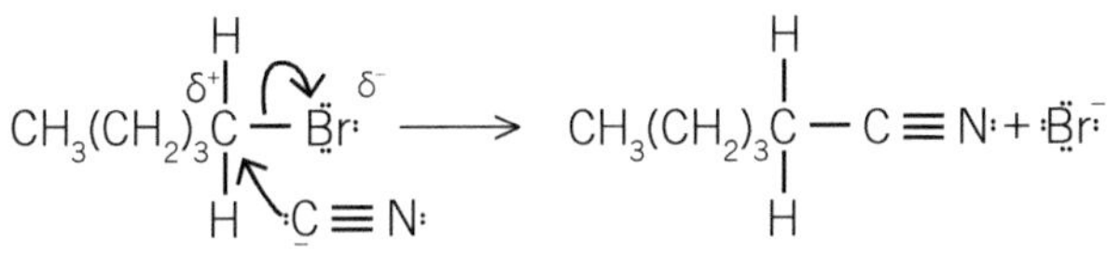

Chlorobenzene will not react with cyanide ions as the ring carbon atoms are not susceptible to nucleophilic attack.

Top tip

When writing the formula of the cyanide ion, make sure that the negative charge is located on the carbon atom. This means that the carbon atom has a lone pair of electrons and has used three electrons in the triple bond to the nitrogen atom.

Formation of hydroxynitriles from aldehydes and ketones

An aldehyde and some simple ketones will react with hydrogen cyanide in the presence of sodium or potassium cyanide to produce a hydroxynitrile.

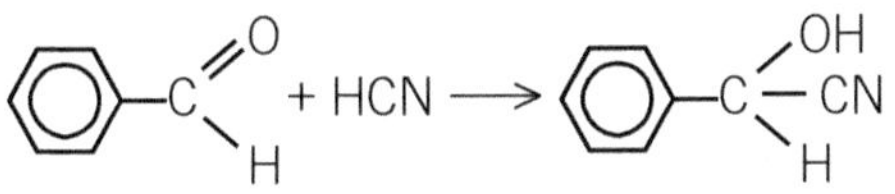

The mechanism of this reaction has been discussed under aldehydes and ketones.

Link

Nucleophilic addition reactions of aldehydes and ketones pages 131–132.

Hydrolysis of nitriles and amides

The hydrolysis of nitriles

When heated under reflux with a dilute acid, a nitrile is hydrolysed, giving a carboxylic acid.

$$CH_3CH_2CN \xrightarrow{H_2SO_4(aq)} CH_3CH_2COOH$$

The nitrogen of the cyanide group becomes an ammonium group. If aqueous sulfuric acid is used, the nitrogen-containing product is a solution of ammonium sulfate. The hydrolysis can also be carried out by heating the nitrile under reflux with an alkali, for example sodium hydroxide, giving the anion of the carboxylic acid. Acidification of the mixture produces the carboxylic acid.

Knowledge check 12

Complete the equation for the reaction of this acid chloride with water.

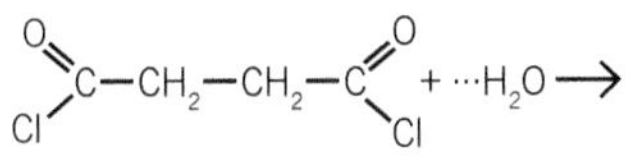

The hydrolysis of amides

When an amide is heated under reflux with a base such as aqueous sodium hydroxide, the amide is hydrolysed as the carbon to nitrogen bond is broken, producing ammonia gas.

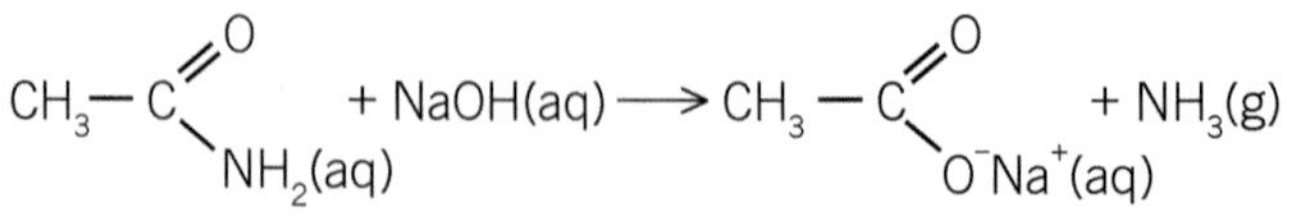

$$CH_3-C\overset{O}{\underset{NH_2(aq)}{\diagdown}} + NaOH(aq) \longrightarrow CH_3-C\overset{O}{\underset{O^-Na^+(aq)}{\diagdown}} + NH_3(g)$$

sodium ethanoate

A similar reaction occurs if an N-substituted amide is hydrolysed, giving an amine as one of the products.

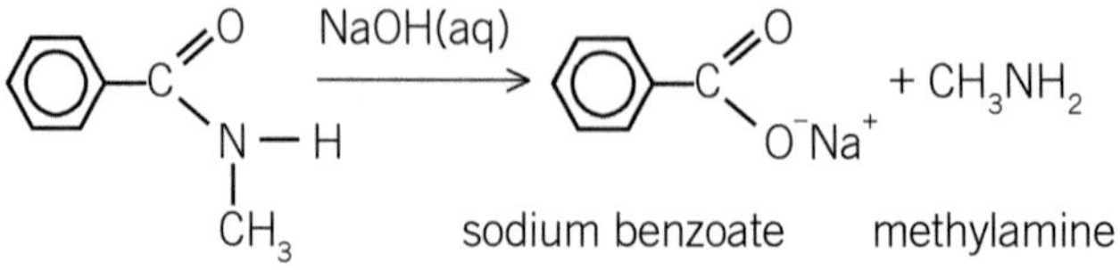

sodium benzoate   methylamine

Reduction of nitriles

Warming a nitrile with a solution of lithium tetrahydridoaluminate(III) in ethoxyethane will reduce a nitrile to a primary amine.

$$CH_3CH_2C\equiv N \xrightarrow[\text{ethoxyethane}]{\text{LiAlH}_4} CH_3CH_2CH_2NH_2$$

propylamine

Other reducing agents that can be used include hydrogen and a nickel catalyst, and sodium metal/ethanol. The formation of a nitrile, followed by its reduction, produces compounds with longer carbon chains.

The reduction of a nitrile with lithium aluminium hydride is generally the preferred method as yields are higher.

Aromatic nitriles such as benzonitrile are similarly reduced by lithium aluminium hydride.

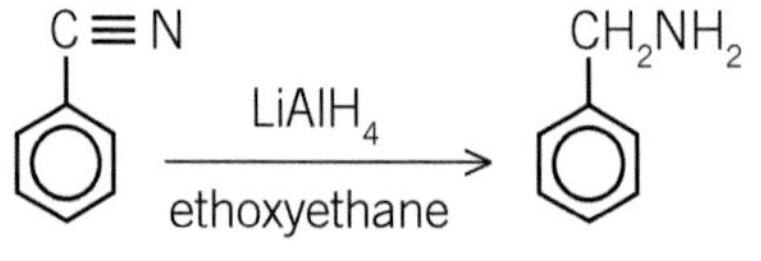

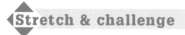

benzonitrile   phenylmethylamine

Hydroxynitriles (see page 131) can be similarly reduced by lithium aluminium hydride. For example, 2-hydroxypropanenitrile is reduced to 1-aminopropan-2-ol,

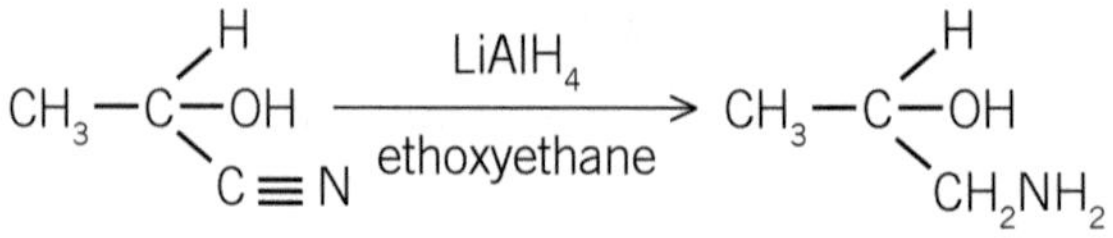

$$CH_3-\overset{H}{\underset{C\equiv N}{\overset{|}{C}}}-OH \xrightarrow[\text{ethoxyethane}]{\text{LiAlH}_4} CH_3-\overset{H}{\underset{CH_2NH_2}{\overset{|}{C}}}-OH$$

2-hydroxy-2-methylpropanenitrile

which contains a chiral centre (see page 110). 2-Hydroxy-2-methylpropanenitrile is similarly reduced to 1-amino-2-methylpropan-2-ol, which does not contain a chiral centre.

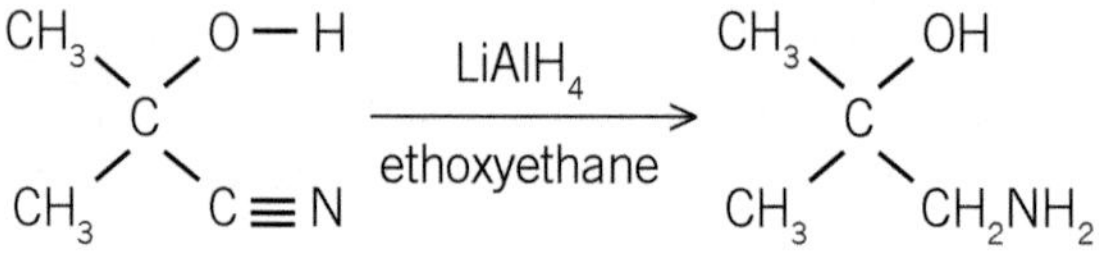

$$\overset{CH_3}{\underset{CH_3}{\diagup}}C\overset{O-H}{\underset{C\equiv N}{\diagdown}} \xrightarrow[\text{ethoxyethane}]{\text{LiAlH}_4} \overset{CH_3}{\underset{CH_3}{\diagup}}C\overset{OH}{\underset{CH_2NH_2}{\diagdown}}$$

Test yourself

1. Many carboxylic acids have common names.

   Give the structures of: [5]

   (a) Formic acid (methanoic acid)

   (b) Lactic acid (2-hydroxypropanoic acid)

   (c) Succinic acid (butane-1,4-dioic acid)

   (d) Caproic acid (hexanoic acid)

   (e) Adipic acid (hexane-1,6-dioic acid)

2. Give the systematic names of these carboxylic acids: [5]

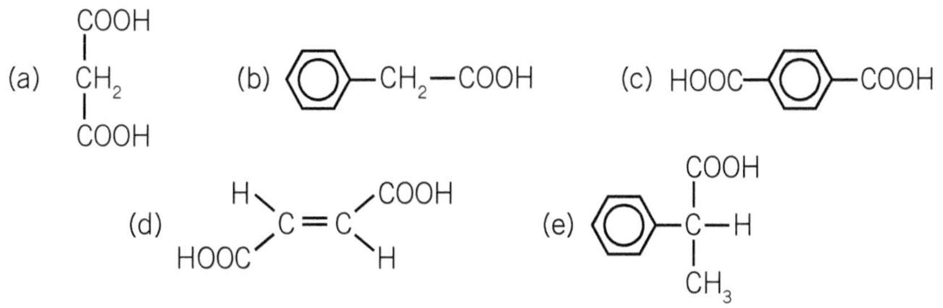

   (a) CH_2 with COOH above and COOH below

   (b) phenyl$-CH_2-COOH$

   (c) $HOOC-$phenyl$-COOH$

   (d) H and $HOOC$ on one carbon, $C=C$, $COOH$ and H on other carbon

   (e) phenyl$-\overset{\displaystyle COOH}{\underset{\displaystyle CH_3}{C}}-H$

3. Give the equation for the reduction of phenylethanoic acid by $LiAlH_4$ to the corresponding alcohol. Represent the reducing agent by [H]. [1]

4. A carboxylic acid, $HOOC–R–CH_3$, contains 31.3% of oxygen by mass. Suggest a formula for the acid, showing your working. [3]

5. A kettle descaler contains an aqueous solution of lactic acid, $(CH_3CH(OH)COOH)$. The acid has an M_r of 90. This solution was diluted ten times with water. The concentration of this diluted solution was found by using an acid–base titration. $25.00\ cm^3$ of the diluted lactic acid solution required $47.20\ cm^3$ of aqueous sodium hydroxide of concentration $0.500\ mol\ dm^{-3}$ for complete neutralisation.

   Show that the concentration of the original lactic acid solution was $850\ g\ dm^{-3}$. [5]

6. Sodium butanoate is heated with soda lime (represented as NaOH). Give the equation for this reaction and name the alkane formed. [2]

7. Calcium butanoate, $(CH_3(CH_2)_2COO)_2Ca$, is heated in the absence of soda lime, giving a ketone and calcium carbonate. Deduce the formula of the ketone produced and name it. [2]

8. State the names of the reagents **R** and **S** seen in the reaction sequence below: [2]

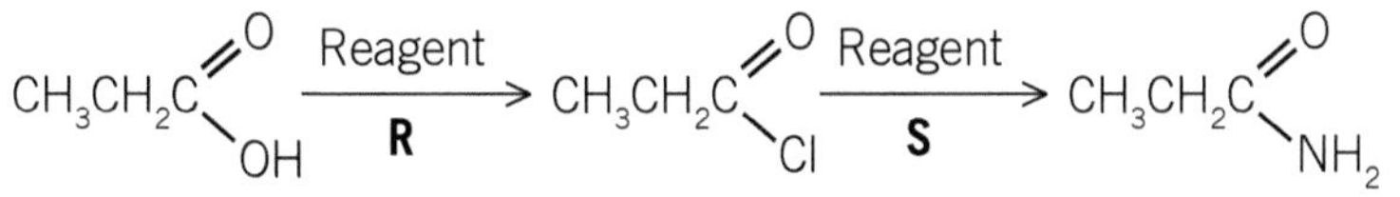

   $CH_3CH_2C\overset{\displaystyle O}{\underset{\displaystyle OH}{<}}$ $\xrightarrow[\textbf{R}]{Reagent}$ $CH_3CH_2C\overset{\displaystyle O}{\underset{\displaystyle Cl}{<}}$ $\xrightarrow[\textbf{S}]{Reagent}$ $CH_3CH_2C\overset{\displaystyle O}{\underset{\displaystyle NH_2}{<}}$

9. 5.84 g of an amide **T** is hydrolysed by heating it with aqueous sodium hydroxide:

   $R-C\overset{\displaystyle O}{\underset{\displaystyle NH_2}{<}} + NaOH \longrightarrow R-C\overset{\displaystyle O}{\underset{\displaystyle O^-Na^+}{<}} + NH_3$

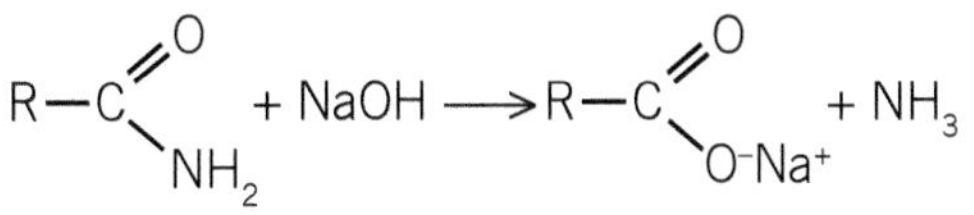

   The reaction produces 1.36 g of ammonia.

   Find the relative molecular mass of the amide and hence the formula of the R group. [3]

Amines

Primary amines are organic compounds that contain an $-NH_2$ group bonded directly to an alkyl or aryl group, giving $R-NH_2$. They can be considered as derivatives of ammonia, NH_3, where one of the hydrogen atoms has been substituted by an $-R$ group. Substitution of the remaining hydrogen atoms gives a secondary amine, R_2NH and a tertiary amine, R_3N. The nitrogen atom in amines retains the lone pair of electrons, allowing amines to react as bases by electron pair donation. Amines are very reactive compounds and can take part in reactions where the C–N bond is broken to give, for example, alcohols and phenols. They can also react where the C–N bond is retained, giving, for example, compounds with peptide linkages and azo dyes.

Topic contents

You should be able to demonstrate and apply knowledge and understanding of:

- The formation of primary aliphatic amines from halogenoalkanes and nitriles.

- The formation of aromatic amines from nitrobenzenes.

- The basicity of amines.

- The ethanoylation of primary amines using ethanoyl chloride.

- The reaction of primary amines (aliphatic and aromatic) with cold nitric(III) acid.

- The coupling of benzenediazonium salts with phenols and aromatic amines.

- The role of the –N=N– chromophore in azo dyes.

- The origin of colour in terms of the wavelengths of visible light absorbed.

Maths skills ≫

- Calculate percentage yields
- Construct and/or balance equations using ratios
- Carry out structured and unstructured mole calculations

Structure and naming of amines

All amines contain a nitrogen atom bonded directly to a carbon atom of an alkyl or aryl group. If this nitrogen atom is bonded to only one carbon atom, the compound is called a primary amine. Secondary amines have the nitrogen atom bonded directly to two carbon atoms and if the nitrogen atom is bonded directly to three carbon atoms then it is a tertiary amine. The longest carbon chain is used to name the amine, with the number of carbon atoms indicated in the usual way and -amine used as the suffix.

$CH_3CH_2CH_2CH_2NH_2$

butylamine

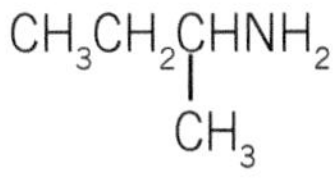

1-methylpropylamine

Aromatic amines have the nitrogen atom of the –NH_2 group bonded directly to the benzene ring.

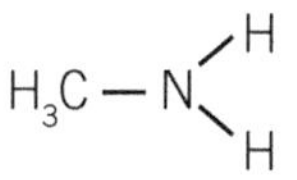

methylamine

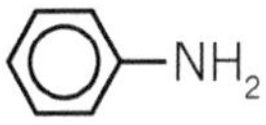

phenylamine

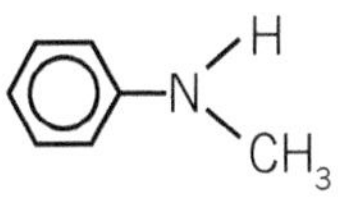

N-methylphenylamine

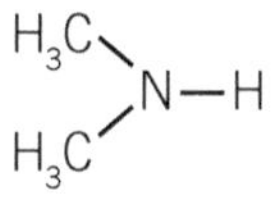

dimethylamine

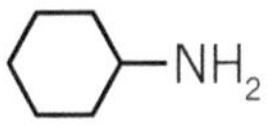

cyclohexylamine

$HO—CH_2—CH_2—NH_2$

2-aminoethanol

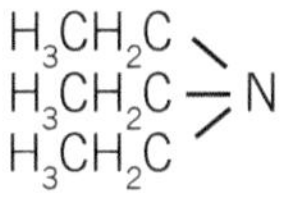

triethylamine

$H_2N—CH_2(CH_2)_4CH_2—NH_2$

hexane-1,6-diamine

Knowledge check 1

Give the systematic names of:

(a) $CH_3(CH_2)_4NH_2$

(b)

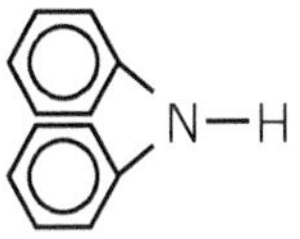

(c) $H_2N—CH_2—CH_2—NH_2$

> **Key term**
>
> An **aromatic amine** must have the –NH_2 group directly bonded to the benzene ring.

Knowledge check 2

Write the displayed formula of:
(a) 4-ethylphenylamine
(b) 2-methylpropylamine.

◄Stretch & challenge▼

Write the displayed formula of a compound $C_7H_{10}N_2$ that contains both aliphatic and aromatic primary amine groups.

Formation of primary aliphatic amines

Making primary aliphatic amines from halogenoalkanes

The reaction of a halogenoalkane with ammonia using a water/ethanol solvent produces an amine.

$$CH_3CH_2CH_2Br + NH_3 \longrightarrow CH_3CH_2CH_2NH_2 + HBr$$

This is a simplified picture as, if the mixture is warmed, ammonia gas is lost from the mixture, reducing the yield. Some methods of preparation use the reactants in a sealed tube, which is gently heated. Generally, an excess of ammonia is used. This is to react with the acidic gas hydrogen bromide, to give ammonium bromide.

The type of mechanism for the formation of the amine is nucleophilic substitution.

$$CH_3CH_2\overset{H}{\underset{H}{C}}\overset{\delta+}{-}\overset{\delta-}{Br} \longrightarrow CH_3CH_2\overset{H}{\underset{H}{C}}-\overset{H}{\underset{H}{\overset{\cdot\cdot}{N}}}^{H} + H^+ + Br^-$$

Depending on the conditions used, the usual product is not the free amine but its salt.

$$CH_3CH_2CH_2NH_2 + HBr \longrightarrow CH_3CH_2CH_2{}^+NH_3\,Br^-$$
propylammonium bromide

Propylammonium bromide is a substituted ammonium bromide (NH_4Br), with one of the hydrogen atoms replaced by a propyl group. If an ammonium salt is heated with a base, ammonia is produced. Similarly, when propylammonium bromide is heated with aqueous sodium hydroxide, propylamine is obtained.

$$CH_3CH_2CH_2{}^+NH_3Br^- + NaOH \longrightarrow CH_3CH_2CH_2NH_2 + NaBr + H_2O$$

Making primary aliphatic amines from nitriles

Nitriles can be reduced with a suitable reducing agent. The usual reducing agent is lithium tetrahydridoaluminate(III) (represented as [H] in the equation) dissolved in ethoxyethane as solvent.

$$CH_3CH_2\underset{CH_3}{\overset{|}{C}}HC \equiv N + 4[H] \longrightarrow CH_3CH_2\underset{CH_3}{\overset{|}{C}}HCH_2NH_2$$
2-methylbutylamine

This reaction has been discussed in the section on carboxylic acids.

Formation of aromatic amines from nitrobenzenes

Benzene is not easily attacked by nucleophiles such as ammonia, as the ring electrons repel nucleophiles but are attracted to electrophiles. Phenylamine, $C_6H_5NH_2$, is made by the reduction of nitrobenzene. A general equation for this reduction is

$$\bigcirc\!\!\!\!-NO_2 + 6[H] \longrightarrow \bigcirc\!\!\!\!-NH_2 + 2H_2O$$

The traditional reducing agent for this reaction is tin metal and hydrochloric acid. Making phenylamine commercially requires the use of more economical reducing agents. These include hydrogen and a nickel catalyst, or iron and hydrochloric acid. Derivatives of nitrobenzene can be reduced in a similar way.

Basicity of amines

Amines, like ammonia, have a lone pair of electrons on the nitrogen atom. The lone pair can be used to accept a proton by means of a coordinate bond.

methylammonium ion

The basic nature of amines can be shown by the reaction of 'smaller' amines with water.

$$CH_3NH_2 + H_2O \rightleftharpoons \left[CH_3NH_3 \right]^+ + {}^-OH$$

Amines are weak bases and the equilibrium position of this equation is well to the left – a strong smell of 'fishy ammonia' indicates that free methylamine is present in the solution. The pH of an aqueous solution of methylamine of concentration 0.1 mol dm^{-3} is about 11.8, whereas an ammonia solution of the same concentration has a pH of about 11.1. Methylamine and other alkylamines are stronger bases than ammonia because the alkyl groups 'push' electrons slightly towards the nitrogen atom, making it more δ– when compared with ammonia. Phenylamine is a much weaker base than ammonia or alkylamines because the nitrogen lone pair of electrons becomes, to a certain extent, part of the delocalised ϖ system, and this makes the nitrogen relatively less δ–. Phenylamine is, however, more susceptible to electrophilic ring substitution than benzene because of this nitrogen lone pair effect. Although the –NH$_2$ group of amines can hydrogen bond with water, phenylamine is only slightly soluble in water because of the hydrophobic effect of the benzene ring. Phenylamine, like alkylamines, will react with acids to form salts.

phenylammonium chloride

Ethanoylation of primary amines

The nitrogen lone pair enables amines to react as nucleophiles. They attack the δ+ carbon atom of the carbonyl group in an acid chloride.

N-methylethanamide

The organic product is N-methylethanamide. The letter 'N' means that the methyl group is bonded to the nitrogen atom. This compound is an N-substituted derivative of ethanamide, CH$_3$CONH$_2$. Phenylamine reacts in a similar way to produce N-phenylethanamide.

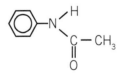

N-phenylethanamide

The — N — C — is known as a peptide linkage or peptide bond and is also present in peptides, polypeptides and proteins. Polyamides such as nylon also contain this linkage. A familiar compound that contains the peptide linkage is the painkiller paracetamol.

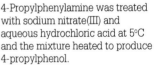

Paracetamol
(4-acetamidophenol)

Knowledge check 7

Give the empirical formula of N,N'-diacetyl-1,4-phenylenediamine:

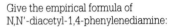

Study point

The reaction of a primary amine with an acid chloride is a method of producing a peptide linkage in a molecule.

Study point

Nitric(III) acid is unstable and made when needed, It can be written as HNO$_2$ or HONO.

Knowledge check 8

4-Propylphenylamine was treated with sodium nitrate(III) and aqueous hydrochloric acid at 5°C and the mixture heated to produce 4-propylphenol.

$$CH_3CH_2CH_2 - \bigcirc - NH_2 \rightarrow$$

$$CH_3CH_2CH_2 - \bigcirc - OH$$

The yield of 4-propylphenol was 0.14 mole, which represented a yield of 70%. Calculate the mass of 4-propylphenylamine used in this experiment.

Reaction of primary amines with cold nitric(III) acid

Nitric(III) acid (or nitrous acid), HNO_2, is an unstable compound and is made when required by the reaction of a dilute acid (eg HCl) on sodium nitrate(III), $NaNO_2$ (also called sodium nitrite).

$$NaNO_2 + HCl \longrightarrow HNO_2 + NaCl$$

A primary aliphatic amine reacts with nitric(III) acid with the evolution of nitrogen gas. One equation for this reaction is:

$$R-NH_2 + HNO_2 \longrightarrow R-OH + N_2 + H_2O$$

This looks as if it provides a useful route for preparing a primary alcohol from a primary amine. Although the yield of nitrogen is quantitative, the yield of the alcohol is poor. The reaction may initially form an alkyl diazonium ion, $R-N^+\equiv N$⁝, which then loses nitrogen to give the primary carbocation.

The corresponding reaction with a primary aromatic amine, for example phenylamine, proceeds via a diazonium intermediate but this is more stable than an alkyldiazonium ion and, if the temperature is between 0°C and 10°C, a solution containing the benzenediazonium ion is produced.

benzenediazonium chloride

If aqueous sulfuric acid is used in place of hydrochloric acid, the intermediate is benzenediazonium hydrogensulfate rather than benzenediazonium chloride. Above 10°C, decomposition of the benzenediazonium compound occurs, giving phenol.

If the reaction is performed at temperatures below 0°C the production of the benzenediazonium compound is too slow.

Coupling reaction of benzenediazonium salts

Benzenediazonium compounds are very reactive compounds and can be used as intermediates in the formation of many other compounds. In the formation of phenol from benzenediazonium chloride, the nitrogen atoms are lost as nitrogen gas as the unstable ion decomposes. Below 10°C benzenediazonium compounds react with phenols and aromatic amines to produce compounds where the –N=N– azo group is retained. The benzenediazonium ion is a weak electrophile and will bond, via electrophilic substitution, with aromatic compounds where the ring has been activated by the presence of –OH or –NH_2 groups. The resulting compound is highly coloured, generally yellow, orange or red, and is called an **azo dye**. This reaction is often called a coupling reaction and is carried out in alkaline solution. Coupling often occurs at the 4-position but can occur at the 2-position relative to the –OH or NH_2 group.

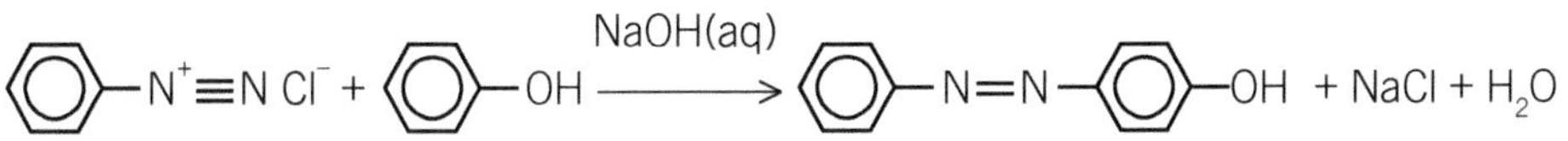

$$\text{C}_6\text{H}_5\text{-N}^+\equiv\text{N Cl}^- + \text{C}_6\text{H}_5\text{-OH} \xrightarrow{\text{NaOH(aq)}} \text{C}_6\text{H}_5\text{-N=N-C}_6\text{H}_4\text{-OH} + \text{NaCl} + \text{H}_2\text{O}$$

(4-phenylazo)phenol

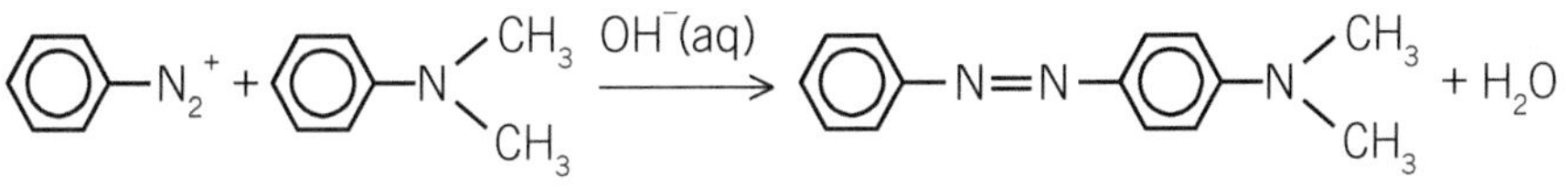

$$\text{C}_6\text{H}_5\text{-N}_2^+ + \text{C}_6\text{H}_5\text{-N(CH}_3)_2 \xrightarrow{\text{OH}^-\text{(aq)}} \text{C}_6\text{H}_5\text{-N=N-C}_6\text{H}_4\text{-N(CH}_3)_2 + \text{H}_2\text{O}$$

Coupling can also occur with other aromatic systems, for example with naphthalen-2-ol.

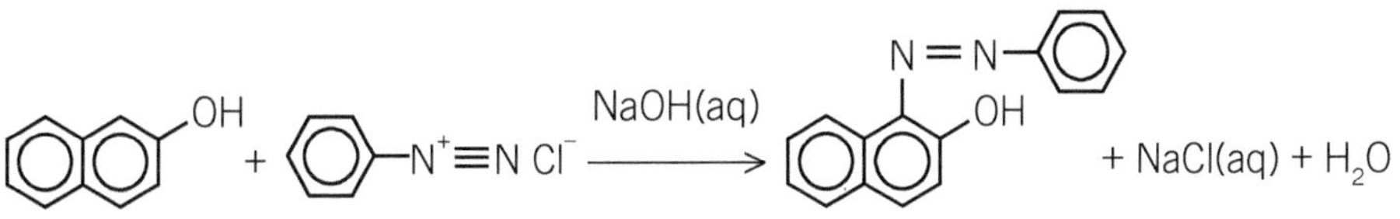

$$\text{naphthalen-2-ol} + \text{C}_6\text{H}_5\text{-N}^+\equiv\text{N Cl}^- \xrightarrow{\text{NaOH(aq)}} \text{(1-phenylazo)naphthalen-2-ol} + \text{NaCl(aq)} + \text{H}_2\text{O}$$

naphthalen-2-ol   (1-phenylazo)naphthalen-2-ol

This azo dye forms as a red solid and is used under the name Sudan 1. It has been used to colour curry powder but its use for this purpose is now banned in the United Kingdom.

Role of the –N=N– chromophore in azo dyes and the origin of their colour

A **chromophore** is a structural unit in a molecule that is primarily responsible for the absorption of radiation of a certain wavelength. If the absorption is in the visible region, the colour that is *not* absorbed is the colour observed. The actual wavelengths of light absorbed depend on other groups present in the molecule.

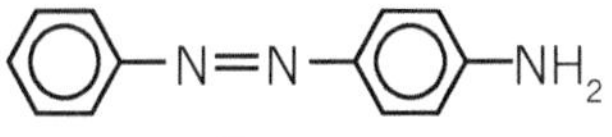

yellow

λ_{max} 384 nm

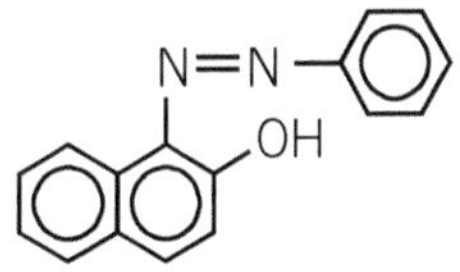

orange–red

λ_{max} 476 nm

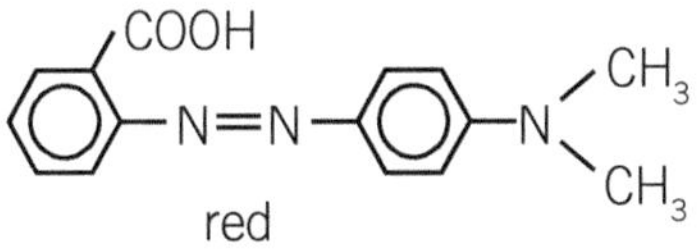

red

λ_{max} 520 nm

The intensity of the colour that is observed in solution depends on the concentration of the compound present, and this is the basis of colorimetry. The colour hexagon (see next page) gives an indication of the colour absorbed and the colour transmitted (not absorbed) when light is shone through the sample.

Knowledge check

An azo dye has the formula:

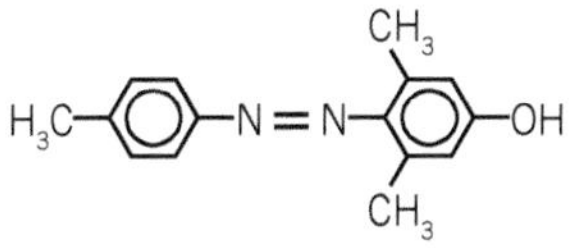

Give the systematic names of the aromatic amine and the phenol used to produce this azo dye in a diazotisation reaction.

>>> **Key term**

A **chromophore** is a structural unit in a molecule that is primarily responsible for the absorption of radiation a certain wavelength, generally in the visible or ultraviolet region of the electromagnetic spectrum.

Knowledge check 10

The acid–base indicator phenolphthalein in a strongly alkaline solution absorbs in the visible region of the electromagnetic spectrum.

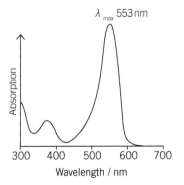

Use the information in this section to show that this alkaline solution of phenolphthalein has a red/violet (magenta) colour.

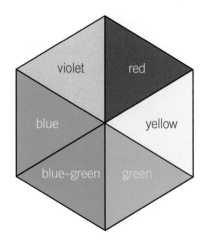

Colour	Wavelength/nm
Violet	380–430
Blue	430–490
Green	490–560
Yellow	560–580
Orange	580–620
Red	620–750

🏠 **11 Knowledge check**

Curcumin is a yellow dye found in curry powder. Use the colour hexagon to suggest a wavelength range for the maximum absorption of a solution containing curcumin, explaining your answer.

The colour of a solution of Sudan 1 is red. Using the colour hexagon, the opposite colour to red is blue-green, which indicates that Sudan 1 has its maximum absorption in the blue-green region. This occurs at 476 nm.

The visible spectrum of methylene blue is shown below.

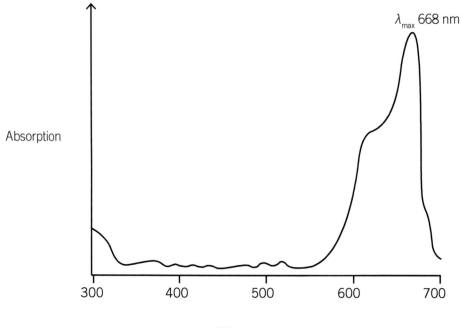

λ_{max} 668 nm

Absorption

Wavelength / nm

The spectrum shows that its maximum absorption is at 668 nm, which is in the red region of the spectrum. The bluish colour of methylene blue is opposite red in the colour hexagon.

Test yourself

1. Write the displayed formulae of the two primary amines whose molecular formula is C_3H_9N. [2]

2. The formula of piperidine is:

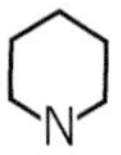

   Explain why piperidine is **not** a primary amine. [1]

3. Bromoethane reacts with ammonia to give ethylamine.

   State why ammonia acts as a nucleophile in this reaction and why it is a substitution reaction. [2]

4. Cyanogen has the formula N≡C–C≡N. Suggest the structure of the amine produced if cyanogen is reduced by lithium tetrahydridoaluminate(III). [1]

5. (a) Benzene-1,3-diamine can be obtained by the reduction of 1,3-dinitrobenzene.

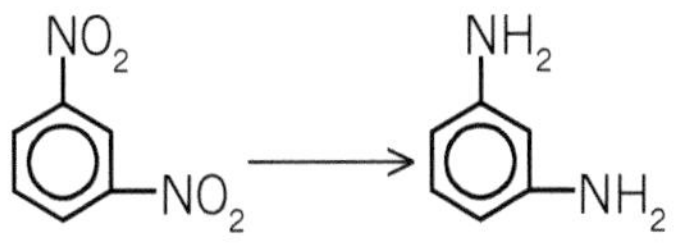

Suggest a reagent(s) for this reaction. [1]

(b) The azo dye Chrysoidine G (present as salt) has the formula:

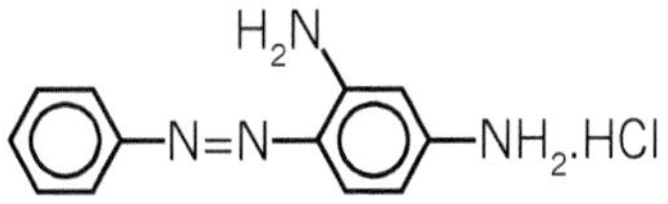

(i) State the name of the aromatic amine that will couple with benzene-1,3-diamine to give this dye. [1]

(ii) Chrysoidine G has a maximum absorption (λ_{max}) in the visible region of the electromagnetic spectrum at 449 nm.

State the colour of Chrysoidine G in white light, giving a reason for your answer. [2]

6. Give the structure of the organic compound formed when benzoyl chloride (C_6H_5COCl) reacts with phenylamine. [1]

7. Ethanoyl chloride reacts with ethylamine to give N-ethylethanamide. Deduce the structure of this compound and explain its name. [2]

8. 0.0150 mol of phenylmethylamine, $C_6H_5CH_2NH_2$, reacts with an excess of nitric(III) acid to produce nitrogen gas. Calculate the maximum volume of nitrogen gas obtained, measured at 298 K and 1 atmosphere pressure. Give your answer in cm³. [2]

9. 2-Methylphenol can be formed from 2-methylphenylamine by diazotisation and then warming the mixture obtained.

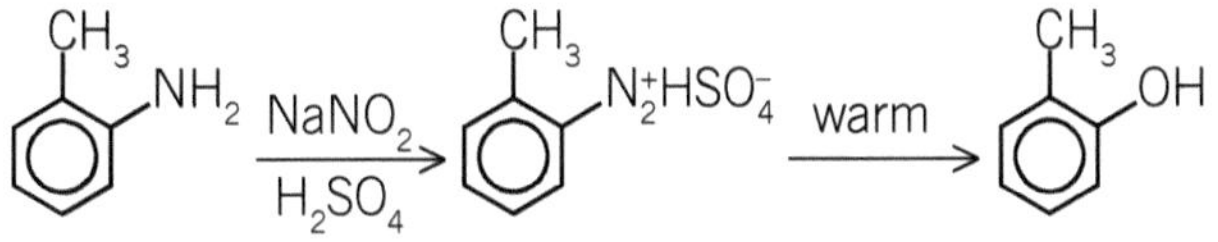

(a) State a suitable temperature used to make the intermediate diazonium compound. [1]

(b) In an experiment 4.81 g of 2-methylphenol was obtained. This was a yield of 89% based on the starting amine.

(i) Calculate the mass of 2-methylphenylamine used in the first stage. [2]

(ii) The density of 2-methylphenylamine is 1.01 g cm⁻³ at 25°C. Calculate the volume of 2-methylphenylamine that should be used in this reaction, giving your answer to an appropriate number of significant figures. [2]

4.7

Amino acids, peptides and proteins

Amino acids, peptides and proteins are naturally occurring nitrogen compounds that form the basis of all living organisms. Amino acids are carboxylic acids that also contain the amino functional group, $-NH_2$. The condensation of two amino acid molecules gives a dipeptide. Further condensations between amino acid molecules lead to polypeptides and to proteins. This topic also considers the basic ideas behind the primary, secondary and tertiary structure of proteins.

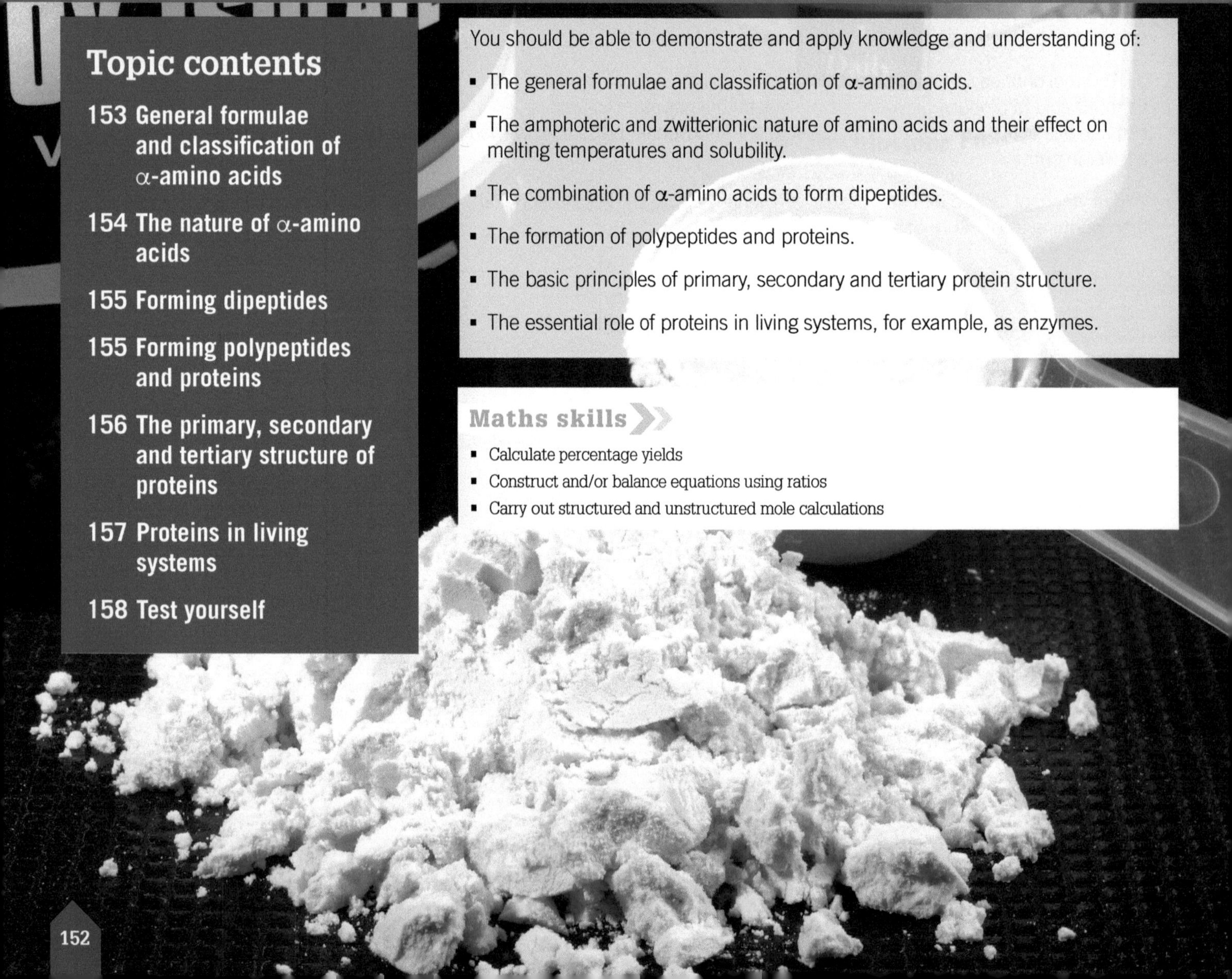

Topic contents

You should be able to demonstrate and apply knowledge and understanding of:

- The general formulae and classification of α-amino acids.

- The amphoteric and zwitterionic nature of amino acids and their effect on melting temperatures and solubility.

- The combination of α-amino acids to form dipeptides.

- The formation of polypeptides and proteins.

- The basic principles of primary, secondary and tertiary protein structure.

- The essential role of proteins in living systems, for example, as enzymes.

Maths skills ⟫

- Calculate percentage yields
- Construct and/or balance equations using ratios
- Carry out structured and unstructured mole calculations

General formulae and classification of α-amino acids

An α-amino acid has the –NH$_2$ group bonded to the carbon atom that is next to the carboxylic acid group. This carbon atom is called the α-carbon atom. The general formula of these α-amino acids is:

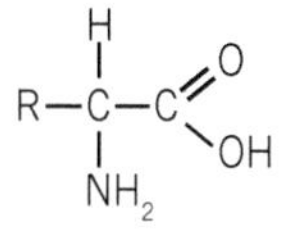

The simplest α-amino acid is aminoethanoic acid, H$_2$NCH$_2$COOH where R = H. R can also be a simple alkyl group or a carbon-containing group that may also contain an –OH, –SH or another –NH$_2$ or –COOH group. Common α-amino acids have traditional names that continue to be used, especially in biology and biochemistry.

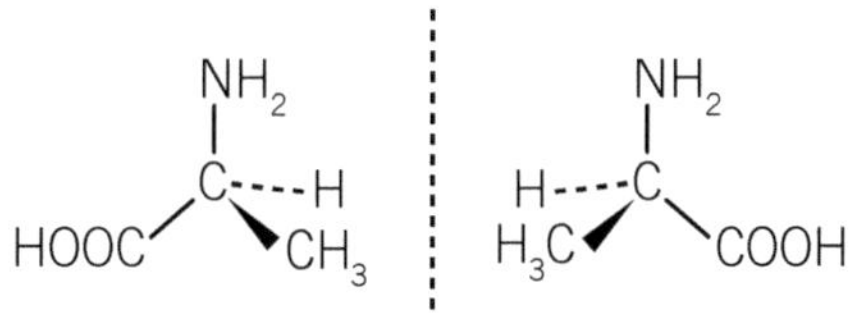

Knowledge check **1**

Give the displayed formula of 2-amino-3-methylpentanoic acid.

H—C—COOH (H, NH$_2$)
aminoethanoic acid
(glycine)

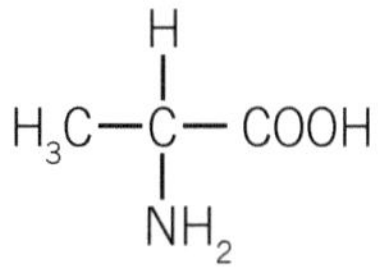

H$_3$C—C—COOH (H, NH$_2$)
2-aminopropanoic acid
(alanine)

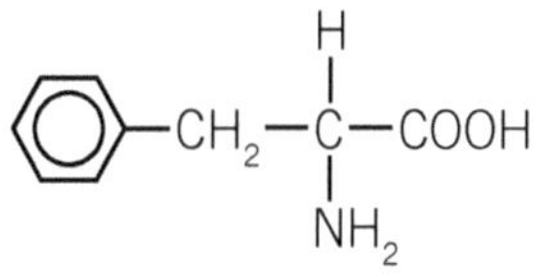

◀**Stretch & challenge**

Give the molecular formula of the α-amino acid, tryptophan.

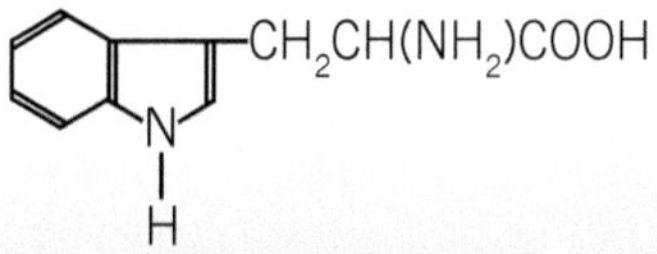

—CH$_2$—C—COOH (H, NH$_2$)
2-amino-3-phenylpropanoic acid
(phenylalanine)

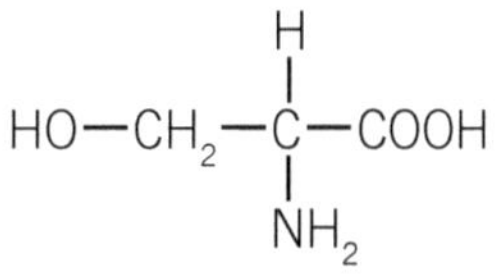

HO—CH$_2$—C—COOH (H, NH$_2$)
2-amino-3-hydroxypropanoic acid
(serine)

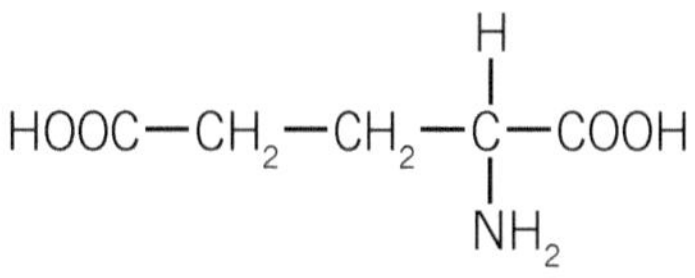

Knowledge check **2**

Cystine has the formula

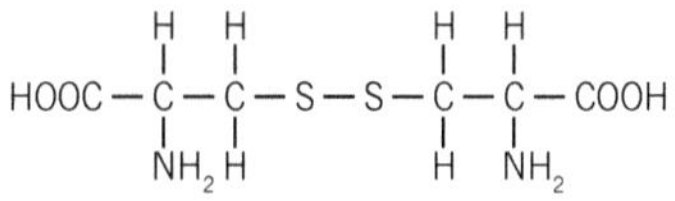

HOOC — C — C — S — S — C — C — COOH (H H, NH$_2$ H, H NH$_2$)

Indicate any chiral centres in the formula of cysteine using an asterisk (*).

HOOC—CH$_2$—CH$_2$—C—COOH (H, NH$_2$)
2-aminopentanedioic acid
(glutamic acid)

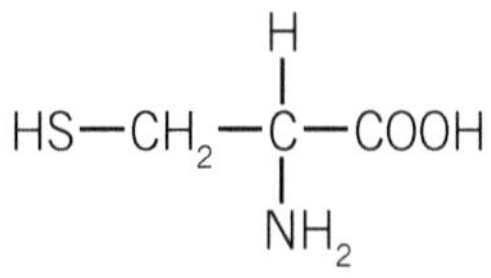

HS—CH$_2$—C—COOH (H, NH$_2$)
2-amino-3-sulfhydrylpropanoic acid
(cysteine)

Apart from aminoethanoic acid, all α-amino acids contain a chiral centre(s) (shown as *).

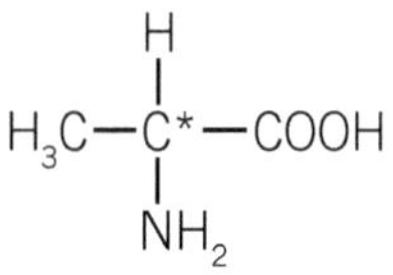

H$_3$C—C*—COOH (H, NH$_2$)
2-aminopropanoic acid
(alanine)

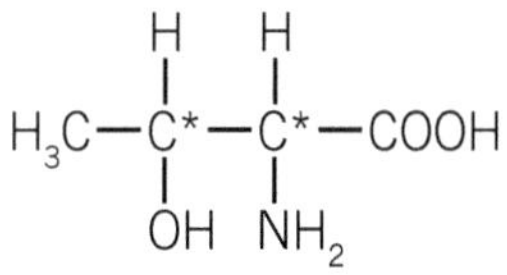

H$_3$C—C*—C*—COOH (H H, OH NH$_2$)
2-amino-3-hydroxybutanoic acid
(threonine)

Link

Some terms used in optical isomerism page 110.

The two optical isomers of 2-aminopropanoic acid (alanine) can be shown as mirror image forms.

Knowledge check **3**

Write the formulae of the two mirror image forms of serine (2-amino-3-hydroxypropanoic acid).

The nature of α-amino acids

α-Amino acids exist as solids at room temperature, whereas similar sized molecules are often liquids or occur as solids with a much lower melting temperature.

aminoethanoic acid
melting temperature 240°C

hydroxyethanoic acid
melting temperature 75–80°C

methoxyethanoic acid
melting temperature 8°C
boiling temperature 203°C

The melting temperatures of α-amino acids are much higher than might be expected because they exist as **zwitterions**. A hydrogen ion (H^+) is lost from the carboxylic acid group and gained by the nitrogen atom of the amino group, by use of its lone pair of electrons. The zwitterion formula of aminoethanoic acid is

The ionic nature of zwitterions means that there are strong ionic forces between positive and negative ions and more energy is needed to overcome these forces. As a result, α-amino acids have higher melting temperatures than related covalently bonded compounds. Aminoethanoic acid is described as a neutral amino acid, as the positive and negative charges are balanced out. The zwitterion dipolar form of the amino acid suggests that α-amino acids are soluble in water. The solubility of aminoethanoic acid at 25°C is 25 g in 100 cm³ of water, and the solubility of the larger molecule 2-amino-3-phenyl-propanoic acid is 3 g in 100 cm³ of water. Aqueous solutions of α-amino acids contain zwitterions but the compounds only exist as the zwitterions themselves at a certain pH, called the isoelectric point. This pH varies depending on the amino acid. The isoelectric point of aminoethanoic acid is at pH 6.0. If the pH of the solution is below the isoelectric point then the amino acid as a base by accepting a hydrogen ion.

In solutions where the pH is greater than the isoelectric point the amino acid acts as an acid, losing a proton.

4 Knowledge check

The isoelectric point of proline is 6.3.

Write the displayed formula of the species formed from proline at a pH of 9.0.

Key term

A **zwitterion** is a dipolar form of an amino acid where the carboxylic acid group loses a proton, becoming COO^- and the amino group gains the proton, becoming $^+NH_3$.

Stretch & challenge

Horses' urine contains hippuric acid.

A sample of hippuric acid is hydrolysed by heating it with aqueous sodium hydroxide. Give the displayed formulae of the two products obtained from this reaction.

5 Knowledge check

Explain how the zwitterion form of glycine suggests that the nitrogen atom cannot exist as a nucleophile.

Forming dipeptides

A dipeptide is formed when two amino acid molecules join together in a condensation reaction, for example, using two molecules of aminoethanoic acid:

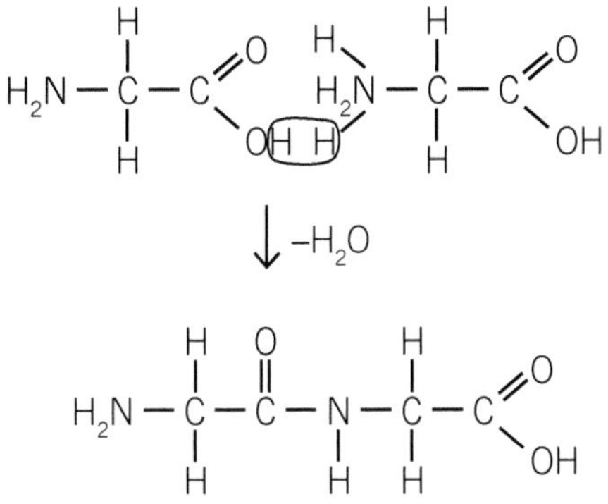

$$\downarrow -H_2O$$

Dipeptides are formed by the reaction of a COOH group in one amino acid with the NH_2 group of another amino acid and vice versa. For example, if phenylalanine and glycine are used then the two dipeptides are

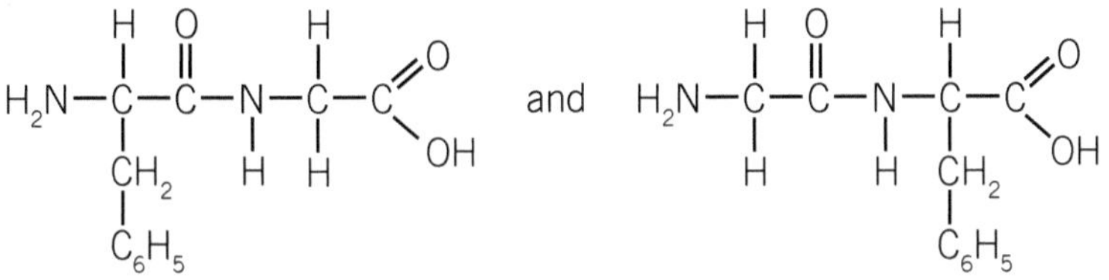

and

The formation of a dipeptide introduces a peptide linkage (also called a peptide bond or amide linkage) into the condensation product.

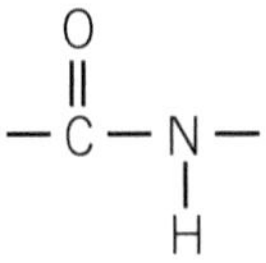

One of the artificial sweeteners used in 'diet' carbonated sugar-free drinks is aspartame, which is the methyl ester of the condensation product between phenylalanine and aspartic acid, $HOOC–CH_2–CH(NH_2)COOH$.

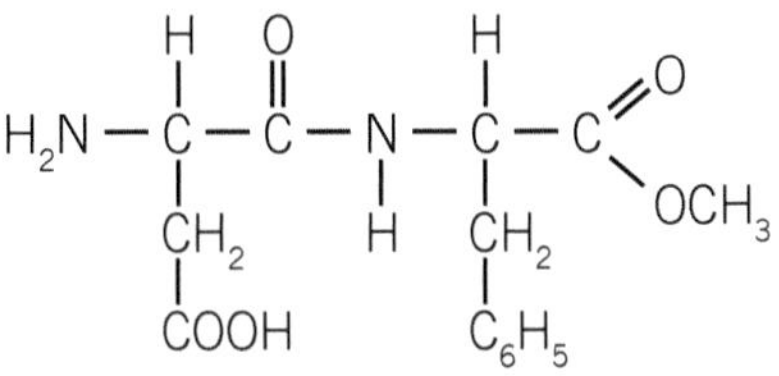

aspartame

Forming polypeptides and proteins

A polypeptide is a long chain of condensed amino acids joined together by peptide linkages. Part of a polypeptide chain is:

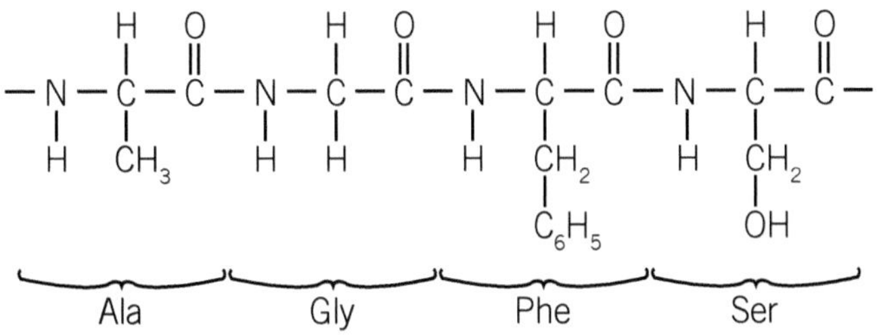

Ala   Gly   Phe   Ser

Knowledge check 6

The diagram shows the formula of a dipeptide formed between valine, $(CH_3)_2CHCH(NH_2)COOH$ and cysteine, $HSCH_2CH(NH_2)COOH$.

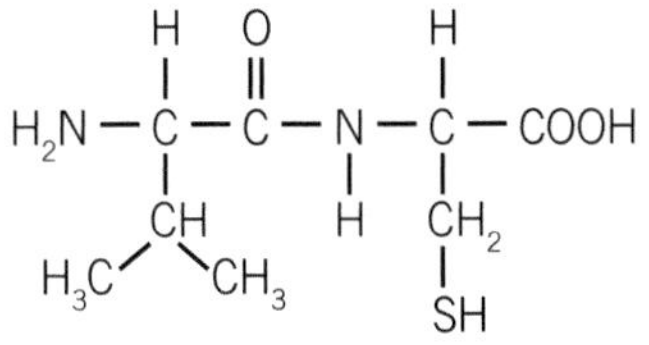

Write the displayed formula of the other dipeptide formed between molecules of these two α-amino acids.

Knowledge check 7

Calculate the maximum mass of the dipeptide (M_r 132) that can be made from 30.0g of aminoethanoic acid (M_r 75).

Stretch & challenge

Some α-amino acids contain more than one basic $–NH_2$ or acidic –COOH group. One of these is aspartic acid, $HOOC–CH_2–CH(NH_2)–COOH$. At the isoelectric point the zwitterion of aspartic acid has the formula $^-OOC–CH_2–CH(^+NH_3)COOH$.

Three letter abbreviations are sometimes used for simplicity when looking at the formula of a polypeptide.

	name	systematic name
Ala	alanine	2-aminopropanoic acid
Gly	glycine	aminoethanoic acid
Phe	phenylalanine	2-amino-3-phenylpropanoic acid
Ser	serine	2-amino-3-hydroxypropanoic acid

Proteins are polypeptides formed by condensing many amino acids together. Human insulin is a protein that consists of two amino acid chains, joined by disulfide bridges (–S–S–). People who have diabetes are not able to produce enough insulin and, as a result, sugar can build up in the blood. If the diabetes cannot be controlled by diet then insulin needs to provided externally, by the use of injections.

Haemoglobin is the oxygen-carrying protein in the blood. The replacement of just one amino acid in its structure can have a profound effect. In sickle cell anaemia, a molecule of glutamic acid (2-aminopentane-1,5-dioic acid) is replaced in haemoglobin by a molecule of valine (2-amino-3-methylbutanoic acid). This mutation reduces the oxygen-carrying capacity of haemoglobin but it does provide an increased resistance to malaria.

▲ Human insulin

The primary, secondary and tertiary nature of proteins

There are several levels to consider when describing the structure of proteins. The primary structure of a protein has already been discussed – it is the order of amino acids in the chain(s). There are twenty amino acids that can make up polypeptide chains. We obtain, or make in the body, twelve of these but the remaining eight cannot be made in the body and we need to obtain these from the food that we eat. These eight amino acids are called **essential amino acids** and include valine and lysine.

Key term

Essential amino acids are those α-amino acids that cannot be synthesised in the body and must be supplied through the diet.

lysine

8 Knowledge check

The shortened formula of lysine is shown in the text. Give the systematic name of lysine.

The number of possible peptides from these twenty amino acids is enormous. For example, there are 400 possible dipeptides, and 8000 possible tripeptides. The secondary structure of proteins is concerned with how the amino acid chains are arranged. The two commonest arrangements are as an α-helix or as a β-pleated sheet. In an α-helix the polypeptide chain is coiled into a spiral and this shape is maintained by hydrogen bonds between the N–H hydrogen atom of an amide group and the C=O carbonyl oxygen atom of another amide group.

9 Knowledge check

Explain why hydrogen bonding occurs between a carbonyl group oxygen atom and an amino group hydrogen atom in the α-helix of a protein.

$$C=O \cdots H-N$$

Proteins that have the α-helix structure are found in muscle and in wool. When wool fibres are stretched, the hydrogen bonds are broken but the strong disulfide bonds remain. When the stretching force is removed this hydrogen bonding is restored and the wool resumes its original shape. Keratin is another protein that has the α-helix structure.

The other common arrangement is a β-pleated sheet. The structure is again maintained by hydrogen bonds as before, but the C=O and N—H groups are in different chains. Van der Waals forces are responsible for producing a pleated sheet rather than a flat arrangement.

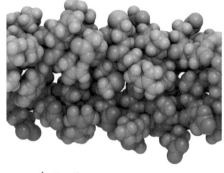

▲ Keratin

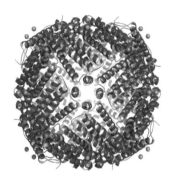

▲ Ferritin

The tertiary structure of proteins is concerned with the way in which the protein chain is folded. Fibrous proteins have a chain length that is many times its diameter and this type of protein tends to be insoluble in water. Other proteins, including most enzymes, operate in an aqueous environment. Some of these are water soluble but most are colloidal. These proteins are called globular proteins and are roughly spherical in shape. In water these proteins often take up a shape where the polar groups are on the surface of the protein, and lipophilic (fat loving) groups are towards the interior of the structure.

Proteins in living systems

Enzymes are compounds that catalyse chemical reactions. They can be described as macromolecular biological catalysts and they catalyse over 5000 biochemical reactions. They work like other catalysts, by lowering the activation energy that is necessary for the reaction to take place. Most enzymes are proteins and they work through their unique 3-dimensional structures. The role of enzymes in the body is essential to maintain life. Without them reactions in the body would be rather slow! Amylase is an enzyme that catalyses the hydrolysis of starch into sugars. It is present in human saliva. Rice and potatoes, which are mainly starch, may taste slightly sweet in the mouth as smaller molecule sugars start to be formed from starch.

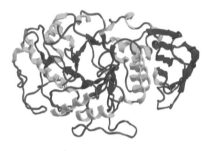

▲ Amylase

The use of enzymes in commerce is an important developing area of research. Some examples of the commercial use of enzymes are given in the table.

Application	Type of enzyme	Mode of action
Laundry	Amylases Lipases Proteases	Removing the stains from clothes caused by starch, fats or proteins
Brewing	Amylases Glucanases Proteases	Hydrolysing polysaccharides and proteins to smaller molecules
Dairy	Rennin	Hydrolysing protein in cheese making

Knowledge check 10

In the brewing industry, glucose ($C_6H_{12}O_6$) molecules are broken down by the enzymes from yeast into ethanol and carbon dioxide. Give the equation for this reaction.

About 150 enzymes have commercial uses. They can operate in aqueous conditions of mild acidity or alkalinity. Enzymes are used in detergents to remove stains from clothes; however, a number of consumers suffer from allergies when handling these detergents or when wearing clothes that have been washed in detergents containing enzymes. In laundry work, amylases degrade starches to water-soluble sugars, lipases hydrolyse fats, and proteases digest proteins. Enzymes also have important uses in the dairy industry where rennin is used to degrade the protein κ-casein in cheese making. The hydrophobic product from this degradation is the main component of curd from which cheese is made.

▲ Traditional cheese making

Test yourself

1. Give the displayed formula of norleucine (2-aminohexanoic acid). [1]

2. (a) Give the systematic name of the amino acid valine. [1]

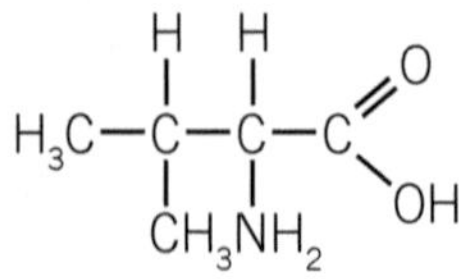

   (b) Draw the structure of the ion formed when valine is dissolved in a strongly alkaline aqueous solution. [1]

   (c) Drawn below is the structure of one dipeptide formed from valine and glycine:

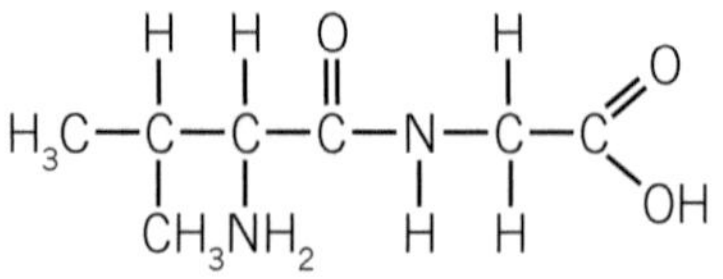

   Draw the structure of the other dipeptide formed from these two amino acids. [1]

3. The formula of a dipeptide formed from two molecules of the same amino acid is shown below:

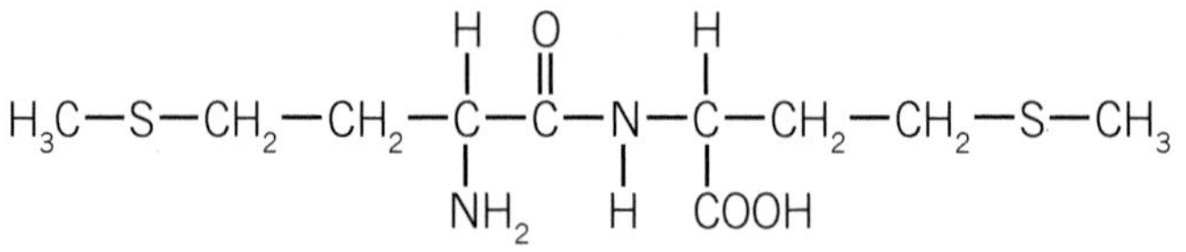

   Draw the structure of the starting amino acid. [1]

4. This section in the book uses a number of terms. Describe the meaning of each of these terms.

   (a) α-amino acid [1]

   (b) protein [1]

   (c) enzyme [1]

   (d) hydrolysis [1]

   (e) hydrophobic [1]

5. Monosodium glutamate (MSG) is used in restaurants as a flavour enhancer for foods.

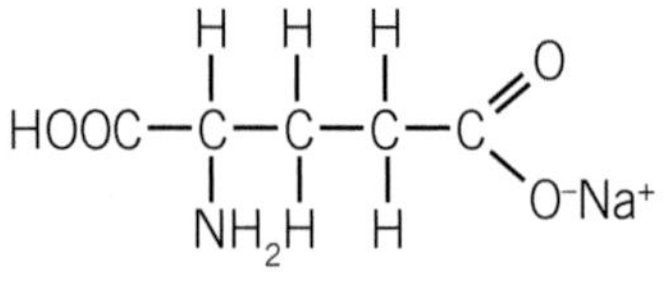

   Write the displayed formula of the species formed from MSG in an aqueous solution of pH 14. [1]

Organic synthesis and analysis

The study of organic chemistry should not just be the learning of the reactions of a number of different compounds. It is also attaining the ability to synthesise a given compound from suitable starting materials, often via a number of stages. When choosing a pathway to the compound, consideration should be given to a number of factors, which include the availability of the reactants, health and safety, cost and yield. Increasingly, attention is being given to 'green' synthetic methods. The evaluation of the chosen method to make the compound is also very important. Another important feature of the work is the qualitative and quantitative analysis of the reactants and products. Since the 1970s, instrumental methods of analysis, particularly NMR and mass spectrometry, have become increasingly important and with the continuing sophistication of these instrumental methods, there seems to less need for the traditional 'wet' methods of analysis.

You should be able to demonstrate and apply knowledge and understanding of:

- The synthesis of organic compounds by a sequence of reactions.

- The principles of the techniques of manipulation, separation and purification.

- The distinction between condensation polymerisation and addition polymerisation.

- How polyesters and polyamides are formed.

- The use of melting temperature as a determination of purity.

- The use of high-resolution ^{1}H NMR spectra in the elucidation of the structure of organic molecules.

- The use of chromatographic data from TLC/paper chromatography, GC and HPLC in finding the composition of mixtures.

Maths skills ≫

- Carry out structured and unstructured mole calculations
- Calculate percentage yields
- Carry out calculations using numbers in standard and ordinary form

Topic contents

furfural

methyl furan

MTHF

H⁺, T

C5 sugars

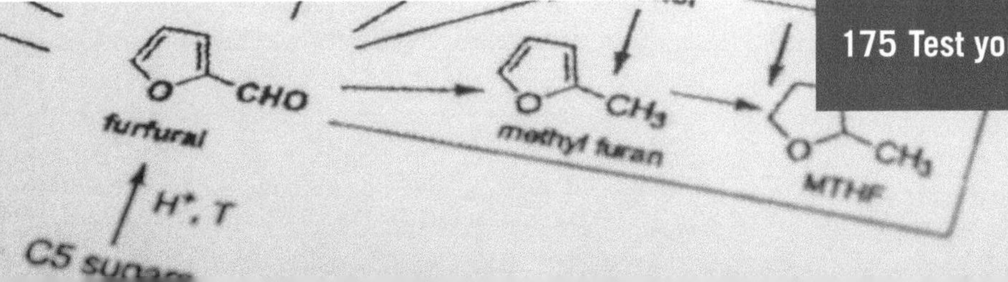

Synthesis of organic compounds

Making an organic compound by a series of steps is a topic that is challenging for many students at 'A' level. This may be because good problem-solving skills and a sound knowledge of organic chemistry are required. An important point to note when devising a reaction sequence is to see if there is a difference in the number of carbon atoms between the starting compound and the compound required. If there is a difference, then one or more steps must involve a reaction where the length of the carbon chain is altered.

Increase in chain length

The reaction of potassium cyanide with a halogenoalkane produces a nitrile, which can be hydrolysed to produce a carboxylic acid. Alternatively, the nitrile can be reduced by lithium tetrahydridoaluminate(III) to give the amine.

$$CH_3CH_2CH_2CH_2Br \xrightarrow{\text{KCN}} CH_3CH_2CH_2CH_2C{\equiv}N$$

1-bromobutane

warm with dilute acid

$LiAlH_4$

$CH_3CH_2CH_2CH_2CH_2NH_2$

pentylamine

$CH_3CH_2CH_2CH_2COOH$

pentanoic acid

Another method of lengthening the carbon chain is by the addition of hydrogen cyanide to an aldehyde or ketone, followed by hydrolysis of the hydroxynitrile.

$$\underset{H}{\overset{CH_3CH_2}{>}}C{=}O \xrightarrow{\text{HCN}} \underset{H}{\overset{CH_3CH_2}{>}}\underset{CN}{\overset{OH}{C}} \xrightarrow[\text{dilute acid}]{\text{warm with}} \underset{H}{\overset{CH_3CH_2}{>}}\underset{COOH}{\overset{OH}{C}}$$

propanal

2-hydroxybutanoic acid

In aromatic systems, a Friedel–Crafts alkylation or acylation reaction can be used to introduce a carbon-containing side chain to a benzene ring.

$CH_3CH_2CH_2Cl$

$AlCl_3$

⬡—$CH_2CH_2CH_3$

propylbenzene

$AlCl_3$

$CH_3C{\overset{O}{<}}{Cl}$

⬡—$C{\overset{O}{<}}CH_3$

phenylethanone

1 Knowledge check

Show how 3-phenylpropanoic acid, $C_6H_5CH_2CH_2COOH$, can be made in two steps from 2-bromoethylbenzene, $C_6H_5CH_2CH_2Br$.

2 Knowledge check

State the reagents, lettered **A**, **B** and **C**, that are needed to produce benzene from methylbenzene.

⬡—$CH_3 \xrightarrow{A}$ ⬡—COO^-Na^+

↓ **B**

⬡ $\xleftarrow{C}$ ⬡—$COOH$

◀Stretch & challenge

There are many types of Friedel-Crafts reaction. One is used in industry to react together benzene and ethene in the presence of a suitable acidic catalyst to produce ethylbenzene.

⬡ $+ CH_2{=}CH_2 \rightarrow$ ⬡$\overset{CH_2-CH_3}{}$
(g)   (g)   (g)

The product is then dehydrogenated to give phenylethene, $C_6H_5CH{=}CH_2$.

Decrease in chain length

Heating an acid or its salt with soda lime (in a decarboxylation reaction) produces an alkane that contains fewer carbon atoms than the starting material.

$$CH_3C \underset{O^-Na^+}{\overset{O}{\diagup\diagdown}} \xrightarrow[\text{(NaOH)}]{\text{soda lime}} CH_4 + Na_2CO_3$$

Flow charts of reactions are a useful way of learning about reaction sequences. A sample one, centred on ethene, is shown. Only the reagents are given, essential conditions have been omitted for clarity.

Worked example

State the reagents required, and any necessary conditions, to prepare benzene from (chloromethyl)benzene, $C_6H_5CH_2Cl$.

Answer

This reaction cannot be carried out in one step. It requires a reduction in the length of the carbon chain to give a hydrocarbon as the product. This suggests a decarboxylation reaction, perhaps from benzenecarboxylic acid, C_6H_5COOH, by use of soda lime. A carboxylic acid can be obtained by the oxidation of a primary alcohol. To obtain a primary alcohol from a halogenoalkane requires a hydrolysis reaction using an alkali in aqueous solution. A suggested route is

>> **Study point**

Constructing flow charts for the reactions of the main groups of organic compounds is a very useful way of learning reaction sequences.

Knowledge check 3

State the name of a compound used, together with a catalyst, in a Friedel–Crafts reaction to prepare 1,4-diethylbenzene from ethylbenzene.

Knowledge check 4

Give the displayed formula of the aldehyde that reacts with hydrogen cyanide/sodium cyanide to give a hydroxynitrile that is then hydrolysed to give compound **P**.

$$CH_3CH_2CH(OH)COOH$$
compound P

Practical check

A two-step synthesis, including purification and determination of the melting temperature of the product, is **a specified practical task**.

Practical check

Planning a sequence of tests to identify organic compounds from a given list is a **specified practical task**.

Knowledge check 5

Give the name of the compound produced by the decarboxylation of $Na^+ {}^-OOC-CH_2-CH_2-CH_2-CH_2-COO^- {}^+Na$

6 Knowledge check

Deduce which one of these compounds will undergo the triiodomethane reaction:

(a) ⬡—CH₂—CH₂OH

(b) ⬡—CH(OH)—CH₃

(c) ⬡—CH₂—CH₂—CH₂—OH

(d) ⬡—CH(OH)—CH₂—CH₃

 Key term

Miscible liquids are completely soluble in each other at all concentrations. An example of this is ethanol and water.

 7 Knowledge check

State the reagents, D, E and F, needed for this reaction sequence:

⬡—CH(OH)—CH₂Br $\xrightarrow{D}$ ⬡—CH(OH)—CH₂CN $\xrightarrow{E}$

⬡—CH(OH)—CH₂COOH $\xrightarrow{F}$ ⬡—C(O)—CH₂COOH

 **Stretch & challenge**

Devise a reaction scheme to make 2,4,6-tribromophenol starting from phenylamine.

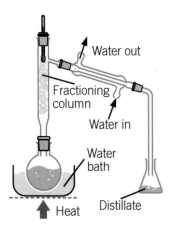

Principles of the techniques

Many organic reactions do not go to completion as required and other products may be produced, while some unreacted starting products may remain. An important part of any organic preparation is the separation and purification of the products. When ethyl ethanoate is made from ethanol and ethanoic acid, the reaction products include the required ester, together with water and some unreacted ethanol and ethanoic acid.

$$CH_3CH_2OH + CH_3COOH \rightleftharpoons CH_3COOCH_2CH_3 + H_2O$$

These four compounds are liquids that are **miscible** with each other, but that is not necessarily the case with other reactions. The products may be an insoluble solid, be present in solution or as immiscible liquids. Each of these mixtures of products needs a different method of separation.

Separating miscible liquids

If the product does not decompose at or below its boiling temperature and the boiling temperature is not too high, then distillation can be used to separate the product from the other substances present in the reaction mixture. Simple distillation can be used if the boiling temperature of the product differs by a reasonable amount (20°C or more) from the boiling temperatures of the other compounds present. It can also be used to separate a volatile liquid from other substances in the mixture that are not volatile.

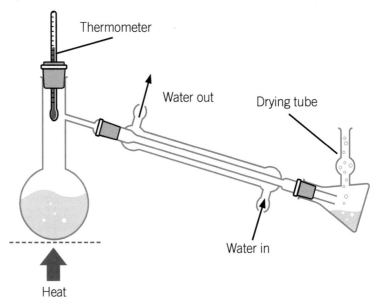

An example is the distillation of 2-chloro-2-methylpropane from its mixture with 2-methylpropan-2-ol and hydrochloric acid.

$$(CH_3)_3COH + HCl \longrightarrow (CH_3)_3CCl + H_2O$$

boiling temperatures: 82°C, 51°C, 100°C

Fractional distillation (shown in the margin) is used if the boiling temperatures are closer together. A fractionating column is used, which enables a more efficient separation of the products to occur.

The separation of ethanol obtained from the fermentation of sugars is more effective if fractional distillation is used. The primary separation of the products present in crude oil (petroleum) is also carried out by fractional distillation. For compounds that decompose just before or at their boiling temperatures, or have very high boiling temperatures, distillation can be carried out under reduced pressures (vacuum distillation) or steam distillation can be used. The reduction in pressure used in vacuum distillation enables compounds to boil at lower temperatures than when distillation occurs at atmospheric pressure. For example, the alkane dodecane, $C_{12}H_{26}$, boils at 216°C under atmospheric pressure (~ 101 kPa) but at 92°C if the pressure is reduced to 1.3 kPa.

Separating immiscible liquids

Steam distillation is a very important method in the perfumery industry where essential oils extracted from plants may decompose if heated to their boiling temperatures at atmospheric pressure.

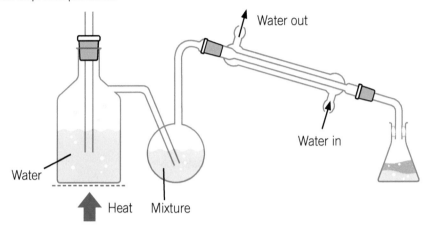

Steam is passed into the reaction mixture and the volatile compounds present pass over with the steam and condense in the receiving flask. Rose oil is one of the most widely used oils in the perfume industry. Steam distillation of rose petals gives an oil (containing a number of different compounds) and water from the condensed vapours.

Solvent extraction

This is a method that depends on the differing solubility of a compound in two immiscible solvents. For example, iodine is about 90 times more soluble in tetrachloromethane than in water. If tetrachloromethane is added to an aqueous solution of iodine and the mixture shaken, most of the iodine is extracted into the tetrachloromethane layer. The two layers can then be separated using a separating funnel.

Insoluble solid separation

Filtration is used to separate a solid from the liquid present. This can be carried using a filter paper and funnel. The use of a fluted filter paper is quicker than the traditional method as the filtrate only needs to travel through one layer of filter paper and the paper only touches the funnel at the folds.

Alternatively, filtration using a Buchner funnel can be used (vacuum filtration).

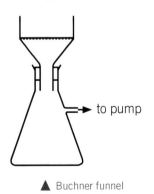

▲ Buchner funnel

▲ Fluted filter paper

Once the solid is in the funnel it needs to be washed with an appropriate solvent and dried – in the air, or in a drying oven at a temperature below its melting temperature.

◀ Practical check

The synthesis of a liquid organic product, including separation using a separating funnel is a **specified practical task**.

Knowledge check 8 ◀

Explain why heating under reflux cannot be used to concentrate a solution.

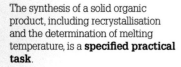
Soluble solids from solution

If the solid is present as a solute in solution then the products are obtained by crystallisation. If the solution should be colourless but is coloured due to the presence of impurities then decolourising charcoal can be used. The solution is boiled with a little decolourising charcoal to remove the colour, and then filtered hot to remove the charcoal that contains the absorbed colour. The filtrate is then concentrated by boiling and then cooled. If the solution has been sufficiently concentrated then crystals of the solute appear on cooling. These are filtered off and dried. If no crystals appear on cooling then the solution needs to be made more concentrated. Extra care needs to be taken when concentrating the solution if the solvent is flammable – generally a water bath or some method of electrical heating, for example a hot plate, is used. The solute obtained by crystallisation may not be pure and it needs to be recrystallised. The essential steps for this are:

- Dissolve the solute in a minimum volume of hot solvent
- Filter hot, if necessary, to remove insoluble impurities
- Allow to cool
- Filter
- Wash the solid with a small amount of an appropriate solvent
- Dry at a temperature below its melting temperature.

Distinguishing between condensation polymerisation and addition polymerisation

Polymerisation is the joining together of a large number of **monomer** molecules. Alkenes undergo addition polymerisation when the –C=C– double bond is used to join the monomer units together giving a polymer that now contains only single bonds between carbon atoms in the chain.

$$n \quad \begin{array}{c} H_3C \\ H \end{array} C = C \begin{array}{c} H \\ H \end{array} \longrightarrow \left[\begin{array}{cc} CH_3 & H \\ | & | \\ C - C \\ | & | \\ H & H \end{array} \right]_n$$

propene

poly(propene)

Condensation polymerisation occurs when a large number of monomer molecules join together with the loss of small molecules (often water or hydrogen chloride). If two different monomers are used, each one having different functional groups, then small molecules are lost when bonding occurs between them. For example, a condensation polymer is formed when a dicarboxylic acid bonds with a diol.

$$\underset{(OH)}{\overset{O}{\|}}C - (CH_2)_n - \underset{(OH)}{\overset{O}{\|}}C \quad HO - \underset{H}{\overset{H}{|}}C - (CH_2)_n - \underset{H}{\overset{H}{|}}C - O(H)$$

$$\downarrow -H_2O$$

$$\cdots \underset{}{\overset{O}{\|}}C - (CH_2)_n - \underset{}{\overset{O}{\|}}C - O - \underset{H}{\overset{H}{|}}C - (CH_2)_n - \underset{H}{\overset{H}{|}}C - O \cdots$$

Condensation polymerisation can also occur between two different functional groups present in just one type of monomer molecule. For example, using an α,ω-amino acid (i.e. with an amino group –NH_2 at one end of the molecule and a carboxylic acid group –COOH at the end of the carbon chain).

$$\underset{H}{\overset{H}{N}}-(CH_2)_n-\overset{O}{\underset{}{C}}\!\!\overset{}{(O-H} \quad \underset{}{H)}\overset{H}{N}-(CH_2)_n-\overset{O}{\underset{}{C}}\!\!(OH)$$

$$\downarrow -H_2O$$

$$\cdots N-(CH_2)_n-\overset{O}{\underset{}{C}}-\underset{H}{N}-(CH_2)_n-\overset{O}{\underset{}{C}}\cdots$$

It is useful to have some rules for distinguishing between these two types of polymerisation. If:

- The monomer is an alkene, then addition polymerisation occurs.

- No small molecule is lost and the polymer is the only product, then addition polymerisation has occurred.

- The monomer(s) contain functional groups such as –NH_2, –COOH or –OH then condensation polymerisation occurs.

- The chain contains the amide link –C(O)N(H)– or an ester linkage –OC(O)– then condensation polymerisation has occurred.

- The chain only consists of carbon atoms then addition polymerisation has occurred.

Forming polyesters and polyamides

Polyesters are very important materials that have extensive uses in the production of clothing, packaging and plastic bottles.

The most common polyester is PET, this an abbreviated name of polyethylene terephthalate. The monomer or monomers from which a polyester is made need to have a functional group at each end of the molecule. PET is produced from ethane-1,2-diol and benzene-1,4-dicarboxylic acid (terephthalic acid).

$$\cdots$$

Knowledge check 12

A section of a polymer is

$$\left[\!\!-O-CH_2-(CH_2)_6-\overset{O}{\underset{}{C}}-\!\!\right]_n$$

State whether addition polymerisation or condensation polymerisation occurred when this polymer was made from its monomer(s), giving a reason for your answer.

Stretch & challenge

Give the structure of a repeating section of the addition polymer 'polyacrylonitrile' that is made by polymerising propenenitrile.

Knowledge check 13

Give the displayed formula of the monomer that can be used to make the polyamide, Nylon 8.

Key term

A **polyester** is a compound made up of repeating units that each contain the

group.

PET itself is made into synthetic fibres such as Terylene, either then used by itself or together with natural fibres such as cotton. The polyester is a good insulator and its fibres can be used in the making of blankets and as a filling material for duvets. PET does not easily biodegrade and there is a need for polyesters that will degrade quickly in landfill. One of these is poly(lactic acid), PLA. This polyester has the added advantage that it can be derived from renewable resources such as corn starch or sugar cane. The equation for the polymerisation from 2-hydroxypropanoic acid (lactic acid) is shown below.

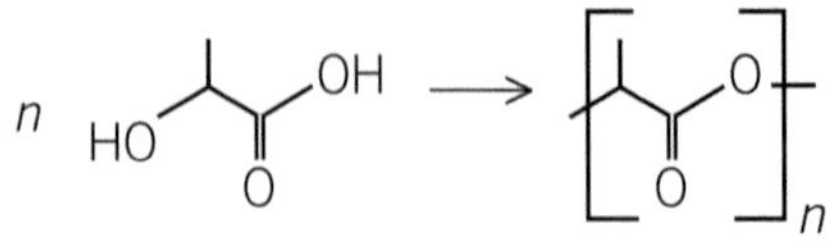

Study point

When finding the formula of a polymer, given the formula of the starting material(s), ensure that the number of carbon atoms or CH_2 groups does not change.

PLA is a polyester that is made from one type of monomer molecule whereas PET requires two different types of monomer molecules.

Polyamides

Polyamides are also a product of condensation polymerisation. The first polyamide, Nylon 6,6 was made in 1935. The starting material can be benzene, which is reduced to cyclohexane and this product is then oxidised, giving cyclohexanol and cyclohexanone. This mixture is then itself oxidised to hexanedioic acid.

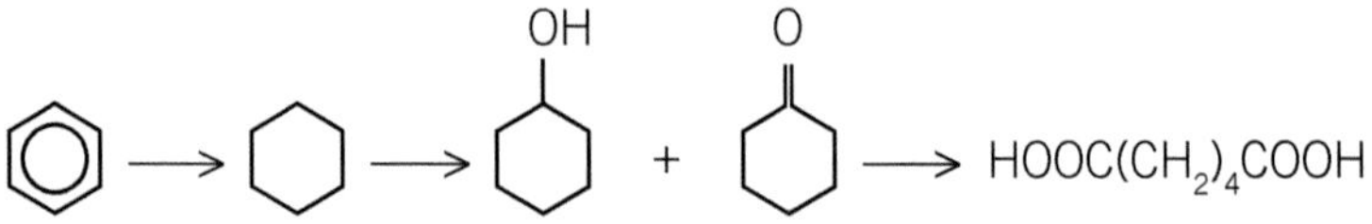

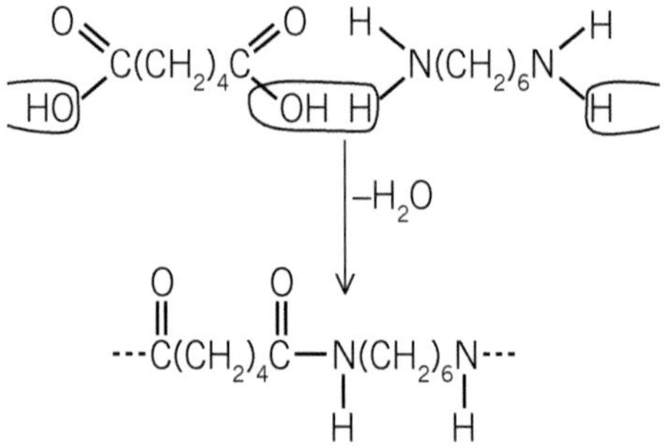

Stretch & challenge

The starting material for making melamine resins is melamine itself. This is made by heating carbamide (urea).

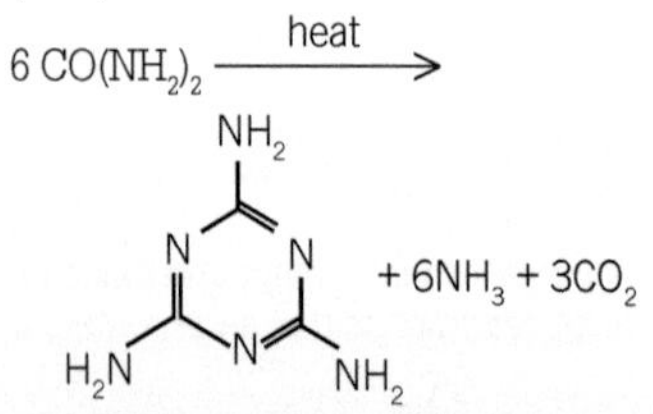

(a) Give the empirical formula of melamine

(b) Calculate the atom economy of this reaction, assuming that melamine is the only useful product.

Some of the hexanedioic acid is converted to hexane-1,6-diamine, which is combined with hexanedioic acid.

The product is named Nylon 6,6 because each of the two monomers has six carbon atoms. Another polyamide, Nylon 6, was developed in Germany in 1939. Nylon 6 is derived from the six carbon-containing acid, 6-aminohexanoic acid,

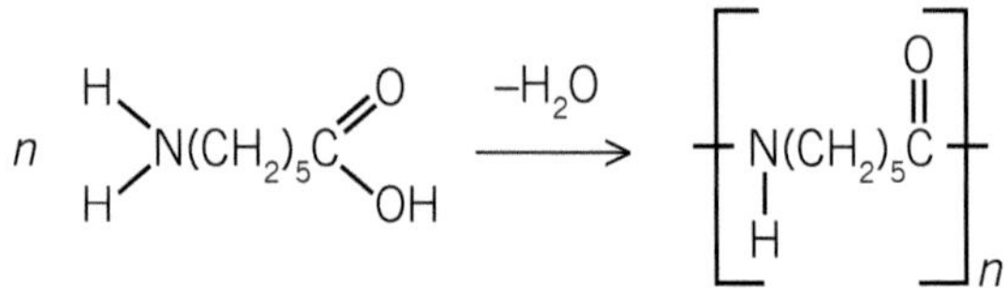

It is not produced directly from the acid itself. Instead it is produced from caprolactam, which itself is made in several stages from benzene. On treating with water caprolactam ring opens and polymerises giving Nylon 6.

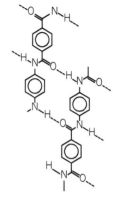

Another important polyamide is Kevlar. This can be produced from benzene-1,4-dioic acid and benzene-1.4-diamine.

Kevlar has good fire retardant properties and is five times stronger than steel. The compound can be produced as a fibre which is spun into bullet proof vests.

Using melting temperature to determine purity

The melting temperature of a solid is the temperature at which the solid begins to change into a liquid. For many pure substances, the temperature at which this change from a solid to a liquid occurs is quite sharp (within 1°C) and the figure obtained is useful for identification purposes. The melting temperature is affected by the presence of impurities and the values obtained give an indication of the compound's purity. The presence of impurities lowers the expected melting temperature and the compound melts over a range of temperature rather than at a fixed value. For example, a sample of a compound has a sharp melting temperature of 122°C and is suspected to be benzenecarboxylic acid. A little pure benzenecarboxylic acid is mixed with the sample and the melting temperature again taken. If the melting temperature remains at the same temperature, it is likely that the compound is as suggested. However, if the melting temperature is now lower and not sharp, then the original substance was not benzenecarboxylic acid. Aldehydes and ketones often exist as liquids or low melting temperature solids and it is sometimes difficult to obtain an accurate melting temperature for them. The aldehyde or ketone is reacted with 2,4-dinitrophenylhydrazine to give a derivative. These derivatives (2,4-dinitrophenylhydrazones) are usually orange-red solids that have a melting temperature that is easier to measure. The melting temperature of the 2,4-dinitrophenylhydrazone can be compared with a table of melting temperatures to identify the starting aldehyde or ketone.

The melting temperature of the compound is usually found using an electrical heating method.

> **Study point**

Kevlar chains are relatively rigid and tend to form planar sheets.

The Kevlar chains lock together by hydrogen bonding between the carbonyl group oxygen of one chain and the N–H hydrogen atom of another chain. This leads to a relatively rigid sheet structure with a high tensile strength.

▲ Electrical melting point apparatus

Alternatively, the melting temperature can be found using a heating bath method.

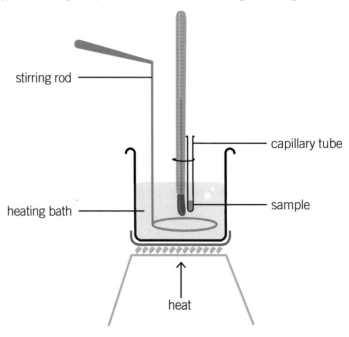

A 2–3 mm depth sample is placed in a capillary tube that is attached to a thermometer by a rubber band. The thermometer and capillary tube are placed into a suitable heating bath. Water is used if the melting temperature is likely to be below 100°C. For melting temperatures above 100°C, silicone oil or other non-flammable liquid is used. The mixture is gently heated and stirred.

Using high resolution ¹H NMR in finding the structure of organic molecules

During the first year of this course, you will have studied the use of low-resolution proton magnetic resonance spectroscopy (¹H NMR) in the identification of chemical structures.

ethane   chloroethane

In ethane all the hydrogen protons are in equivalent **environments** and only one signal is seen. In chloroethane, one hydrogen atom is replaced by a chlorine atom. The hydrogen atoms are now in two environments, the three CH_3 protons are identical but the two CH_2 protons are in a different environment. This gives a low-resolution NMR spectrum showing two peaks with peak areas 3:2, reflecting the protons of the CH_3 and the CH_2 hydrogen atoms.

Knowledge check

14 The melting temperature of a pure compound is 158 °C. It is thought that it is either 2-hydroxybenzenecarboxylic acid or 3-chlorobenzenecarboxylic acid, both of which also have the melting temperature of 158 °C. State how you would find out which acid is present as the unknown compound, using a melting temperature method.

Key term

In NMR spectroscopy the **environment** means the nature of the surrounding atoms or groups in the molecule.

Study point

If the compound contains more than two carbon atoms, the ¹H NMR spectrum can become more complicated. In propane ($CH_3CH_2CH_3$) there are eight hydrogen protons. The six CH_3 protons are all in the same environment and the two CH_2 protons are equivalent to each other but in a different environment to the methyl protons. There are two peaks in a low resolution ¹H NMR spectrum, reflecting the CH_3 and CH_2 hydrogen protons. These are in a relative peak area ratio of 6:2 (i.e. 3:1) respectively.

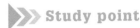

Dichloroethane has two isomers, 1,2-dichloroethane and 1,1-dichloroethane.

1,2-dichloroethane

1,1-dichloroethane

In 1,2-dichloroethane all the hydrogen protons are in an identical environment and the ^1H NMR spectrum shows a single peak (a singlet) at 3.7δ. However, in 1,1-dichloroethane the four hydrogen protons are in two different environments, one signal is seen at 5.9δ for the $CHCl_2$ hydrogen proton and the other is seen at 2.1δ for the CH_3 hydrogen protons. The peak areas for these two signals occur as a 1:3 ratio respectively.

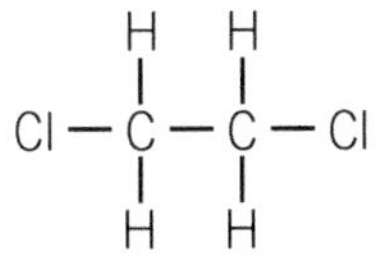

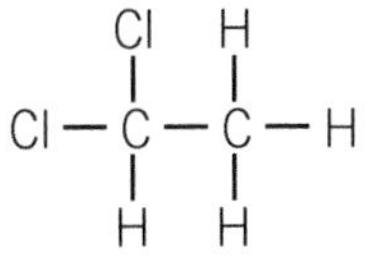

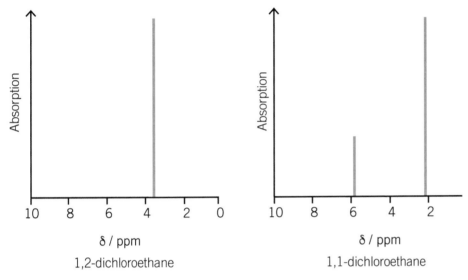

1,2-dichloroethane

1,1-dichloroethane

If the ^1H NMR spectrum of 1,1-dichloroethane is measured using a high resolution spectrometer the two peaks are seen to be split. This splitting occurs because the magnetic environment of a proton or protons in one group is affected by the magnetic environment of neighbouring groups. The high resolution ^1H NMR spectrum of 1,1-dichloroethane shows that both signals are split – one into four peaks (a quartet) and the other into two peaks (a doublet). This process of signal splitting is called **spin-spin coupling**. The process is only considered for hydrogen protons on neighbouring atoms – usually carbon, nitrogen or oxygen. Thus in 1,1,1,2-tetrachloropropane the hydrogen proton on carbon two is affected by the hydrogen protons on carbon three and vice versa.

1,1,1,2-tetrachloropropane

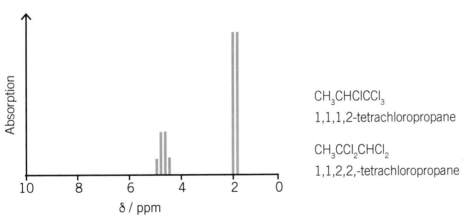

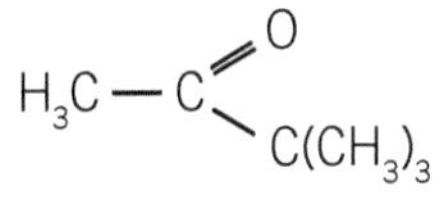

$CH_3CHClCCl_3$
1,1,1,2-tetrachloropropane

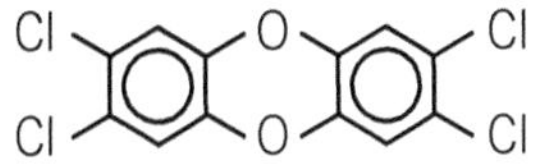

$CH_3CCl_2CHCl_2$
1,1,2,2,-tetrachloropropane

However, in 1,1,2,2-tetrachloropropane the hydrogen atoms are not bonded to adjacent carbon atoms and the high resolution ^1H NMR spectrum shows two signals that show no splitting.

Study point

A very common spin-spin splitting pattern at this level is a quartet and a triplet. This is given by an ethyl group

$CH_3 — CH_2 —$
a **b**

where the hydrogen protons at **a** are split to give a triplet and the hydrogen protons at **b** are split to give a quartet.

Knowledge check 15

Pinacolone has the formula

Describe its ^1H NMR spectrum, including relative peak areas.

Knowledge check 16

The compound whose formula is shown below is a very toxic tetrachlorodioxin.

Explain why (a) its mass spectrum shows a number of molecular ion signals that range from *m*/*z* 320 to 328 and (b) its ^{13}C NMR spectrum shows three signals.

Stretch & challenge

The formula shown below is of an ester of a dicarboxylic acid, where R is an alkyl group.

Its low resolution ^1H NMR consists of two signals having a relative peak area of 3:2. Deduce a possible displayed formula for the ester.

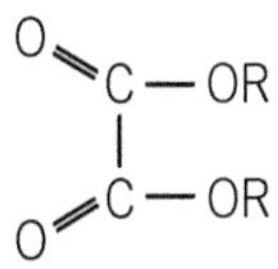

17 Knowledge check

The ^{1}H high-resolution NMR spectrum of a methyl ketone consists of a singlet, a doublet and a triplet. The ^{13}C NMR spectrum indicates that there are four distinct environments for the carbon atoms. Suggest a displayed formula for this ketone.

18 Knowledge check

Butanedione has the formula:

$$H_3C - \overset{\overset{\displaystyle O}{\|}}{C} - \overset{\overset{\displaystyle O}{\|}}{C} - CH_3$$

State what you would see in its ^{1}H high-resolution NMR spectrum.

Study point

In questions involving NMR spectra, you should assume that the magnetic field around ^{13}C nuclei has no effect on the ^{1}H NMR spectrum of a compound.

If a hydrogen proton bonded to a carbon, nitrogen or oxygen atom has n hydrogen protons bonded to an adjacent carbon, nitrogen or oxygen atom then its single peak will be split into $(n + 1)$ smaller peaks. The table shows some examples of this rule in practice.

compound	hydrogen(s) a		hydrogen(s) b	
	splitting pattern	relative peak area	splitting pattern	relative peak area
$\overset{H^a}{\underset{F}{>}}C=C\overset{H^b}{\underset{Br}{<}}$	doublet	1	doublet	1
$\overset{a}{C}H_3\overset{b}{C}H_2C \equiv N$	triplet	3	quartet	2
$ClCH_2^a - \overset{\overset{\displaystyle O}{\|}}{C} - \overset{b}{C}H_3$	singlet	2	singlet	3
$\overset{a}{C}H_3 - C\overset{\nearrow O}{\underset{\searrow H^b}{}}$	doublet	3	quartet	1

Questions may be set where candidates are asked to find the structure of a compound by the use of its ^{13}C NMR spectrum, and its high-resolution ^{1}H NMR spectrum, as well as its mass spectrum and its infrared absorption spectrum. In ^{1}H NMR spectra the peak height gives information about the number of protons. In ^{13}C NMR spectra the peak height does not provide any information.

Worked example

A candidate was given the following information about compound **M** and was asked to deduce its displayed formula.

- The infrared spectrum showed an absorption peak at 1718 cm^{-1} but did not show a peak at ~2800 cm^{-1} indicating the C–H bond of an aldehyde group, nor an absorption at 1000 – 1300 cm^{-1} that is characteristic of a C–O single bond

- The mass spectrum showed a molecular ion at m/z 86 and significant fragmentation peaks at m/z 71 and 43

- The ^{13}C NMR spectrum showed four distinct environments for carbon atoms

- The high resolution ^{1}H NMR spectrum of compound **M** gave the spectrum below

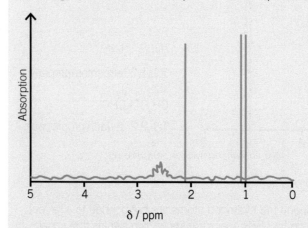

S/ppm	Splitting pattern	Relative peak area
1.11	doublet	6
2.14	singlet	3
2.55	heptet	1

Answer

The infrared spectrum shows an absorption peak at 1718 cm^{-1}. This suggests that the compound contains a C=O double bond and therefore could be an aldehyde, ketone, ester or carboxylic acid. The information indicates that it cannot be an aldehyde as there is no absorption at ~2800 cm^{-1}. It is not an ester or a carboxylic acid as the absorption due to a C–O single bond is absent. The compound must therefore be a ketone of formula R$_1$–C(O)–R$_2$ where R$_1$ and R$_2$ could be the same group or different groups. The relative molecular mass is 86 as compound **M** has a molecular ion at m/z 86. The fragmentation peak at m/z 71 shows a reduction of 15, indicating the possible loss of a methyl group. The fragment at m/z 43 could indicate a CH$_3$C=O$^+$ or C$_3$H$_7^+$ ion. The ^{13}C NMR spectrum indicates that the carbon atoms are in four different environments. One peak in the ^{1}H NMR spectrum is a singlet – suggesting that there are no hydrogen atoms bonded to the adjacent carbon atom. It is possible that there is a –C(O)CH$_3$ group present in the molecule. This group has M_r 43 and therefore the other alkyl group must also have M_r 43, corresponding to C$_3$H$_7$. A doublet in the ^{1}H NMR spectrum suggests a single hydrogen atom on an adjacent carbon atom. This removal of CH leaves 2 carbon atoms and 6 hydrogen atoms. The ^{1}H NMR spectrum shows a signal at 2.6δ as a heptet. This suggests that there are 6 equivalent hydrogen protons on the adjacent two carbon atoms. This is a characteristic pattern for a 2-propyl group, –CH(CH$_3$)$_2$. This evidence suggests that the compound could be 3-methylbutan-2-one,

which has the molecular formula C$_5$H$_{10}$O and relative molecular mass 86.

Using chromatography to find the composition of mixtures

Chromatography is a technique that is used to separate substances from a mixture by their slow movement, at different rates, through or over a stationary phase. This technique of separation was developed early in the 20th century and was initially used to separate plant pigments by using a column containing powdered calcium carbonate. Since its discovery, chromatography has been extensively developed so that mixtures of substances can be separated using a number of different methods, which are appropriate to the number and nature of the components present. Although chromatography was originally a qualitative method, it can now be used in a quantitative way and is often used in conjunction with mass spectrometry to identify individual components that are present in the mixture. The emphasis in this topic is finding the composition of mixtures rather than the theory and principle of this technique. A description of the two mechanisms for the separation process – partition and adsorption – is not required.

Top tip

Questions involving gas chromatography often show peak areas. These figures are relative values and not necessarily percentages, unless the sum of the peak areas adds up to 100.

Study point

When deducing a structure from the fragmentation pattern of a mass spectrum, a signal at m/z 29 often indicates a C$_2$H$_5^+$ ion and a signal at m/z 43 may indicate C$_3$H$_7^+$ or CH$_3$CO$^+$ ions.

Stretch & challenge

The equation shows a dehydration reaction.

(a) State how the infrared red absorption spectrum of the distinguishing peaks for the two compounds differ from each other.

(b) Another product of the dehydration is

Comment on how the ^{13}C NMR spectrum of the two **products** would differ. It is not necessary to consider the position of the signals in the spectrum.

(c) State whether either of the two products could exist as E–Z isomers, explaining your answer.

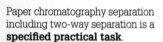

Top tip

A common mistake when finding the R_f value in paper chromatography and TLC is to measure the distances from the bottom of the paper/plate rather than from the start line.

19 Knowledge check

(a) A TLC was taken of a sample of some carboxylic acids. The solvent front distance was measured as 6.9 cm and the spot for benzenecarboxylic acid was measured at 4.0 cm from the start line. Calculate the R_f value for this acid under these conditions.

(b) A spot was also seen around 5.2 cm that appeared to consist of two spots close together. Suggest what should be done to separate the spots caused by these two compounds.

Practical check

Paper chromatography separation including two-way separation is a **specified practical task**.

Paper chromatography / Thin layer chromatography (TLC)

In paper chromatography the stationary phase is water trapped in the cellulose fibres of the paper, whereas in TLC the stationary phase is a layer of silica (SiO_2) or aluminium oxide (Al_2O_3) coated onto a plastic or glass plate. The techniques for paper chromatography and TLC are similar. Spots of the starting materials in a suitable solution are placed at the bottom of a piece of chromatography paper or TLC plate, which is then placed in a suitable solvent with the initial solvent level below the spots. The solvent front then rises up the paper/plate, separating the mixture into a series of spots. When the solvent front has risen to a suitable level, the paper/plate is removed and dried. The position of the separated spots and the solvent front are noted and the distance that these have risen from the starting line is measured.

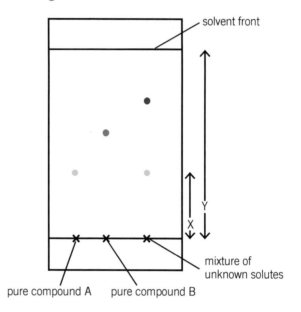

The position of the spots for a known and the unknown solute are then compared, to see if they have travelled the same distance. The R_f value can also be calculated.

$$R_f = \frac{\text{distance moved by spot (X)}}{\text{distance move by the solvent front (Y)}}$$

Sometimes the use of a particular solvent does not completely separate the spots. This may be overcome by rotating the dried chromatogram through 90° and then using a different solvent. This technique is called two-way separation.

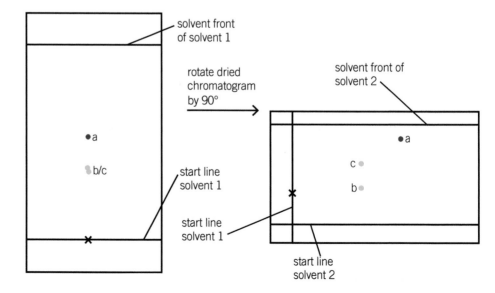

The identification of spots is relatively easy if they are coloured. Sometimes solutes produce colourless spots and the chromatogram is sprayed with a suitable reagent that causes the spots to become coloured. An example of this is to dry a chromatogram containing spots of different amino acids and then spray it with a solution of ninhydrin and then gently warm the paper. Blue-purple spots appear that show the position of the different amino acids. Another method that can be used to show the position of colourless spots is to shine UV light onto the plate. If the separated compounds are fluorescent then they show a colour. Alternatively the plate itself can have a coating of fluorescent materials and then the plate is exposed to UV light. The spots then show up as dark spots on a fluorescent background. Advantages of TLC over paper chromatography are that it is faster and that the thin layer on the plates can be made from a variety of materials. TLC continues to have important uses in forensic science.

Gas chromatography

The most common type of gas chromatography is gas–liquid chromatography (GLC) where a gaseous mixture is passed 'through' liquid particles supported on an unreactive (inert) solid. The gaseous mixture is swept into the column by a carrier gas, which could be hydrogen, helium or argon. The column itself contains fine solid material or is a hollow column whose walls are coated with a solid on which there is a liquid stationary phase. The retention time is the time taken from the sample entering the injection port until it reaches the detector. The efficient separation of the compounds in the mixture depends on a number of factors – these include the volatility of the compound itself, the column temperature, the length of the column and the flow rate of the carrier gas. For similar compounds (for example those in the same homologous series) an important factor in separating the compounds by GLC is their boiling temperatures. Retention times vary enormously because of the factors outlined above and identification solely by retention time depends on the conditions being exactly the same. Very often, the separated components of the mixture are passed into a mass spectrometer, where a positive identification can be made. This important technique is abbreviated to GC–MS.

The ABE fermentation process is a method of bacterial fermentation that produces propanone (acetone), butan-1-ol and ethanol from starch. The chromatogram shows a typical GC chromatogram of the products from the ABE process.

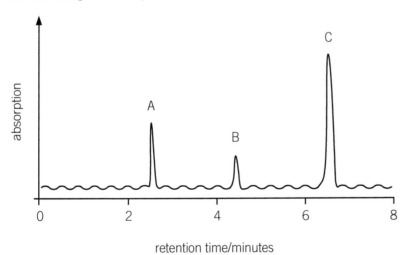

Peak **A** represents propanone and peaks **B** and **C** represent ethanol and butan-1-ol respectively. Propanone is less polar than the two alcohols and its lower boiling temperature of 56°C suggests that it will come off the column first. This is followed by ethanol and then butan-1-ol, which have boiling temperatures of 78°C and 117°C respectively.

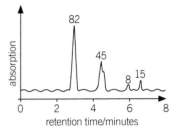

▶▶▶ Study point

HPLC is a form of liquid chromatography. A simpler form of this chromatography is column chromatography where solutes dissolved in a suitable solvent are passed down a column, leading to separation. Another solvent (the eluting solvent) is then added to the column and the separated components are run off separately. Removal of the eluting solvent then gives the individual solute. Silica (SiO$_2$), and alumina (Al$_2$O$_3$) are often used as the stationary phase.

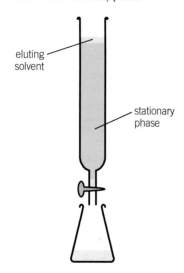

◀ Stretch & challenge

There have been a number of HPLC studies to separate, and then determine the amounts of, caffeine and theobromine in beverages. Worries about the addictive and stimulant nature of caffeine have led to the development of decaffeinated colas, coffee and tea.

(a) A 330 cm^3 can of cola contains 35 mg of caffeine (M_r 194). Calculate the concentration of caffeine in the cola in mol dm^{-3}.

(b) Both caffeine and theobromine are colourless substances in solution. Suggest how scientists could find the HPLC retention times for these two compounds.

High performance liquid chromatography (HPLC)

In this technique, the column is packed with solid particles (of uniform size) and the mixture sample is dissolved in a suitable solvent. This solution is then forced through the column at a high pressure. High performance liquid chromatography (previously high-pressure liquid chromatography) is generally shortened to HPLC and is a very important method of separation that can be used for compounds that vaporise at high temperatures where they may start to decompose. This method has many applications, for example in testing urine samples of athletes for the presence of banned substances. Another application is in food chemistry, where antioxidants are added to fatty food products, such as margarine and cream cheese to help prevent oxidation. The most common antioxidants include BHA, BHT and various esters of gallic acid. In BHA the C(CH$_3$)$_3$ group can be in the 2- or 5- positions in the ring.

The presence of these antioxidants in food can be detected by HPLC:

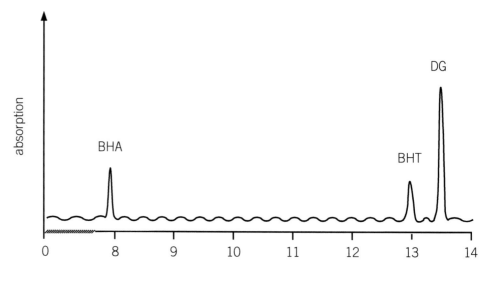

Test yourself

1. Identify reagents **L**, **M** and **N** in this reaction sequence. [3]

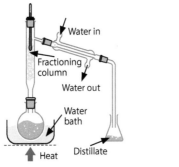

2. Identify compounds **0** and **P** in this reaction sequence. [2]

3. Devise a route to produce methyl ethanoate from bromomethane. [5]

4. Identify the two things that are wrong with this diagram of distillation. [2]

5. Identify compounds **R** and **S** in this reaction sequence. [2]

6. 4.25 g of butane-1,4-dioic acid (M_r 118) was dissolved in water and an equal volume of the solvent ethoxyethane (immiscible with water) was added and the mixture shaken. The two layers were allowed to separate. The aqueous layer now contained 6×10^{-3} mol of the acid. Calculate how many more times the acid is soluble in ethoxyethane under these conditions. [3]

7. Iodine (I_2) can be extracted from its aqueous solution by the use of tetrachloromethane (CCl_4), which is immiscible with water. I_2 is 90 times more soluble in CCl_4 than it is in water. 100 cm³ of an aqueous solution of iodine contained 0.030g of dissolved iodine. An equal volume of CCl_4 was added and the mixture shaken. The mixture was allowed to separate and the CCl_4 layer removed. Calculate approximately how much iodine remained in the aqueous layer. [2]

8. Two addition polymers have the repeating structures A and B.

(a) Give the name of the monomer that produces polymer A. [1]

(b) State the names of the two functional groups present in the monomer that produces polymer B. [2]

9. Octane-1,8-dioic acid reacts with 1,3-diaminopropane to give a polyamide.
   Draw the repeating structure of this polyamide. [2]

10. 1,6-Diaminohexane is made from hexanedioic acid by the three steps outlined below.

    (a) Hexanedioic acid is neutralised with ammonia to give the diammonium salt of the acid. Complete the equation for this reaction: [2]

    $HOOC-(CH_2)_4-COOH \quad + \quad \underline{\quad} \ NH_3 \ \rightarrow \ \underline{\hspace{5cm}}$

    (b) The diammonium salt is then heated to give hexane-1,6-dinitrile. Give the equation for this reaction. [2]

    (c) The dinitrile is then reduced by hydrogen to produce 1,6-diaminohexane. State a suitable catalyst for this reduction. [1]

11. Complete the table below, which describes the splitting patterns for the ^1H NMR spectra of three compounds. The first row is completed for you. [4]

compound	hydrogen(s) a		hydrogen(s) b	
	splitting pattern	relative peak area	splitting pattern	relative peak area
H^a $\diagdown$ $\diagup$ H^b $C=C$ F $\diagup$ $\diagdown$ Br	doublet	1	doublet	1
$C^aH_3-\overset{\overset{\displaystyle Cl}{\mid}}{\underset{\underset{\displaystyle Cl}{\mid}}{C}}-\overset{\overset{\displaystyle O}{\parallel}}{C}-C^bH_2CCl_3$				
$F-\!\!\bigcirc\!\!-C^aH_2-C^bH_3$ with F F at top, F F at bottom				

12. (a) The gas chromatogram of a mixture of chloroalkanes of formula C_4H_9Cl shows that it contains three compounds. The numbers given are relative peak areas. Calculate the percentage by mass of the largest peak. [2]

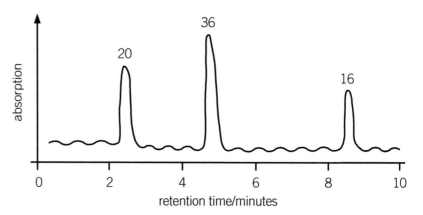

(b) Further work showed that one of the compounds in this mixture is 1-chloro-2-methylpropane.

    (i) Draw the displayed formula for this compound. [1]

    (ii) State how many signals would be present in the ^{13}C NMR spectrum of this compound, explaining your answer. [2]

Unit 4

As with all examination questions, it is essential to read the questions very carefully, and be certain to answer exactly what is needed. The command words and the mark allocation can be used as a guide to the depth of response needed. Many questions increase in complexity through the different question sections. The command words in the questions below are shown in **bold**.

‹ Link ›

Command words are explained on page 8.

(a) (i) **Draw** the displayed formula of (Z)-butenedioic acid. [1]

(ii) **Give** the empirical formula of (Z)-butenedioic acid. [1]

(iii) On heating with an aqueous acid, this (Z)-isomer is converted into the corresponding (E)-isomer. The (Z)-isomer is soluble in water, whereas the (E)-isomer is only slightly soluble in water and precipitates as white crystals.

Explain how you would obtain pure dry crystals of the (E)-isomer from this precipitate. [3]

*In part (i) you should be clear that if a displayed formula is requested then all the bonds should be given, and in part (ii) you should be looking for the simplest possible whole number ratio of the atoms present. In part (iii) more detail is required as the word **explain** is the command word. Many candidates do not describe how the product could be dried.*

(b) (E)-butenedioic acid reacts with hydrogen bromide to give 2-bromobutanedioic acid.

(i) **Draw** the structure of 2-bromobutanedioic acid and indicate the chiral centre on your formula by use of an asterisk (*). [2]

(ii) **Draw** the two mirror image forms of 2-bromobutanedioic acid. [2]

Part (i) for asks for a structure of the product. Not all bonds need to be drawn and as long as the formula given is unambiguous and correct, then it will gain credit. In part (ii) the answer given must be a 3-D representation of the mirror image forms.

(c) 5.22g of (E)-butenedioic acid reacted with hydrogen bromide to give an 80% yield of 2-bromobutanedioic acid.

Calculate the mass of 2-bromobutanedioic acid that was produced. [3]

In this mole calculation you will firstly need to calculate the M_r values of both acids, and hence the number of moles of (E)-butenedioic acid that was used. In calculations you should show all stages of your working as some credit can then be given, even if you later make a mistake when working through the question.

1. (a) The analysis of a compound shows that its formula is C_4H_7Br.

The high resolution ^1H NMR spectrum of the compound is shown below:

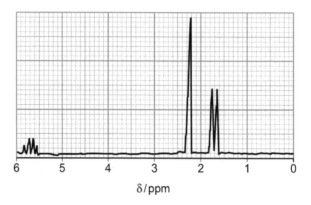

δ/ppm

(i) A student suggested that the compound might be one of the following:

compound **L** $BrCH_2 — CH_2 — CH = CH_2$

compound **M** $CH_3 — CH_2 — C = CH_2$
                                            |
                                            Br

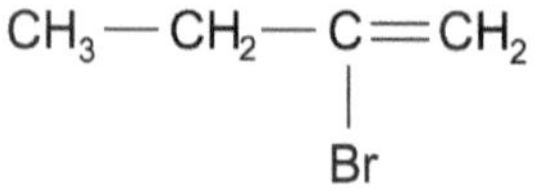

compound **N**

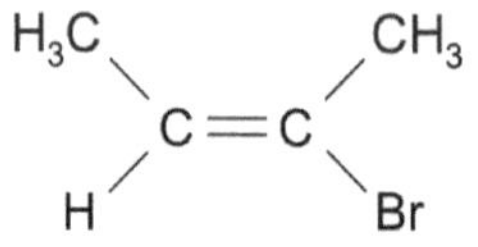

Discuss the splitting pattern seen in the ^1H NMR spectrum to decide whether the compound is **L**, **M** or **N**. Give reasons for your answer.   [4]

(ii) Another student suggested that the compound was bromocyclobutane:

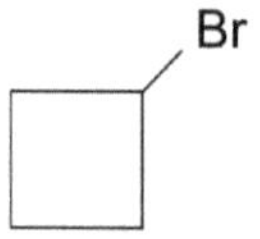

I. Describe a chemical test that would show that the compound was either **L**, **M** or **N** and not bromocyclobutane. You should state the result of your test with each compound.   [1]

II. Explain how the ^{13}C NMR spectrum would show that the compound was **L**, **M** or **N** and not bromocyclobutane.   [2]

(b) The bromination of methylbenzene using electrophilic substitution gives a mixture of 1-bromo-2-methylbenzene and 1-bromo-4-methylbenzene.

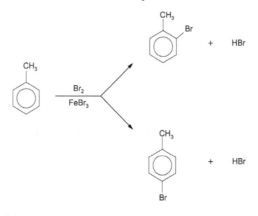

Give the mechanism of the reaction that produces 1-bromo-4-methylbenzene.   [3]

(c) A reaction sequence to produce 4-bromophenylethanoic acid is shown below:

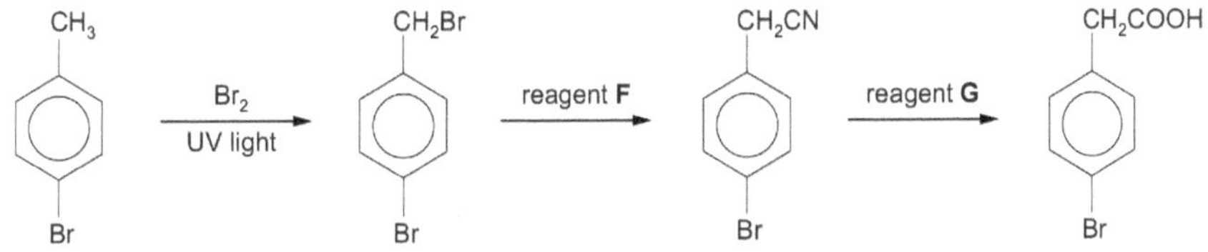

(i) The bromination of 1-bromo-4-methylbenzene is a free radical process.

Give the equation for this reaction and then use the equation to calculate the minimum mass (in g) of bromine needed to convert 0.15 mol of 1-bromo-4-methylbenzene to 4-bromo-1-(bromomethyl)benzene in the first stage of this reaction sequence. [2]

(ii) State the name of reagent **F**. [1]

(iii) State the name of reagent **G**. [1]

(Total 14 marks)

[*WJEC Unit 4 2017 Q13*]

2. (a) When ethene is passed into aqueous chlorine one of the products is 2-chloroethanol:

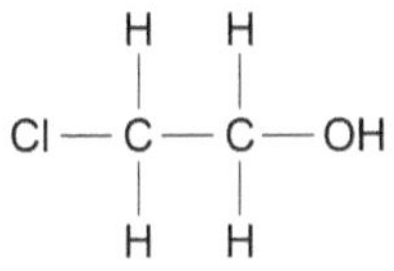

(i) The presence of a C–Cl bond in this compound can be shown by a simple test tube reaction. Outline the practical steps that are used in this method. You can assume that 2-chloroethanol is soluble in water. [4]

(ii) The oxidation of 2-chloroethanol (M_r 80.6) produces chloroethanoic acid together with a little chloroethanal.

5.80 g of chloroethanol were oxidised and the chloroethanoic acid in the resulting mixture reacted with 0.0600 mol of sodium hydroxide in a 1:1 stoichiometric ratio.

Show that the percentage conversion to chloroethanoic acid was 83%. [3]

(b) DDT is an effective insecticide but its use in recent years has become restricted because of its persistence in the environment. The DDT that is sold is a mixture of closely related compounds that includes DDT, DDE and DDD.

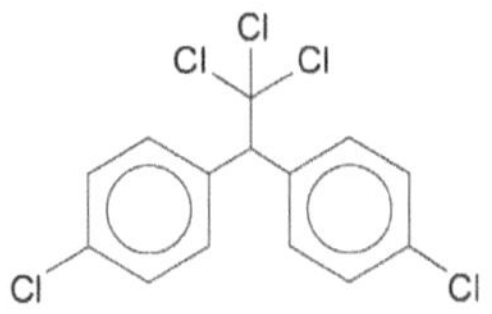

DDT

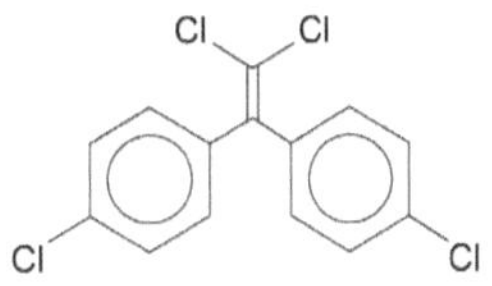

DDE

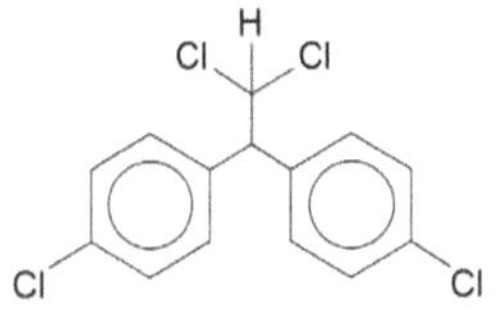

DDD

(i)   Suggest why DDT is largely insoluble in water.   [1]

(ii) Suggest suitable reagent(s) that can be used to convert DDE into DDD.   [1]

(iii) Many areas around former DDT manufacturing sites remain contaminated.

The soil from one contaminated area was analysed by gas chromatography.

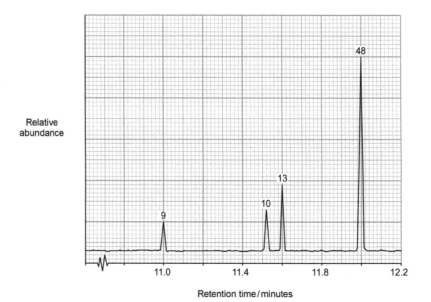

Retention time / minutes

Calibration of the chromatogram showed that the concentration of DDT in the sample (shown as the largest peak) was 0.018 mol kg^{-1}.

Calculate the total concentration of all the contaminants shown in the chromatogram in mol kg^{-1}.   [2]

(c)   Over the last thirty years the production and use of CFCs has declined, mainly because of their adverse environmental effects. Using fluorotrichloromethane as your example, outline how the compound causes a reduction in the amount of ozone present in the upper atmosphere and state two problems that can be caused by this reduction.   [6 QER]

(d)   The formula of chloropentafluoroethane is shown below:

$$\text{F}-\overset{\displaystyle \text{F}}{\underset{\displaystyle \text{F}}{\text{C}}}-\overset{\displaystyle \text{F}}{\underset{\displaystyle \text{F}}{\text{C}}}-\text{Cl}$$

State and explain what you would expect to see in the molecular ion region of its mass spectrum and in its ^{13}C spectrum.   [4]

(Total 21 marks)

[Eduqas Component 2 2017 Q12]

3. (a) Explain why amino acids are amphoteric compounds. [1]

(b) If an amino acid is treated with methanal, the resulting compound can be titrated against sodium hydroxide solution in a 1:1 ratio.

4.95 g of an amino acid was treated with methanal and the resulting solution made up to 250 cm³. 25.0 cm³ of this solution was then titrated with sodium hydroxide of concentration 0.105 mol dm⁻³. The results are shown in the table below.

Titration	1	2	3	4	5
NaOH(aq) used / cm³	38.73	35.90	36.00	32.00	36.10

(i) Suggest a practical reason why the reading for titration 1 was too high. [1]

(ii) Use appropriate titration values to calculate the relative molecular mass (M_r) of the amino acid. [5]

(iii) Assuming that the amino acid in part (ii) is a straight chain aliphatic α-amino acid, deduce its structure. [2]

(c) The formulae of three amino acids are shown below:

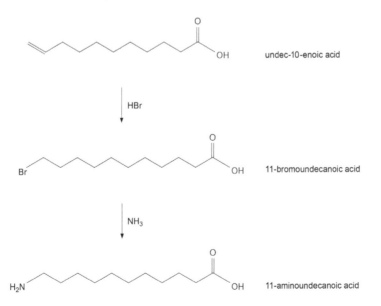

compound **R**   compound **S**   compound **T**

(i) State which of the three compounds could be identified by its ability to rotate the plane of plane polarised light. Give a reason for your answer. [1]

(ii) State how the infrared absorption spectrum of compounds **S** and **T** would differ from each other in their significant functional group absorption(s). [1]

(iii) State which of these amino acids would **not** be able to form two **different** dipeptides with either of the other two amino acids. Explain your answer. [1]

(d) Nylon-11 is a bio-sourced polyamide which is made from castor oil. Undec-10-enoic acid is produced as an intermediate compound. This acid is reacted, under suitable conditions, to give 11-bromoundecanoic acid, which is then treated with ammonia to produce 11-aminoundecanoic acid. Polymerisation of this product gives nylon-11.

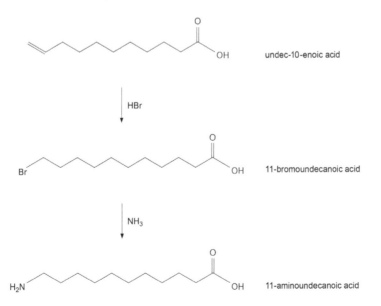

undec-10-enoic acid

HBr

11-bromoundecanoic acid

NH₃

11-aminoundecanoic acid

(i) The addition of hydrogen bromide to undec-10-enoic acid would give 10-bromoundecanoic acid as the major product. Explain why this is the case. [1]

(ii) The bromo-compound reacts with ammonia to produce 11-aminoundecanoic acid.

One step in the mechanism for this reaction is shown below:

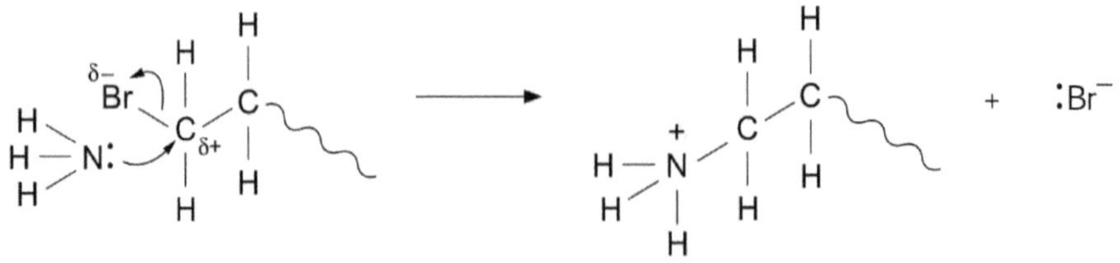

I. Explain how partial charges ($\delta+$ and $\delta-$) arise on the carbon and bromine atoms. [1]

II. State the role of ammonia in this reaction. [1]

(iii) Draw the structure of nylon-11, indicating the repeating unit present.

The formula of 11-aminoundecanoic acid is shown below. [1]

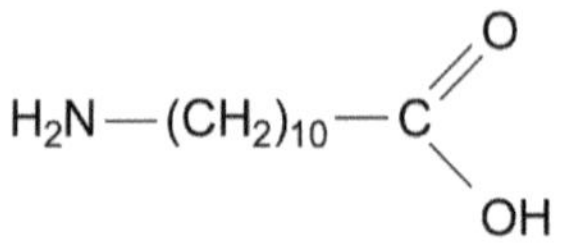

(Total 16 marks)

[Eduqas Component 2 2018 Q13]

Practical work

Practical work throughout the course

To ensure that you have developed a sound understanding of the practical elements of chemistry, you are expected to have experienced a range of practical techniques throughout the two years of the course. In most cases these will be the practical tasks set out in the WJEC lab book, but your teachers may have substituted alternative tasks that cover the same skills. When performing these tasks, it is important that you do not treat them as methods to follow, but as part of developing your understanding of chemistry. For each step in the method, you should be considering why you are doing it in a particular way. For example:

- Why do I need an excess of acid?

- Why do I need to cool or heat the mixture? Why am I adding anti-bumping granules when heating?

- Why must the temperature not exceed a certain level?

- What is the function of the substance I'm adding?

- Why am I using this particular indicator rather than another one?

- Why am I using this apparatus to measure rather than another option? Is there a better option?

- Why do I need to make so many measurements over time or so many repeat measurements?

- To what precision should I record my results? My stop clock says 44.27 seconds – are all these figures valid?

You should also assess the quality of your results:

- Do my readings make sense? Is the pattern plausible and does it agree with expectations?

- Do my repeat readings agree? Why is one result so far from the others? Do I need to take more repeat readings?

- Why is my melting temperature low?

- What is the accuracy of my measurements? What is causing the greatest problems with accuracy?

- Do my results lead to sensible calculated values, e.g. are all percentages below 100%?

> **Top tip**
>
> One type of AO3 question requires you to improve a practical method. Often one of the easiest practical improvements is to use more precise measuring devices – a 4-decimal place balance in place of a 2-decimal place balance or to use a burette or graduated pipette in place of a measuring cylinder.

Types of practical skills

Practical work includes a range of skills such as preparation and purification; qualitative and quantitative analysis; and physical chemistry measurements.

Preparation and purification

When a chemist needs to prepare a pure sample of a new substance they will prepare the substance, purify it, and then test to see if the substance is pure.

When preparing a substance, the amounts of each reactant need to be considered carefully. In a few cases, stoichiometric amounts are used – that is, exactly the correct numbers of moles of all reactants. In most cases, one substance is the limiting reagent, and all of this will react. The remaining substances are in excess, and there will be small amounts of these left at the end of the reaction. When choosing which substance to be in excess, the choice is often based on which substance is easiest to remove. In the reaction between solid copper carbonate and acid, it is usually the solid copper carbonate that is in excess as this can be removed by filtration.

Reactions usually need to be performed under specific conditions. This frequently includes heating to a specific temperature – heating too much can lead to a different reaction, and heating too little leads to a slow reaction. Heating flammable substances needs to be done safely, and this can be done using electrical heating. If electrical heating is not possible then using a water bath that is heated by a Bunsen burner (if the temperature needed is below 100°C) or an oil bath (for temperatures above 100°C) can be used.

A common method is to heat to reflux, where the solvent is heated until it boils, but then passes into a condenser where it condenses and falls back into the reaction vessel. In this method the reaction mixture stays at a constant temperature, and this is the boiling temperature of the solvent. Small fragments of an inert solid, called anti-bumping granules, are often added to the reaction mixture as this helps the solution to boil evenly.

Purification of the final product can use a range of techniques, including recrystallisation, distillation or use of a separating funnel. Once this is done, the purity of the substance can be tested. One common method is melting temperature analysis, as a pure substance has a sharp melting temperature which matches the literature value. Another is to use chromatography, where a pure substance gives one spot in paper chromatography or one peak in gas chromatography. If a substance is not pure, then it is purified again until it is pure.

Experiments in this section include:

- Preparation of a soluble salt by titration
- Preparation of an ester and separation by distillation
- Nucleophilic substitution reaction
- Synthesis of a liquid organic product
- Synthesis of a solid organic product
- Two-step synthesis
- Paper chromatography, including two-way separation.

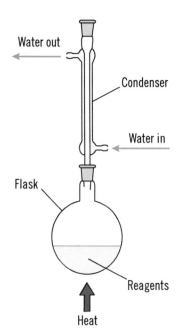

Water out

Condenser

Water in

Flask

Reagents

Heat

▲ Heating to reflux

‹ Link ›

Purification techniques are outlined in Chapter 4.8.

1 Knowledge check

A sample of the ester methyl benzoate is treated with a mixture of nitric acid and sulfuric acid to nitrate the benzene ring. The main product is methyl 2,4-dinitrobenzoate.

(a) The final product can be purified by recrystallisation using ethanol. Outline the steps needed to recrystallise the product.

(b) Suggest two methods of showing the final product after recrystallisation is pure. Outline the expected observations (i) if the sample is pure and (ii) if it is not.

Qualitative analysis

Qualitative analysis includes methods that show what is present in a substance, but not the amount that is present. In inorganic chemistry there are a range of specific tests for ions, many of which you have seen throughout the course. These include tests for the following ions:

- From year 12 work: Li^+, Na^+, K^+, Mg^{2+}, Ca^{2+}, Sr^{2+}, Ba^{2+} (by flame test); Ba^{2+}, Cl^-, Br^-, I^-, SO_4^{2-} (by precipitation), CO_3^{2-} (by reaction with acid)
- From year 13 work: Al^{3+}, Pb^{2+}, Cu^{2+}, Fe^{2+}, Fe^{3+}, Cr^{3+} (by precipitation and the effect of excess hydroxide ions)

Qualitative inorganic analysis is not limited to these formal tests – any observation linked to a chemical rection can be used to analyse a substance. For example, adding concentrated sulfuric acid to a solid gives a smell of rotten eggs if iodide ions are present, mixing $Cu^{2+}(aq)$ and $I^-(aq)$ gives a white solid in a brown solution.

In qualitative organic analysis, tests indicate specific functional groups. A molecule may include many functional groups, and tests for each will give positive results. The formal tests covered include:

- From year 12 work: Br_2 (for alkenes); MnO_4^- (for alkenes); hydrolysis and acidification followed by $AgNO_3(aq)$ (for chloro-, bromo- and iodo-alkanes); $NaHCO_3$ (for carboxylic acids); acidified $K_2Cr_2O_7$ (for primary and secondary alcohols).
- From year 13 work: Br_2 (for phenols); $FeCl_3$ (for phenols); Tollens' reagent (for aldehydes); Fehling's reagent (for aldehydes); I_2/OH^- (iodoform test for $CH_3CH(OH)$ or CH_3CO); 2,4-dinitrophenylhydrazine or 2,4-DNPH (for carbonyl compounds); using cold HNO_2 (from HCl and $NaNO_2$) for amines.

Qualitative organic analysis is not limited to these formal tests, and the observations seen for any reaction studied can be used to gain information about a molecule. An example is hydrolysis of an amide, which produces ammonia gas, and so turns moist red litmus paper blue.

Experiments in this section include:

- Identification of unknown solutions by qualitative analysis
- Planning a sequence of tests to identify organic compounds.

Knowledge check 2

Describe chemical tests that would allow you to identify the following compounds. Your tests should identify both the cation and anion present in each compound:

(a) barium chloride;

(b) copper(II) sulfate;

(c) lead(II) carbonate.

Top tip

When testing for amines using cold HNO_2 you need to make sure you distinguish between aromatic amines (ones attached to benzene rings) and aliphatic amines (ones attached to saturated hydrocarbons) as they react in different ways.

Worked example

A student has access to the following four solutions: $Br_2(aq)$; acidified $MnO_4^-(aq)$; $NaHCO_3(aq)$; an alkaline solution of iodide ions ($I^-/OH^-(aq)$).

He needs to use these solutions to distinguish between the following four pairs of substances. He can use each solution to distinguish between one pair only.

A. 2-hydroxypropanoic acid and 3-hydroxypropanoic acid

B. Propane and propene

C. Phenol and benzoic acid

D. 2,4-dimethylphenol and 2,4-dimethylbenzene

Identify which solution should be used to distinguish each pair, giving the expected observations with each compound.

Continued ▶

Answer

Looking at each test, we can identify which pairs would be distinguished by these solutions.

- $Br_2(aq)$ can test for alkenes and also for phenols, so would work for B, C and D.
- Acidified $MnO_4^-(aq)$ can test for alkenes so would work for B only.
- $NaHCO_3(aq)$ can test for carboxylic acids so would work for C only. It cannot distinguish the pair in A as both are carboxylic acids.
- An alkaline solution of iodide ions ($I^-/OH^-(aq)$) can test for $CH_3CH(OH)$ or $CH_3C=O$, so it will distinguish the pair in A only.

The correct answers are therefore:

A. 2-hydroxypropanoic acid and 3-hydroxypropanoic acid: Alkaline iodine will give a yellow precipitate with 2-hydroxypropanoic acid but no reaction with 3-hydroxypropanoic acid.

B. Propane and propene: Acidified $MnO_4^-(aq)$ would turn from purple to colourless with propene but no change with propane.

C. Phenol and benzoic acid: $NaHCO_3(aq)$ will give effervescence with benzoic acid but not with phenol.

D. 2,4-dimethylphenol and 2,4-dimethylbenzene: $Br_2(aq)$ will give a white precipitate with 2,4-dimethylphenol but not with 2,4-dimethylbenzene.

Quantitative analysis

In quantitative analysis, the amounts of each substance present can be measured. Quantitative methods rely on measurements of the number of moles of substance needed for reaction or produced in a reaction. Most **quantitative** analysis can be classed as either **gravimetric**, which relies on measuring masses, or **volumetric**, which relies on measuring concentrations and volumes.

In gravimetric analysis, the amount of an ion present in a substance can be measured by forming a precipitate and drying this to constant mass to ensure **all** water has been removed. The mass of the precipitate formed can be used to find the amount of the ion in the original substance. Similar methods can be used to find the water of crystallisation in a solid by drying to constant mass, which removes all water.

In volumetric analysis, standard solutions are prepared in volumetric flasks and titration is used to measure the volume of standard solution needed for complete reaction.

When comparing gravimetric and volumetric techniques, it is important to assess the precision of the measurements taken. In titrations, most measurements have four significant figures, e.g. 25.45 cm³ with the final digit being 0 (on the line of the burette) or 5 (between the lines of the burette). In gravimetric analysis, the number of significant figures often relies on the number of decimal places on the balance. To get the best possible results, it is helpful to set up the experiment so the measurements are larger, as this reduces the percentage error. In gravimetric analysis, this means using ions that give a precipitate with a large relative formula mass; whilst in volumetric analysis, this means using similar concentrations for both solutions, so the volume measured is approximately 25 cm³.

The titration method can be applied to a wide range of problems. The simple calculations involved in an acid–base titration can be extended to problems such as double titrations (where there are two substances in a mixture) or back titrations (where titration measures the amount of acid remaining after a reaction). It is important that you are familiar with the calculations in each case and understand why these calculations are used. This will allow you to apply the skills to unfamiliar problems.

Knowledge check

The amount of silver ions dissolved in a solution can be measured by precipitating the silver as silver chromate (Ag_2CrO_4), filtering, washing with water then heating to constant mass.

(a) Explain why the sample is washed with water after filtration.

(b) Explain why the sample is heated to constant mass.

(c) A 100 cm³ sample of $AgNO_3(aq)$ was treated with excess $CrO_4^{2-}(aq)$. A mass of 0.3310 g of silver chromate was precipitated. Calculate the concentration of the $AgNO_3(aq)$ in mol dm⁻³.

Link

The detailed method for making a standard solution and titration is covered in Chapter 1.7 of the AS book.

Link

Redox titrations for finding the amounts of ions such as Fe^{2+} or Cu^{2+} are also covered in Chapter 3.2 of this book.

Worked example

A fungicide contains $CuSO_4.5H_2O$ as its active ingredient, alongside other compounds. Measure precisely approximately 6 g of the fungicide using a 3 decimal place balance and use this to make precisely 250.0 cm³ of a solution in deionised water.

Take a 25.0 cm³ sample of the solution, place in a conical flask and add a spatula of solid KI and mix thoroughly. Titrate the sample against a standard solution of sodium thiosulfate $(Na_2S_2O_3)$ of concentration 0.100 mol dm⁻³. Add a starch indicator as you approach the equivalence point.

Repeat the titration at least three times or until concordant results are obtained. Use your results to calculate the percentage of $CuSO_4.5H_2O$ in the fungicide.

Results

- Mass of fungicide used = 6.027 g
- Mean volume of $Na_2S_2O_3$(aq) used for complete reaction = 17.35 cm³

Answer

Moles of $Na_2S_2O_3$(aq) used = $17.35 \times 10^{-3} \times 0.100 = 1.735 \times 10^{-3}$ mol

Moles of Cu^{2+} in 25 cm³ = Moles of $Na_2S_2O_3$ = 1.735×10^{-3} mol

Moles of Cu^{2+} in 250 cm³ = $1.735 \times 10^{-3} \times 10 = 1.735 \times 10^{-2}$ mol

M_r of $CuSO_4.5H_2O$ = 249.7
therefore mass $CuSO_4.5H_2O$ = $249.7 \times 1.735 \times 10^{-2} = 4.332$ g

Percentage $CuSO_4.5H_2O$ = $4.33 \times 100 \div 6.027 = 71.9\%$

Top tip

It is common to be told to measure an approximate mass precisely. This may sound odd, but it simply means that you need to know the exact mass, but it doesn't matter if this matches the mass given. Measuring approximately 10 g precisely means that 10.224 g or 9.956 g are acceptable but 10 g with no decimal places is not.

Experiments in this section include:

- Gravimetric analysis
- Standardisation of an acid solution
- Back titration
- Double titration
- Titration using a pH probe
- Simple redox titration
- Estimation of copper in copper(II) salts
- Determination of an equilibrium constant.

Physical chemistry measurements

Physical chemistry includes energy changes, equilibria, rates and electrochemistry. All experiments involve numerical measurements. When performing experiments in these areas it is important to assess the precision required and how much data needs to be collected.

In energetics experiments, temperature changes can be relatively small, with changes of 10°C being common. A 1°C thermometer would lead to a 10% error in measurements, so any thermometer used needs be more precise. Frequent temperature

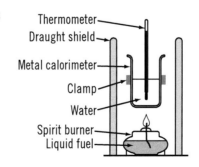

Thermometer
Draught shield
Metal calorimeter
Clamp
Water
Spirit burner
Liquid fuel

Link

The details of the practical methods used to measure temperature changes can be found in the AS book in Chapter 2.1.

‹ Link ›

Electrochemistry experiments are described in Chapter 3.1 of this book.

Top tip ››

You should be able to apply the ideas of standard electrode potentials to predict or explain any cell measurements and understand the function of each part of the electrochemical cell.

measurements are taken to ensure the temperature is stable before the experiment, and temperature changes can account for transfer of heat between the reaction and the environment.

In rate experiments, the time taken for a colour change to occur (iodine clock) or for a volume of gas to be released (gas collection) is used to calculate the rate. The devices used frequently give a measurement to the nearest hundredth of a second; however, they rely on a person to press the start and end button to get this measurement. Human reaction times are not fast enough to make this precision possible and recording to the nearest second is more appropriate.

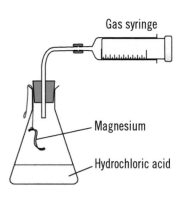

Experiments in this section include:

- Indirect determination of an enthalpy change of reaction
- Determination of an enthalpy change of combustion
- Investigation of a rate of reaction by a gas collection method
- Study of an 'iodine clock' reaction
- Determination of the order of a reaction
- Construction of electrochemical cells and measurement of E_{cell}.

Test yourself

1. Benzenediazonium chloride, $C_6H_5N_2Cl$, decomposes over time releasing nitrogen gas, with each benzenediazonium chloride molecule releasing one molecule of $N_2(g)$.

   $100\ cm^3$ solution of benzenediazonium chloride of concentration $0.100\ mol\ dm^{-3}$ was allowed to decompose at 20°C, and the volume of gas released was recorded in the table:

Time / min	0	2	4	6	8	10	12	14	16	18	20	22	24	26	28
Volume of $N_2(g)$ / cm³	0	31	62	89	111	130	147	162	173	184	194	204	211	217	222

   (a) Plot a graph of the volume of nitrogen gas produced against time in minutes. [4]

   (b) Use the graph to determine the initial rate of production of nitrogen gas in dm³ min⁻¹. [3]

   (c) The reaction is first order overall.

       (i) Write a rate equation for the decomposition. [1]

       (ii) Calculate the rate constant for the decomposition, giving its unit. [2]

2. You are provided with six unlabelled bottles that contain the following solid samples: $CuCO_3$, Na_2CO_3, $Pb(NO_3)_2$, KI, $MgSO_4$ and $Al(NO_3)_3$. You also have labelled bottles containing NaOH(aq) and HNO_3(aq).

   Outline a method for identifying the six solid samples. You should include ionic equations for any reactions that occur. [8]

3. Outline a method for finding the enthalpy change of the reaction below: [6]

$$Mg(s) + 2HCl(aq) \rightarrow MgCl_2(aq) + H_2(g)$$

   Explain how the results of the experiment would be used to calculate the enthalpy change.

Unit 5

1. Metals **X** and **Y** are Group 2 metals. Each one could be Mg, Ca, Sr or Ba.

    (a) Metal **X** appears not to react with cold water but reacts rapidly with steam.

    Identify metal **X** and give the equation for its reaction with steam. [1]

    (b) Metal **Y** reacts with cold water according to the following equation:

    $$\textbf{Y}(s) + 2H_2O(l) \longrightarrow \textbf{Y}(OH)_2(aq) + H_2(g)$$

    (i) When 2.27 g of metal **Y** are added to 600 cm³ of water, the concentration of the metal hydroxide solution formed on complete reaction is 0.0431 mol dm⁻³. Identify metal **Y**. [3]

    (ii) When 200 cm³ of the 0.0431 mol dm⁻³ solution of **Y**(OH)₂ formed in part (i) are added to excess sodium carbonate solution, YCO₃ is precipitated as a white solid and separated by filtration. The precipitate is washed with deionised water and heated at 100°C until the mass of YCO₃ remains constant.

    I. Give an equation, including state symbols, for the reaction of aqueous **Y**(OH)₂ with aqueous sodium carbonate. [1]

    II. Calculate the maximum mass of YCO₃ formed, giving your answer to three significant figures. [3]

    (c) Carbonates of Group 2 decompose on heating according to the following general equation:

    $$\textbf{Z}CO_3(s) \longrightarrow \textbf{Z}O(s) + CO_2(g)$$

    Give the trend in thermal stability of the carbonates on going down Group 2 and describe an experiment to show how this pattern could be confirmed.

    Your answer should include:

    - a detailed description of the method used
    - how to collect valid results, identifying the control variable(s)
    - the expected observation(s) [6 QER]

    (Total 14 marks)

    [*Eduqas Chemistry in Practice 2017 Q2*]

2. You are supplied with four unlabelled aqueous solutions containing the following species:

CO_3^{2-}	I^-	Cl_2	$S_2O_3^{2-}$
carbonate	iodide	chlorine	thiosulfate

    You are also provided with the following reagents:

    - dilute H_2SO_4
    - $AgNO_3(aq)$

    Devise a scheme whereby **all** four of the unlabelled solutions could be **positively identified**.

    You should include observations and **ionic** equations for any reactions occurring. [8]

    (Total 8 marks)

    [*WJEC Practical Methods and Analysis Task 2019 Q2*]

3. Aqueous citric acid reacts with sodium hydrogencarbonate according to the following equation:

$$3NaHCO_3(s) + C_6H_8O_7(aq) \rightarrow C_6H_5O_7Na_3(aq) + 3H_2O(l) + 3CO_2(g) \qquad \Delta H^\ominus = +78.8 \text{ kJ mol}^{-1}$$

The following method was used in an experiment to determine the temperature change during the reaction.

- A burette was used to measure 50.0 cm³ of 1.00 mol dm⁻³ citric acid into a polystyrene cup.
- 16.0 g of powdered sodium hydrogencarbonate was weighed.
- The initial temperature of the solution in the polystyrene cup was recorded as 24.4°C.
- The sodium hydrogencarbonate was added and the solution stirred slowly and constantly using the thermometer whilst measuring the temperature.

(a) Using the values given above, show that the sodium hydrogencarbonate was present in excess. [2]

(b) Using the given value of $\Delta H^\ominus$, calculate the expected temperature change and hence the final temperature recorded on carrying out this reaction. [3]

(Total 5 marks)

[WJEC Practical Methods and Analysis Task 2018 Q3]

Maths skills

The weighting for the assessment of maths skills for the A level qualification is a minimum of 20%. This section gives you an explanation and details some of the mathematical concepts that you are likely to meet. The following pages briefly mention those skills that you have used during the AS year and focus in greater detail on the extra maths skills that are tested during the A2 year.

Arithmetic and numerical computation

Standard form

Details of converting from ordinary to standard form and vice versa are given in the AS Student Book and you are likely to be expected to use both types during the A2 year.

Worked examples

Example 1

The dissociation constant, K_a, for butanoic acid at a certain temperature is 0.0000158 mol dm^{-3}. Convert this number to standard form.

Answer

Count how many times the decimal place has moved in order to be directly to the right of the first digit.

$$0.0000158$$

The decimal point has been moved 5 times from its original position; the index must therefore be −5.

In standard form the value of K_a is 1.58×10^{-5} mol dm^{-3}.

Example 2

At a certain temperature the dissociation constant, K_a, for chloric(I) acid is 3.72×10^{-8} mol dm^{-3}. Display this number in ordinary form.

Answer

To obtain the answer move the decimal point 8 places to the left.

$$0.0000000372$$

This gives 0.0000000372 mol dm^{-3}.

Units

At A2 level you may be asked to give the units for an equilibrium constant or for a rate constant.

Worked examples

Example 1

In gas phase reactions partial pressures are often used in place of concentrations. Ammonia is made from nitrogen and hydrogen in the Haber process:

$$N_2(g) + 3H_2(g) \rightleftharpoons 2NH_3(g)$$

Give the expression for the equilibrium constant, K_p, and its units.

Answer

The equilibrium constant, K_p, in terms of partial pressures (pp) (given in kPa) is given by:

$$K_p = (pp_{NH_3})^2 / (pp_{N_2}) \times (pp_{H_2})^3$$

In terms of units

$$K_p = (kPa)^2 / (kPa) \times (kPa)^3$$
$$= 1 / kPa^2$$

The unit of K_p is kPa^{-2}.

Example 2

Benzoyl chloride reacts with phenylamine to give N-phenylbenzamide:

This a first order reaction for each reactant and second order overall.

$$\text{Rate} = k[C_6H_5COCl][C_6H_5NH_2]$$

The rate of the reaction is measured in $mol\ dm^{-3}\ min^{-1}$.

Deduce the unit of the rate constant, k.

Answer

$$k = \text{rate} / [C_6H_5COCl][C_6H_5NH_2]$$
$$= mol\ dm^{-3}\ min^{-1} / mol\ dm^{-3} \times mol\ dm^{-3}$$
$$= dm^3\ min^{-1}\ mol^{-1}$$

Example 3

Find the minimum temperature at which a reaction will occur when the enthalpy change, ΔH, is 75.0 kJ mol^{-1} and the entropy change, ΔS, is 120 J K^{-1} mol^{-1}.

Answer

The temperature at which a reaction becomes feasible is when the value of ΔG changes from positive to negative. When this occurs ΔG is zero and $\Delta H = T\Delta S$.

Using $\Delta G = \Delta H - T\Delta S$

The value of ΔS needs to be in kJ K^{-1} mol^{-1}.

$$T = 75.0 / 120 \times 10^{-3} = 625\ K$$

Exam tip

Sometimes different units are given in equations, and conversion may be necessary. This is common in entropy / enthalpy calculations.

Estimating results

You may be asked to estimate the effect of changing conditions on the value of an equilibrium constant.

Worked examples

Example 1

The table shows the value of the equilibrium constant, K_p, for the reaction:

$$N_2(g) + O_2(g) \rightleftharpoons 2NO(g)$$

Temperature / K	K_p
293	4×10^{-31}
700	5×10^{-13}
1100	4×10^{-8}

Decide whether the forward reaction is exothermic or endothermic.

Answer

$$K_p = pp(NO)^2/pp(N_2) \times pp(O_2)$$

at different temperatures, where 'pp' is partial pressure.

As the temperature rises, the value of K_p increases. The reaction is endothermic in the forward reaction as the reaction is absorbing heat as the temperature rises. This explains why nitrogen and oxygen combine together in the combustion chamber of a motor vehicle but not at room temperature.

Example 2

The dissociation of water molecules into ions is very small.

$$H_2O(l) \rightleftharpoons H^+(aq) + OH^-(aq)$$

Since the amount of dissociation is very small, the concentration of water is assumed to be unchanged and is included into a new constant, K_w, the ionic product of water, where

$$K_w = [H^+(aq)][OH^-(aq)]$$

The table shows the value of K_w at various temperatures:

Temperature / K	K_w / mol^2 dm^{-6}
283	2.92×10^{-15}
313	2.91×10^{-14}
333	9.91×10^{-14}

Deduce whether the dissociation of water is an exothermic or an endothermic process.

Answer

As the temperature increases, the extent of the dissociation of water also increases. Therefore the dissociation of water is an endothermic reaction.

Handling data

Use an appropriate number of significant figures

This has been covered in more detail in the AS Student Book and only a reminder is included here. You may be asked to report a result of a calculation to an appropriate number of significant figures.

Worked example

0.15 mol of ethanol (M_r 46.1) is required for a reaction. Calculate the mass of ethanol needed, to an appropriate number of significant figures.

Answer

Mass of ethanol required = Number of mol $\times M_r$

$$= 0.15 \times 46.1 = 6.915 \text{ g}$$

The lowest number of significant figures given in the question is two. The answer should be given to this lowest number.

Therefore mass of ethanol required = 6.9 g

Finding arithmetic means

This too has been covered in more detail in the AS book and a brief reminder is given here.

Worked example

A sample of gallium consists of 60.0% of ^{69}Ga and 40.0% of ^{71}Ga. Use this data to calculate the relative atomic mass of gallium.

Answer

Relative atomic mass = (60.0 $\times$ 69) + (40.0 $\times$ 71) / 100

$$= 4140 + 2840 / 100$$

$$= 69.8$$

Algebra

Change the subject of an equation

This is an essential part of an 'A' level Chemistry course. Some examples of this procedure are shown in the AS Student book. In A2 we need to extend this to calculating a rate constant from a rate equation.

Worked examples

Example 1

The rate equation for the hydrolysis of 2-chlorobutane under certain conditions is:

Rate = $k\,[CH_3CH_2CHClCH_3]$

At a certain temperature, the initial concentration of 2-chlorobutane is 0.20 mol dm^{-3}. Calculate the value of the rate constant, k, giving its unit, if the initial rate of hydrolysis is 3.0×10^{-4} mol dm^{-3} s^{-1}.

Answer

k = Initial rate / concentration

$= 3.0 \times 10^{-4} / 0.20 = 1.5 \times 10^{-3}$

In terms of units, k = mol dm^{-3} s^{-1} / mol dm^{-3} = s^{-1}

$= 1.5 \times 10^{-3}$ s^{-1}

Example 2

Chlorine(IV) oxide reacts with hydroxide ions in a disproportionation reaction:

$$2ClO_2 + 2OH^- \longrightarrow ClO_3^- + ClO_2^- + H_2O$$

The rate-determining step for this reaction is second order for ClO_2 and first order for OH^- ions.

(a) Write the rate equation for this reaction.

(b) When the starting concentration of both reactants is 0.0500 mol dm^{-3} the initial rate is 2.90×10^{-2} mol dm^{-3} s^{-1}.

Calculate the value of the rate constant, giving its unit.

Answer

(a) Rate = $k\,[ClO_2]^2[OH^-]$

(b) Rate = $k\,[ClO_2]^2[OH^-]$

$2.90 \times 10^{-2} = k \times 0.0500^2 \times 0.0500$

$k = 2.90 \times 10^{-2} / 0.0500^3 = 232$

In terms of units k = mol dm^{-3} s^{-1} / (mol dm^{-3})^3

$=$ dm^6 mol^{-2} s^{-1}

Answer k = 232 dm^6 mol^{-2} s^{-1}

 **Knowledge check**

Iron(II) ions react with cobalt(III) ions in aqueous solution:

Fe^{2+}(aq) + Co^{3+}(aq) $\longrightarrow$ Fe^{3+}(aq) + Co^{2+}(aq)

When both reactants have an initial concentration of 0.050 mol dm^{-3}, the initial rate is 0.025 mol dm^{-3} s^{-1}.

The rate constant for this reaction is 10 dm^3 mol^{-1} s^{-1}.

Show that the rate-determining step is first order for each reactant and second order overall.

As part of the A2 examination you may be asked to carry out equations using the Arrhenius equation:

$$k = Ae^{(-E_a/RT)}$$

Where:

k = rate constant at a certain temperature

A = frequency factor, treated as constant over a limited temperature range

e = mathematical constant, 2.718

E_a = activation energy

R = molar gas constant (8.31 J K^{-1} mol^{-1})

T = temperature in Kelvin

The Arrhenius equation can also be used in its log version:

$\ln k = \ln A - E_a/RT$

Worked examples

Example 1

For the reaction:

$$C_2H_5Br(aq) + OH^-(aq) \rightarrow C_2H_5OH(aq) + Br^-(aq)$$

A has a value of 4.3×10^{11} dm^3 mol^{-1} s^{-1} at 303 K and the value of the rate constant is 1.6×10^{-4} dm^3 mol^{-1} s^{-1}.

Use the Arrhenius equation to find the value of the activation energy for this reaction. Give your answer in kJ mol^{-1}.

Answer

$k = 4.3 \times 10^{11} \times e^{(-E_a/8.31 \times 303)}$

Taking logs of the equation

$\ln k = \ln 4.3 \times 10^{11} + \ln e^{(-E_a/8.31 \times 303)}$

$\ln k = 26.8 + (-E_a/8.31 \times 303)$

$-8.74 = 26.8 - E_a/2518$

$\therefore E_a/2518 = 35.5$

$\therefore E_a \quad = 35.5 \times 2518$

$\quad\quad\quad = 89389$ J mol^{-1}

$\quad\quad\quad = 89.4$ kJ mol^{-1}

The activation energy is 89.4 kJ mol^{-1}

Example 2

Use the value of the activation energy found in example 1 (89.4 kJ mol^{-1}) to calculate the value of the rate constant at 318 K.

Answer

It is often more convenient in calculations to use the Arrhenius equation in its logarithmic form.

Substituting in the logarithmic form of the Arrhenius equation

$\ln k = \ln 4.3 \times 10^{11} - 89.4/8.31 \times 10^{-3} \times 318$

$\ln k = 26.8 - 33.8$

$\ln k = -7.0$

$\therefore k = 9.1 \times 10^{-4}$ dm^3 mol^{-1} s^{-1}

Top tip

Questions on the Arrhenius equation may involve changing the subject of the equation and using logarithms, as well as graphical work.

Knowledge check 2

Use the logarithmic form of the Arrhenius equation to calculate the activation energy for the reaction:

$$C_2H_5I(g) \rightarrow C_2H_4(g) + HI(g)$$

At 660 K the rate constant for this reaction is 7.2×10^{-4} s^{-1} and the value of the factor A is 3.5×10^{18} s^{-1}.

Knowledge check 3

The rate of hydrolysis of 2-chloro-2-methylpropane was measured at two different temperatures and the rate constants calculated:

Temperature / K	Rate constant / s^{-1}
283 (T)	3.57×10^{-3} (k)
293 (T')	15.19×10^{-3} (k')

(a) Use the equation below to calculate the activation energy, E_a, for this reaction:

$\ln k' - \ln k = E_a/R \,(1/T - 1/T')$

(b) Use the value for the activation energy found in (a) to calculate the rate constant at 288 K, using the logarithmic form of the Arrhenius equation. The numerical value of the frequency factor A is 1.02×10^{16}.

Study point

The equation seen in Knowledge check 3(a) is sometimes used to eliminate the value of A in Arrhenius equation calculations. This is done by subtracting the Arrhenius equation for one temperature from the equation at a different temperature.

Calculating the value of an equilibrium constant, K_c

This is a common procedure that is often tested at A2 level.

Worked examples

Example 1

At 500 K, gaseous hydrogen bromide partly dissociates to hydrogen and bromine:

$$2HBr(g) \rightleftharpoons H_2(g) + Br_2(g)$$

At equilibrium, the concentrations of hydrogen bromide and hydrogen are 1.25×10^{-3} mol dm^{-3} and 1.10×10^{-8} mol dm^{-3} respectively. Calculate the value of the equilibrium constant, K_c at this temperature.

Answer

At equilibrium the concentrations of hydrogen and bromine are equal.

$$K_c = [H_2(g)][Br_2(g)] / [HBr(g)]^2$$
$$= (1.10 \times 10^{-8}) \times (1.10 \times 10^{-8}) / (1.25 \times 10^{-3})^2$$
$$= 1.21 \times 10^{-16} / 1.56 \times 10^{-6}$$
$$= 7.76 \times 10^{-11} \quad \text{(no unit)}$$

Example 2

At 500 K the equilibrium constant for the reaction below is 0.031.

$$Br_2(g) + Cl_2(g) \longrightarrow 2BrCl(g)$$

Calculate the equilibrium concentration of bromine monochloride (BrCl) if the initial concentrations of both bromine and chlorine are 0.050 mol dm^{-3}.

Answer

$$Br_2(g) + Cl_2(g) \longrightarrow 2BrCl(g)$$

Initial concentration / mol dm^{-3}

    0.050   0.050   0.000

At equilibrium / mol dm^{-3}

    $0.050 - x$ $0.050 - x$ $2x$

$$K_c = 0.031 = [BrCl(g)]^2 / [Br_2(g)][Cl_2(g)]$$
$$= 4x^2 / (0.050 - x)(0.050 - x)$$

Taking square roots of both sides of the equation:

$$\sqrt{0.031} = 2x / 0.050 - x$$
$$0.176 (0.050 - x) = 2x$$
$$8.80 \times 10^{-3} - 0.176x = 2x$$
$$2.176x = 8.80 \times 10^{-3}$$
$$x = 4.05 \times 10^{-3}$$

The equilibrium concentration of BrCl is 4.05×10^{-3} mol dm^{-3}

▶ 4 Knowledge check

Benzoic acid exists largely as double molecules (dimers) when dissolved in benzene:

$$(C_6H_5COOH)_2 \rightleftharpoons 2C_6H_5COOH$$

12.20 g of benzoic acid (M_r 122) is dissolved in 1 dm^3 of benzene at a certain temperature. At equilibrium 1.29 g of benzoic acid remains as the monomer. Calculate the equilibrium constant for this reaction, giving its units, if any.

Carrying out pH and pK_a calculations

You will have met pH calculations during the first year of this course and this is now extended to the relationship between pH and the dissociation constant of acids, K_a, and its log equivalent, pK_a. As part of this work you will also study buffer solutions.

Worked example

Example 1

The dissociation constant, K_a, for butanoic acid at 298 K is 1.51×10^{-5} mol dm^{-3}. Calculate the pH of an aqueous solution of this acid of concentration 0.15 mol dm^{-3}.

$$CH_3(CH_2)_2COOH(aq) \rightleftharpoons CH_3(CH_2)_2COO^-(aq) + H^+(aq)$$

$$K_a = [CH_3(CH_2)_2COO^-(aq)][H^+(aq)] / [CH_3(CH_2)_2COOH(aq)]$$

Answer

Since the extent of dissociation is very small, the concentration of butanoic acid at equilibrium can be assumed to be 0.15 mol dm^{-3}. The concentrations of the butanoate ion and the hydrogen ion, H$^+$, have the same value.

Therefore $K_a = [H^+]^2 / 0.150$

$[H^+(aq)]^2 = 1.51 \times 10^{-5} \times 0.150 = 2.27 \times 10^{-6}$

$[H^+(aq)] = \sqrt{2.27 \times 10^{-6}} = 1.50 \times 10^{-3}$ mol dm^{-3}

pH $= -\log [H^+] = -\log 1.50 \times 10^{-3} = 2.82$

Example 2

An acidic buffer solution consists of an aqueous solution containing a carboxylic acid and its sodium salt. Work out the pH of a buffer solution that contains butanoic acid of concentration 0.100 mol dm^{-3} and sodium butanoate of concentration 0.120 mol dm^{-3}. The pK_a of butanoic acid is taken as 4.82.

Answer

$$CH_3(CH_2)_2COOH(aq) \rightleftharpoons CH_3(CH_2)_2COO^-(aq) + H^+(aq)$$

$$CH_3(CH_2)_2COO^-Na^+(aq) \rightarrow CH_3(CH_2)_2COO^-(aq) + Na^+(aq)$$

The acid is only slightly dissociated, and we can therefore assume that its concentration remains unchanged. The concentration of the butanoate ions derived from the acid can be neglected as it is very small. The concentration of butanoate ions from the sodium butanoate is 0.120 mol dm^{-3} as the sodium butanoate is fully dissociated in aqueous solution.

Using the Henderson–Hasselbalch equation

pH$_{buffer}$ = pK_a + log $[CH_3(CH_2)_2COO^-]$ / $[CH_3(CH_2)_2COOH]$

        = 4.82 + log (0.120 / 0.100) = 2.82 + log 1.2

        = 4.82 + 0.08 = 4.90

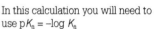

Knowledge check 5

In this calculation you will need to use pK_a = $-\log K_a$

The pK_a values of fluoroethanoic acid, FCH$_2$COOH, and bromoethanoic acid, BrCH$_2$COOH, are 2.66 and 2.90 respectively.

(a) Calculate the dissociation constant, K_a, of each acid.

(b) State, giving a reason, which of these two acids, each present as aqueous solutions of concentration 0.10 mol dm^{-3}, and at the same temperature, will have the lower pH.

Graphs

Another way of using the Arrhenius equation is with the aid of a graph.

The logarithmic form of the equation is:

$$\ln k = \ln A + (-E_a/RT)$$

$$\text{or } \ln k = \ln A - \frac{E_a}{R} \times \frac{1}{T}$$

This is of graphical form:

$$y = c + mx \quad \text{where } m \text{ is the gradient of the graph and is } E_a/R.$$

Worked example

On heating, iodoethane decomposes into ethene and hydrogen iodide:

$$C_2H_5I(g) \longrightarrow C_2H_4(g) + HI(g)$$

Use the data below to plot a graph of ln k against $1/T$ to find the activation energy for this reaction.

T/K	Rate constant k/s^{-1}
650	4.10×10^{-4}
675	1.66×10^{-3}
700	6.10×10^{-3}

Answer

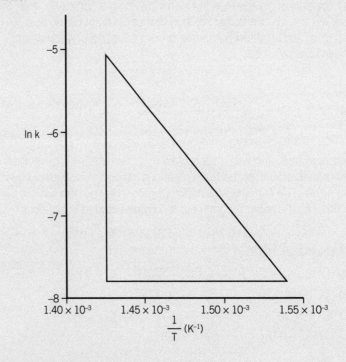

▶ 6 **Knowledge check**

An Arrhenius equation graph of ln k against $1/T$ produces a slope of value −37888.

Calculate the activation energy in kJ mol⁻¹.

The slope of the graph is $2.70 / 0.11 \times 10^{-3}$

This is equal to $-E_a/R$

Therefore $E_a = 2.70 \times 8.31 / 0.11 \times 10^{-3} = 203973$ J

The activation energy is 204 kJ mol⁻¹

Graphs can also be used to find the order of reaction of a compound taking part in a reaction. The order can be found using a concentration–time graph or a rate–concentration graph. The shapes of the graphs are characteristic of the order of reaction.

Zero order

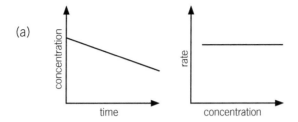

For a zero order reaction the concentration of the reactant decreases at a constant rate as time increases. Since the rate is independent of concentration the line in the rate–concentration graph is horizontal.

First order

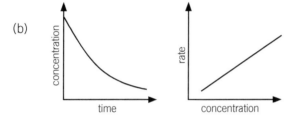

For a first order reaction the concentration of the reactant halves in equal time intervals. Half-life is a constant. Since the order of reaction is one for the reactant, the rate increases linearly with the concentration, giving a straight-line graph.

Second order

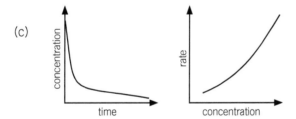

The concentration of the reactant decreases rapidly but the rate then decreases. Half-life is not a constant and increases with time. The rate–concentration graph shows a curve. If the rate is plotted against concentration squared, then a straight-line graph is obtained.

Graphs can also be used to find the rate at a particular time during the reaction. The initial rate can be found by drawing a tangent to the curve at the start, or a tangent drawn later on the curve will provide the rate at that particular time.

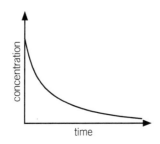

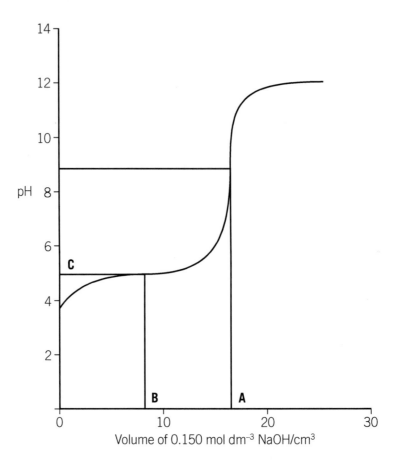

Graphs can similarly be used to find the dissociation constant, K_a, of a weak acid.

The graph shows the titration of 0.150 mol dm⁻³ aqueous sodium hydroxide with 25.0 cm³ of 0.100 mol dm⁻³ aqueous propanoic acid.

$$CH_3CH_2COOH + NaOH \longrightarrow CH_3CH_2COO^-Na^+ + H_2O$$

Number of mol of $CH_3CH_2COOH = 25.0 \times 0.100 / 1000 = 2.50 \times 10^{-3}$
Since the mol ratio of the reactants is 1:1, the number of mol of sodium hydroxide is also 2.50×10^{-3}.

Volume of aqueous NaOH needed $= 2.50 \times 10^{-3} \times 1000 / 0.150$

$= 16.7$ cm³ (point **A** on the graph)

At 'half neutralisation' 8.35 cm³ of aqueous NaOH (point **B** on the graph) have been added and at this point number of moles of CH_3CH_2COOH remaining is equal to the number of moles of CH_3CH_2COOH that have reacted to give $CH_3CH_2COO^-$.

$$CH_3CH_2COOH(aq) \rightleftharpoons CH_3CH_2COO^-(aq) + H^+(aq)$$

$K_a = [CH_3CH_2COO^-(aq)][H^+(aq) / [CH_3CH_2COOH(aq)]$

At 'half neutralisation' $[CH_3CH_2COO^-(aq)] = [CH_3CH_2COOH(aq)]$

$K_a = [H^+(aq)]$

taking negative logs

$- \log K_a = - \log [H^+(aq)]$ or $pK_a = pH$

The graphs show that at half neutralisation the pH is 4.9 (point **C**).

The pK_a of propanoic acid is 4.9 and K_a is 1.3×10^{-5} mol dm⁻³.

Geometry and trigonometry

The work in the A2 year includes a section on stereoisomerism. You will be expected to be able to identify chiral centres in the formulae of molecules drawn either as 2D or 3D representations.

Chiral centres are often identified by the use of an asterisk (*).   Many molecules contain more than one chiral centre.

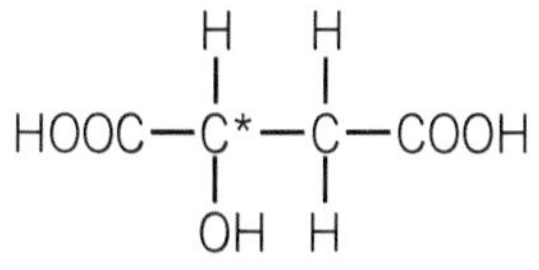

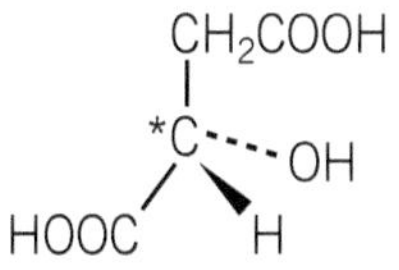

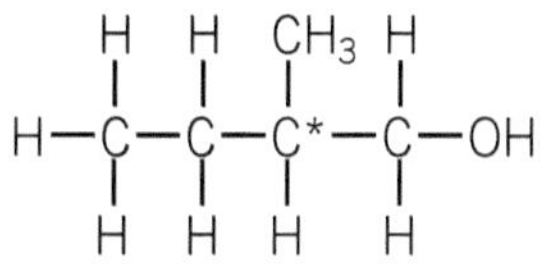

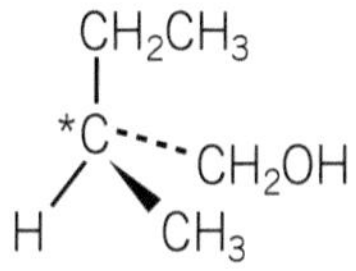

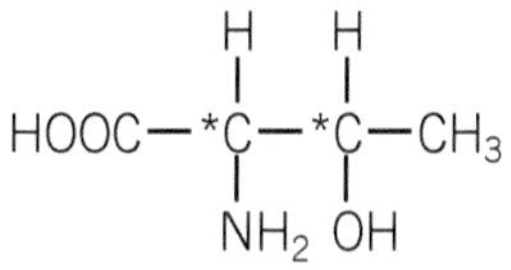

Knowledge check answers

Unit 3

3.1

1

(a) Oxidising agent = Sn^{4+}; Reducing agent = Fe^{2+}
(b) Oxidising agent = Cl_2; Reducing agent = Fe
(c) Oxidising agent = H_2O_2; Reducing agent = V^{3+}

2

(a) Oxidation $Mg(s) \longrightarrow Mg^{2+}(aq) + 2e$
    Reduction $Fe^{2+}(aq) + 2e \longrightarrow Fe(s)$
(b) Oxidation $Zn(s) \longrightarrow Zn^{2+}(aq) + 2e$
    Reduction $2H^+(aq) + 2e \longrightarrow H_2(g)$

3

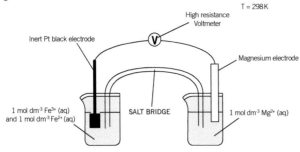

4

(a) $Pt(s) \mid H_2(g) \mid H^+(aq) \parallel Fe^{2+}(aq), Fe^{3+}(aq) \mid Pt(s)$
(b) Diagram of apparatus including: high-resistance voltmeter, salt bridge, platinum electrodes in both cells, and $1 \, mol \, dm^{-3} \, H^+(aq)$ and $1 \, atm \, H_2(g)$ in one cell with $1 \, mol \, dm^{-3} \, Fe^{2+}(aq)$ and $1 \, mol \, dm^{-3} \, Fe^{2+}(aq)$ in the other.

5

(a) (i) Copper electrode   (b) Fe^{2+}/Fe^{3+} electrode
(b) (i) Electrons flow from hydrogen electrode to copper electrode.
    (ii) Electrons flow from magnesium electrode to Fe^{2+}/Fe^{3+} electrode.

6

Oxidising agents: Na^+, I_2, MnO_4^-
Reducing agents: Cu, Cl^-, Mg, H_2
Order of decreasing oxidising power = MnO_4^-, I_2, Na^+.

7

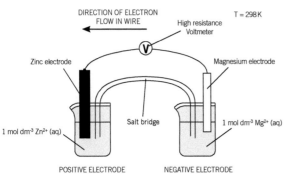

EMF = 1.60 V

8

(a) 0.34 V   (b) 3.13 V   (c) 2.27 V

9

(a) $2Fe^{3+} + Zn \longrightarrow 2Fe^{2+} + Zn^{2+}$
(b) $Cu + Cl_2 \longrightarrow CuCl_2$

10

(a) Yes it is feasible as the EMF for the reaction is +0.76V, and positive values represent feasible reactions.
(b) No it is not feasible as the EMF for the reaction would be negative, which is not feasible.

11

- Temperature is not the standard temperature / 298 K.
- Pressure of gases is not the standard pressure / 1 atm.
- Concentration of H^+ is not the standard concentration / $1 \, mol \, dm^{-3}$.

3.2

1

(a) $2ClO_4^- + 16H^+ + 14e^- \longrightarrow Cl_2 + 8H_2O$
(b) $MnO_4^- + 4H^+ + 3e^- \longrightarrow MnO_2 + 2H_2O$
(c) $C_2O_4^{2-} \longrightarrow 2CO_2 + 2e^-$
(d) $Ta_2O_5 + 10H^+ + 10e^- \longrightarrow 2Ta + 5H_2O$

2

(a) $2BrO_3^- + 12H^+ + 10Fe^{2+} \longrightarrow Br_2 + 6H_2O + 10Fe^{3+}$
(b) $Cr_2O_7^{2-} + 8H^+ + 6HI \longrightarrow 2Cr^{3+} + 7H_2O + 3I_2$
(c) $2MnO_4^- + 6H^+ + 5H_2O_2 \longrightarrow 2Mn^{2+} + 8H_2O + 5O_2$

3

(a) $2MnO_4^- + 5C_2O_4^{2-} + 16H^+ \longrightarrow 2Mn^{2+} + 10CO_2 + 8H_2O$
(b) $5.44 \times 10^{-3} \, mol \, dm^{-3}$

4

Moles manganate(VII) in each titration = $23.30 \times 0.0200 \div 1000$
                                             = 4.66×10^{-4} moles.
Reacting ratio 1:5 so moles iron in the $25.0 \, cm^3$ = 2.33×10^{-3} moles.
Moles iron in original $250 \, cm^3$ = 2.33×10^{-2} moles
Mass of iron = $2.33 \times 10^{-2} \times 55.8 = 1.300 \, g$
Percentage of iron = $1.300 / 1.740 \times 100 = 74.7\%$

5

Moles thiosulfate in $25 \, cm^3$ = $26.45 \times 10^{-3} \times 0.1 = 2.645 \times 10^{-3} \, mol$
Moles Cu^{2+} in $250 \, cm^3$ = $2.645 \times 10^{-3} \times 10 = 2.645 \times 10^{-2} \, mol$
Mass Cu = $2.645 \times 10^{-2} \times 63.5 = 1.680 \, g$
Percentage Cu = $1.680 / 2.741 \times 100 = 61.3\%$

3.3

1

Nitrogen: No d-orbitals in outer shell so it cannot expand its octet to form 5 bonds.

Bismuth: The inert pair effect increases down the group.

2

Acting as a base: $ZnO + 2HCl \longrightarrow ZnCl_2 + H_2O$

Acting as an acid: $ZnO + 2NaOH + H_2O \longrightarrow Na_2[Zn(OH)_4]$

3

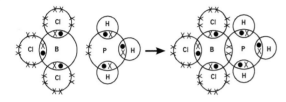

4

The stability of the +2 oxidation state increases down the group due to the inert pair effect so the stable oxidation state for carbon is +4 and for lead is +2. The carbon in CO has an oxidation state of +2 so it will act as a reducing agent and be oxidised to the more stable +4. The lead in PbO_2 has an oxidation state of +4 so it will act as an oxidising agent and be reduced to the more stable +2.

5

The standard electrode potential for Cl_2/Cl^- is much more positive than that for I_2/I^-. This means that chlorine is a stronger oxidising agent than iodine, so chlorine can oxidise iodide to form iodine.

$Cl_2 + 2I^- \longrightarrow 2Cl^- + I_2$

6

$H_2SO_4 + 6HI \longrightarrow S + 3I_2 + 4H_2O$

$H_2SO_4 + 8HI \longrightarrow H_2S + 4I_2 + 4H_2O$

3.4

1

Ti: $1s^2 2s^2 2p^6 3s^2 3p^6 3d^2 4s^2$

V: $1s^2 2s^2 2p^6 3s^2 3p^6 3d^3 4s^2$

Fe: $1s^2 2s^2 2p^6 3s^2 3p^6 3d^6 4s^2$

2

Cr^{3+} : $[Ar]3d^3$

Mn^{2+}: $[Ar]3d^5$

Fe^{2+}: $[Ar]3d^6$

Fe^{3+}: $[Ar]3d^5$

Cu^{2+}: $[Ar]3d^9$

Zn^{2+}: $[Ar]3d^{10}$

3

The blue/pink colour combination suggests Co^{2+} compounds. These exist in the following equilibrium:

$[CoCl_4]^{2-} + 6H_2O \rightleftharpoons [Co(H_2O)_6]^{2+} + 4Cl^-$

The blue solution contains $[CoCl_4]^{2-}$ ions. Adding water shifts the equilibrium towards $[Co(H_2O)_6]^{2+}$ whilst adding concentrated HCl provides chloride ions which shift the equilibrium to the left, according to Le Chatelier's principle.

4

The water ligands cause the d-orbitals to split into three of lower energy and two of higher energy. Electrons can move from lower to higher energy level by absorbing specific frequencies of light. These correspond to the energy gap ($E=hf$). The light not absorbed is the colour seen, so this complex is green as it absorbs all other colours apart from green.

5

The d-orbitals in the Zn^{2+} ion are full, so electrons cannot move between energy levels and therefore cannot absorb light energy.

6

Either $[Cu(H_2O)_6]^{2+}(aq) + 2OH^- \longrightarrow Cu(OH)_2 + 6H_2O$ or

$Cu^{2+} + 2OH^- \longrightarrow Cu(OH)_2$

3.5

1

As a reaction progresses the concentration of reactants decreases so there are fewer particles in any given volume. This leads to fewer collisions in the same time, so successful collisions occur less frequently.

2

Rate = change/time = $(0.84–0.50) / 2 = 0.17$ mol dm^{-3} min^{-1}

(or $\div 60$ to convert to seconds giving 2.83×10^{-3} mol dm^{-3} s^{-1})

3

(a) Second order   units: mol^{-1} dm^3 s^{-1}

(b) First order   units: s^{-1}

(c) First order   units: s^{-1}

(d) First order   units: s^{-1}

(e) Third order

4

Rate = $k[H_2O_2][I^-]$

$k = 2.8 \times 10^{-2}$ mol^{-1} dm^3 s^{-1}

5

(a) $C_2H_4 + Br_2 \longrightarrow$ products

(b) $I_2 \longrightarrow$ products

(c) $H_2O_2 + I^- + H^+ \longrightarrow$ products

6

$A = k \div e^{(-E_a/RT)}$ e.g. $T = \dfrac{E_a}{-R\ln(k/A)}$

7

Rate constant at 300 K = $0.0345 \div 0.100 = 0.345$ s^{-1}

Frequency factor = $A = k / e^{-(E_a/RT)} =$

$0.345 / e^{-(42000/8.314 \times 300)} = 7.10 \times 10^6$

Rate constant at 320 K = 0.989 s^{-1}

Rate = k × [concentration] = $0.989 \times 0.150 = 0.148$ mol dm^{-3} s^{-1}

3.6

1

$-84 - (-32) = -52$ kJ mol^{-1}

2

Methanol is not in its standard state; Water is not in its standard state; The sample is heated so it is not at the standard temperature; the equation is for $2CH_3OH$ when it should be for one.

3

250–800 kJ mol^{-1} as the first ionisation energy is less than the second ionisation energy.

4

Enthalpy of solution

= Enathlpy of lattice breaking +
 Enthalpy of hydration (Ca^{2+} + 2 × Cl$^-$)

= 2237 + (−1650) + (2 × −364)

= −141 kJ mol^{-1}

The calcium chloride will be soluble in water because the process of dissolving is exothermic.

5

(a) $Cu(g) \longrightarrow Cu^+(g)$

(b) $Cl_2(g) \longrightarrow 2Cl(g)$

(c) $\frac{1}{2} O_2(g) \longrightarrow O(g)$

(d) $2Na^+(g) + O^{2-}(g) \longrightarrow Na_2O(s)$

(e) $F(g) + e \longrightarrow F^-(g)$

6

$Cu(s) + F_2(s) \longrightarrow Cu(g) + F_2(g) \longrightarrow Cu^+(g) + 2 F(g) \longrightarrow Cu^{2+}(g) +$
$2 F(g) \longrightarrow Cu^{2+}(g) + 2F^-(g)$
$\longrightarrow Cu^{2+}(g) + 2 F^-(g) \longrightarrow CuF_2(s)$

So energy = (339) + (2 × 79) + (745) + (1960) + (2 × −348) + (−3037) = −531 kJ mol^{-1}

3.7

1

$I_2(s)$, $Br_2(l)$, $Br_2(aq)$, $Cl_2(g)$

2

Above 100°C the water would be a gas, which would have much higher entropy so the entropy change will be much less negative (accept more positive).

3

$C_6H_{14}(l) + 9\frac{1}{2} O_2(g) \longrightarrow 6CO_2(g) + 7H_2O(l)$

Entropy change = 6 × S (CO$_2$) + 7 × S(H$_2$O) − 9.5 × S(O$_2$) − S(hexane)

= 1284 + 490 − 1947.5 − 204

= −377.5 J K^{-1} mol^{-1}

4

(a) $\Delta G = -4 - (323 \times 12/1000) = -7.88$ kJ mol^{-1}

(b) $\Delta G = 24 - (298 \times 42/1000) = +11.5$ kJ mol^{-1}

The reaction is not feasible as the Gibbs free energy change is positive.

5

$T = \Delta H / \Delta S = -37/(-87/1000) = 425$ K

3.8

1

(a): $K_c = \dfrac{[H^+]^2[CO_3^{2-}]}{[H_2CO_3]}$

(b): $K_c = \dfrac{[NH_4OH]}{[NH_4^+][OH^-]}$

(c) $K_c = \dfrac{[HCl][HOCl]}{[H_2O][Cl_2]}$

(d) $K_c = \dfrac{[FeNCS^{2+}]}{[Fe^{3+}][NCS^-]}$

2

(a): $K_p = \dfrac{P_{SO_3}^2}{P_{SO_2}^2 \times P_{O_2}}$

(b): $K_p = \dfrac{P_{CO} \times P_{H_2}^3}{P_{CH_4} \times P_{H_2O}}$

(c): $K_p = \dfrac{P_{HI}^2}{P_{H_2} \times P_{I_2}}$

(d): $K_p = \dfrac{P_{CO} \times P_{Cl_2}}{P_{COCl_2}}$

(e): $K_p = \dfrac{P_{PCl_3} \times P_{Cl_2}}{P_{PCl_5}}$

3

$K_c = [Cl_2][CO]/[COCl_2] = 3.2 \times 10^{-3} \times 1.1 \times 10^{-4} / 3.2 \times 10^{-3}$
$= 1.1 \times 10^{-4}$ mol dm^{-3}

4

	Fe^{3+}(aq)	+ NCS$^-$(aq)	$\rightleftharpoons$	FeNCS^{2+}(aq)
Start (mol dm^{-3})	0.20	0.20		0.00
At equilibrium	0.05	0.05		0.15

To make 0.15 FeNCS^{2+}, 0.15 Fe^{3+} and 0.15 NCS$^-$ are needed, which leaves 0.05 of each of the reactants. (1 mark)

$K_c = \dfrac{[FeNCS^{2+}]}{[Fe^{3+}][NCS^-]}$ (1 mark)

$K_c = \dfrac{0.15}{0.05 \times 0.05} = 60$ mol^{-1} dm^3

5

At the start the partial pressures of H$_2$ and I$_2$ are 50500 Pa each. To form 37500 Pa HI then half this amount of H$_2$ and I$_2$ must react = 18750 Pa, leaving 50500 − 18750 = 31750 Pa of H$_2$ and I$_2$.

$K_p = \dfrac{P_{HI}^2}{P_{H_2} \times P_{I_2}} = 37500^2 / 31750^2 = 1.40$ (no units)

3.9

1

(a) $pH = -\log_{10}(0.2) = 0.70$

(b) $pH = -\log_{10}(0.03) = 1.52$

(c) $pH = -\log_{10}(10^{-9}) = 9$

(d) $pH = -\log_{10}(3 \times 10^{-11}) = 10.52$

2

(a) $[H^+] = 10^{-pH} = 1 \ mol \ dm^{-3}$

(b) $[H^+] = 10^{-pH} = 2 \times 10^{-3} \ mol \ dm^{-3}$

(c) $[H^+] = 10^{-pH} = 5 \times 10^{-7} \ mol \ dm^{-3}$

(d) $[H^+] = 10^{-pH} = 3.16 \times 10^{-11} \ mol \ dm^{-3}$

(e) $[H^+] = 10^{-pH} = 1 \times 10^{-14} \ mol \ dm^{-3}$

3

$K_a = [H^+][ClO^-] / [HClO]$

$K_a = [H^+][CN^-] / [HCN]$

4

As temperature increases, Le Chatelier's principle suggests the equilibrium will shift in the endothermic reaction. If K_w increases this means the equilibrium shifts to the right. The forward reaction must therefore be endothermic, with a positive enthalpy change.

5

So $[H^+]^2 = K_a \times [CH_3COOH]$

(a) (i) $[H^+]^2 = 1.7 \times 10^{-5} \times 0.5 = 8.5 \times 10^{-6}$
  $[H^+] = 2.9 \times 10^{-3}$
  $pH = -\log (2.9 \times 10^{-3}) = 2.54$

  (ii) $[H^+]^2 = 1.7 \times 10^{-5} \times 2 = 3.4 \times 10^{-5}$
  $[H^+] = 5.8 \times 10^{-3}$
  $pH = -\log (2.9 \times 10^{-3}) = 2.23$

  (iii) $[H^+]^2 = 1.7 \times 10^{-5} \times 0.01 = 1.7 \times 10^{-7}$
  $[H^+] = 4.12 \times 10^{-4}$
  $pH = -\log (2.9 \times 10^{-3}) = 3.38$

(b) (i) $[H^+]^2 = 2.9 \times 10^{-8} \times 1 = 2.9 \times 10^{-8}$
  $[H^+] = 1.703 \times 10^{-4}$
  $pH = 3.77$

  (ii) $[H^+]^2 = 2.9 \times 10^{-8} \times 0.5$
  $[H^+] = 1.204 \times 10^{-4}$
  $pH = -\log (1.204 \times 10^{-4}) = 3.92$

  (iii) $[H^+]^2 = 2.9 \times 10^{-8} \times 5 = 1.45 \times 10^{-7}$
  $[H^+] = 3.808 \times 10^{-4}$
  $pH = 3.42$

(c) Ethanoic acid is stronger as it has the larger K_a value (a less negative power indicates a larger number).

6

Using $K_a = \dfrac{[H^+][A^-]}{[HA]}$

where the two terms on the top are equal, we simply need to obtain $[H^+]$ from pH.

(a) For HB, $[H^+] = 10^{-pH} = 0.001258$
  so $K_a = 0.001258^2 \div 0.5 = 3.16 \times 10^{-6}$

(b) For HC, $[H^+] = 10^{-pH} = 6.310 \times 10^{-3}$
  so $K_a = (6.310 \times 10^{-3})^2 \div 1 = 4 \times 10^{-5} \ mol \ dm^{-3}$

(c) For HD, $[H^+] = 10^{-pH} = 3.16 \times 10^{-4}$
  so $K_a = (3.16 \times 10^{-4})^2 \div 0.5 = 2 \times 10^{-7} \ mol \ dm^{-3}$

7

Since $K_w = [H^+] \times [OH^-] = 1 \times 10^{-14} \ mol^2 \ dm^{-6}$,
$[H^+] = (1 \times 10^{-14}) \div [OH^-]$.

(a) $[H^+] = (1 \times 10^{-14}) \div 1 = 1 \times 10^{-14}$
  $pH = -\log (1 \times 10^{-14}) = 14$

(b) $[H^+] = (1 \times 10^{-14}) \div 0.2 = 5 \times 10^{-14}$
  $pH = -\log (5 \times 10^{-14}) = 13.3$

(c) $[H^+] = (1 \times 10^{-14}) \div 0.05 = 2 \times 10^{-13}$
  $pH = -\log (2 \times 10^{-13}) = 12.7$

(d) $[H^+] = (1 \times 10^{-14}) \div 0.003 = 3.33 \times 10^{-12}$
  $pH = -\log (3.33 \times 10^{-12}) = 11.5$

8

To work out the pH of a buffer solution, we can use K_a expression to derive:

$[H^+] = K_a \times [ACID] \div [SALT]$

Then we can use $pH = -\log [H^+]$ to work out the pH of the buffer.

(a) $[H^+] = 1.7 \times 10^{-5} \times [0.20] \div [0.10] = 3.4 \times 10^{-5}$
  $pH = -\log (3.4 \times 10^{-5}) = 4.47$

(b) $[H^+] = 1.7 \times 10^{-5} \times [0.20] \div [0.40] = 8.5 \times 10^{-6}$
  $pH = -\log (8.5 \times 10^{-6}) = 5.07$

(c) $[H^+] = 1.7 \times 10^{-5} \times [1] \div [0.2] = 8.5 \times 10^{-5}$
  $pH = -\log (8.5 \times 10^{-5}) = 4.07$

(d) $[H^+] = 1.6 \times 10^{-2} \times [0.20] \div [0.10] = 0.032$
  $pH = -\log (0.032) = 1.49$

9

Ammonia is a weak base and sulfuric acid is a strong acid so the salt should have an acidic pH e.g. pH 6. This is because the ammonium sets up an equilibrium where H^+ ions are lost increasing $[H^+]$ and decreasing pH.

10

strong acid – strong base titration: Bromophenol blue, Phenol red, Thymol blue

weak acid – strong base titration: Phenol red, Thymol blue

strong acid – weak base titration: Bromophenol blue

All these indicators change colour completely within the vertical regions of the relevant titration curve.

Unit 4

4.1

1

2

Citronellal does not exist as E-Z isomers because one carbon atom of the carbon-to-carbon double bond is bonded to two groups that are the same.

3

Enantiomers have the same molecular formula and hence the same relative molecular mass. Therefore equal 'amounts' will contain the same number of moles of each enantiomer.

4

5

Although each carbon atom of the carbon-to-carbon double bond is bonded to a hydrogen atom, one carbon atom is also bonded to a $CH_3(CH_2)_7-$ group whereas the other carbon atom is bonded to a $-(CH_2)_7COOH$ group.

4.2

1

The bonding is unclear. Bonding in the ring should only link carbon atoms and not involve hydrogen atoms.

2

The correct statement is (b). Each benzene atom is bonded to two carbon atoms and a hydrogen atom, and therefore (a) is wrong. Statement (c) is also wrong as multiple bonds between carbon atoms are shorter than carbon to carbon single bonds. As the value of the resonance energy increases, the stability of the molecule also increases.

3

The student said that the mechanism showed that iron(III) bromide was 'regenerated' at the end of the reaction and that, without the presence of iron(III) bromide, the reaction was very slow.

4

4.3

1

$CH_3CH(OH)CH_2CH_2COOH$

2

Pentane-1,3,5-triol

3

Every time that a ketone group is reduced, the resulting molecule has gained two more hydrogen atoms as a secondary alcohol group is produced. If the product has gained six more hydrogen atoms than the starting ketone, then the original compound had three ketone carbonyl groups.

4

2,3-dimethylbutan-1-ol

5

The relative molecular mass of the 'ethanoate part' of the ester, CH_3COO, is 59. The M_r of the alkyl group is $130 - 59 = 71$; this could be C_5H_{11}. As the question states that it is a primary alcohol, the primary alcohol could be pentan-1-ol, 3-methylbutan-1-ol or 2,2-dimethylpropan-1-ol.

6

Ethyl ethanoate has the formula $CH_3COOCH_2CH_3$. This compound has M_r of 88. The percentage of oxygen in this compound is $32/88 \times 100 = 36.4$.

7

methyl methanoate

8

Salicylic acid contains a phenol group as an OH group is bonded directly to the benzene ring. Aspirin does not contain this phenolic group. If a few drops of iron(III) chloride are added to a solution of aspirin, then a purple colour will be seen if any salicylic acid remains.

4.4

1

$C_{16}H_{24}O$

2

$+ [O] \rightarrow$ $+ H_2O$

3

(a) it reacts with Tollens' reagent to give a silver mirror. It reduces Fehling's solution to an orange-red solid.

(b) for example $HOCH_2CH_2C(O)CH_3$ or other ketone

4

$CH_3CH_2CH(OH)CH_3$

5

$C_2H_5OH \longrightarrow CH_3CHO$
M_r 46   44
Moles of ethanol $= 27.6 / 46 = 0.60$
For 100% yield moles of ethanol is also 0.60
Mass of ethanal for 100% yield $= 0.60 \times 44 = 26.4g$
Percentage yield $= 9.8 \times 100 / 26.4 = 37$

6

Name: butane-1,3-diol

7

$CH_3CH_2CH_2C(O)CH_3$ methyl methanoate

8

2-Methylpentan-3-one

9

(a) Hexan-2-one, this contains a $CH_3C=O$ group and
(c) 1-Phenylethanol, this contains a $CH_3CH(OH)$ group

4.5

1

(a)   (b) OH   (c)

2

Pentanoic acid

3

$$\bigcirc\!\!-CH_2OH + 2[O] \rightarrow \bigcirc\!\!-COOH + H_2O$$

4

The product is benzene-1,4-dioic acid, $C_6H_4(COOH)_2$ M_r 166

% Carbon $= 8 \times 12 \times 100 / 166 = 57.8$

5

Butanoic acid

6

2,2-dimethylpropanenitrile

7

The nucleophile is the cyanide ion, ^-CN. Both potassium cyanide and sodium cyanide produce cyanide ions in solution.

8

$C_3H_5O_2$

9

C_3H_6O

10

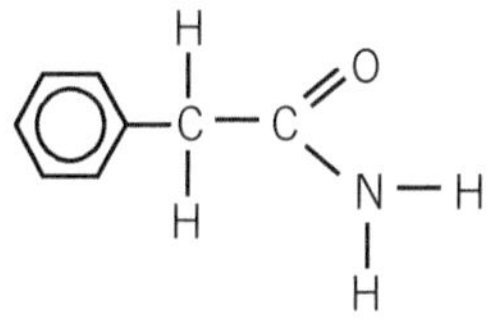

11

$CH_3\,CH_2\,CH_2\,CH_2\,CH_2\,C\!\equiv\!N$

14.4% nitrogen $\longrightarrow A_r$ 14

100% of compound gives $M_r \longrightarrow 14 \times 100 / 14.4 = 97$

CN group has 'M_r' 12 + 14 = 26

'M_r' of the R group = 97 – (12 + 14) = 71

Likely to be 5 carbon atoms (60) and 11 hydrogen atoms (11)

Formula is $CH_3CH_2CH_2CH_2CH_2CN$

12

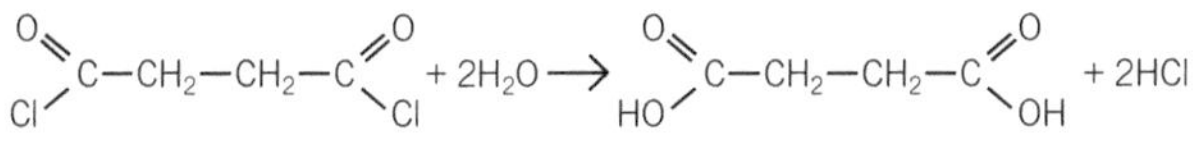

13

2-methylbutanenitrile

4.6

1

(a) Pentylamine   (b) Diphenylamine   (c) Ethane-1,2-diamine

2

(a)   (b)

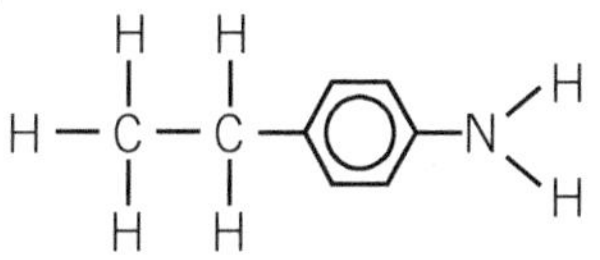

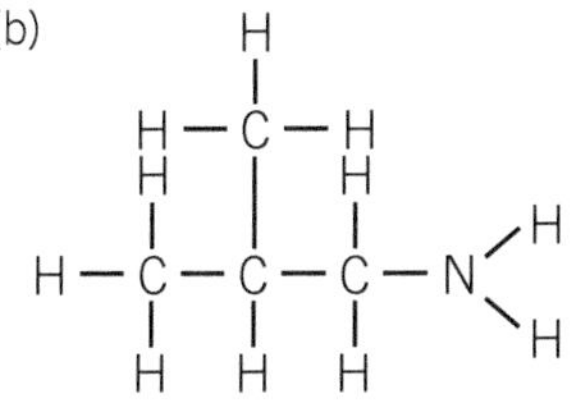

3

The carbon-to-chlorine bond is polarised $C^{\delta+}\!-Cl^{\delta-}$ and the $\delta+$ carbon atom attracts a lone pair of electrons from the nitrogen atom of the amine. This leads to a carbon to nitrogen bond being formed and the loss of chlorine.

4

$CH_3NH_3{}^+Br^- + OH^- \rightarrow CH_3NH_2 + H_2O + Br^-$

5

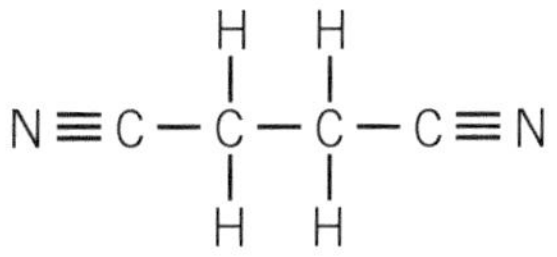

6

(a) $C_2H_5NH_3{}^+ Cl^-$

(b) $C_6H_5NH_3{}^+ HSO_4{}^-$

7

C_5H_6NO

8

27g

9

4-methylphenylamine   and   3,5-dimethylphenol

10

Phenolphthalein in alkaline solution has λ_{max} at 553 nm, in the green region. Using the colour hexagon this is opposite red-violet and this is the colour that is seen.

11

430–490 nm, as blue is opposite to yellow in the colour hexagon

4.7

1

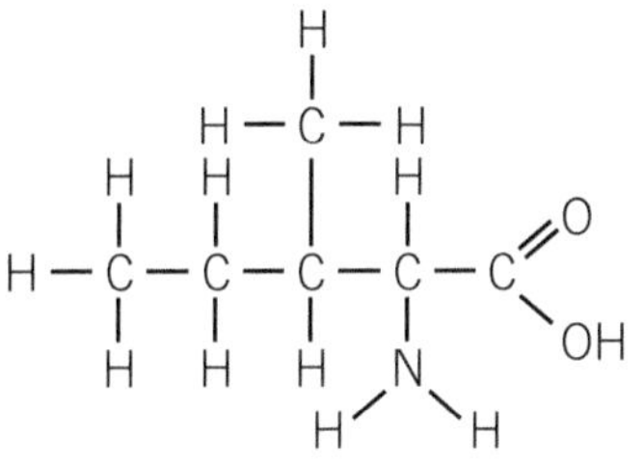

2

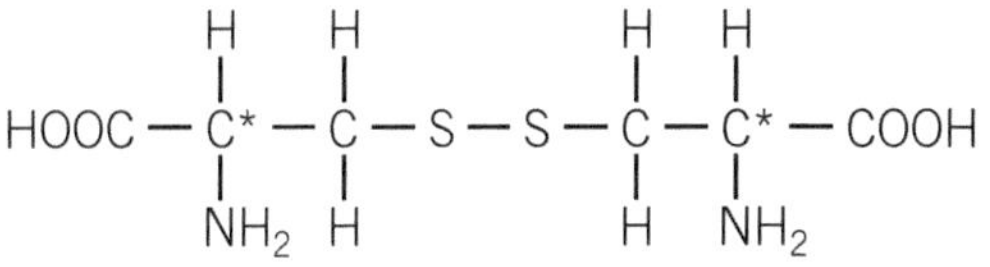

3

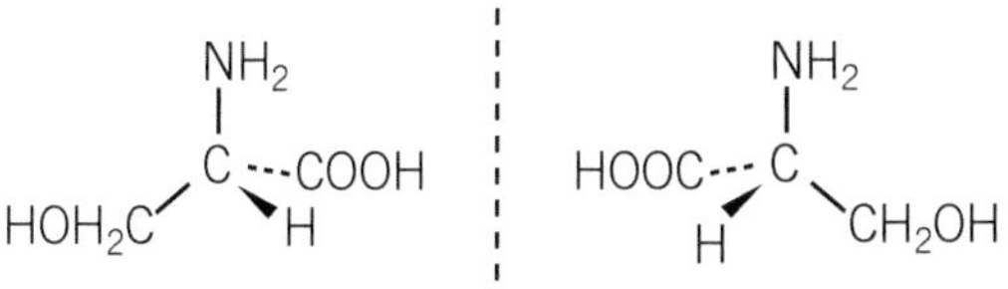

4

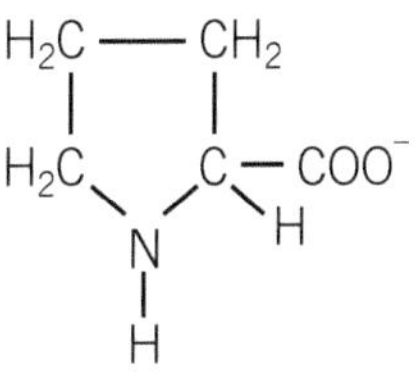

5

The zwitterion form $^+NH_3CH_2COO^-$ has used the lone pair of electrons on the nitrogen atom to bond the hydrogen ion, H^+, and there is now no lone pair of electrons for nucleophilic attack.

6

7

26.4 g

8

2,6-Diaminohexanoic acid

9

There is attraction between the δ– oxygen atom of the $C^{δ+}= O^{δ-}$ group and the δ+ hydrogen atom of the $N^{δ-}- H^{δ+}$ group.

10

$C_6H_{12}O_6 \rightarrow 2C_2H_5OH + 2CO_2$

4.8

1

2

A   Alkaline potassium manganate(VII) solution

B   (Dilute) hydrochloric acid

C   Soda lime

3

Bromoethane/chloroethane/iodoethane

4

5

Butane

6

Compound (b)

7

D   Potassium cyanide or sodium cyanide

E   Aqueous sulfuric acid / hydrochloric acid

F   Acidified potassium dichromate

8

No solvent is lost by evaporation, the concentration remains the same.

9

Simple distillation should suffice, as the boiling temperatures are very far apart.

10

(a)

(b) It may start to decompose if it is distilled at its boiling temperature under normal atmospheric pressure. Consideration should be given as to whether the lower temperature used during vacuum distillation **may** be more cost effective.

11

When the washing water is no longer yellow.

12

The compound is a polyester and it is likely that water has been eliminated during the polymerisation process – it is condensation polymerisation.

13

14

Mix equal quantities of the unknown compound and one of the known compounds. Take the melting point – if it remains at 158°C then it is the known compound used. If the melting temperature is now lower, the unknown compound is the other carboxylic acid.

15

There are two signals, both singlets. The $C(CH_3)_3$ group protons are three times the size of the single methyl group protons.

16

(a) Both ^{35}Cl and ^{37}Cl isotopes contribute to the mass spectrum

(b) There are only three distinct carbon environments.

17

18

A single peak as all its hydrogen atoms are in identical environments.

19

(a) 0.58

(b) Use a different solvent or use two-way chromatography.

20

(a) The total peak area is the sum of the individual values for each component in the mixture. This gives a total of 150 and the percentage of the largest component (82) is therefore $82 \times 100 / 150 = 55$

(b) The proportion of components in the mixture is not affected by running the procedure at a different temperature.

(c) A column that is packed with a different material might well effect a separation of these two compounds.

Unit 5

1

(a) Dissolve in <u>minimum</u> volume of <u>hot</u> ethanol. Filter whilst hot, then allow solution to cool so that crystals can form. Filer when cold then wash crystals with a small volume of cold ethanol. Allow to dry in an oven below the melting temperature of the compound.

(b) Take melting temperature: pure sample will have sharp melting temperature that matches literature value, impure will melt over a range of a few °C and melting temperature will be lower than literature value.

(Thin layer) chromatography: pure sample will have one spot, impure will have more than one. (Also allow gas chromatography and refer to one peak vs several peaks.)

Alternative answer: NMR spectroscopy: pure sample will have peaks for correct compound only, impure will have additional peaks.

2

(a) Barium chloride: Ba^{2+} identified by apple-green flame test OR dissolve and add SO_4^{2-}(aq) which gives a white precipitate. Cl^- identified by dissolving in water then add silver nitrate solution and a white precipitate will form.

(b) Copper(II) sulfate: Cu^{2+} identified by dissolving in water then adding OH^-(aq) to form a pale blue precipitate (also allow adding I^-(aq) to form a white solid in a brown solution); SO_4^{2-} identified by dissolving and adding Ba^{2+}(aq) to form a white precipitate.

(c) Lead(II) carbonate: Pb^{2+} identified <u>by adding nitric acid</u> to dissolve then add I^- to form yellow precipitate OR add excess OH^- and a white precipitate forms then dissolves again in excess; CO_3^{2-} identified by adding any acid and fizzing is seen (can test with lime water to show this is carbon dioxide).

3

(a) To remove <u>soluble</u> impurities.

(b) To remove <u>all</u> water.

(c) M_r (Ag_2CrO_4) = 332

moles of Ag_2CrO_4 produced from
100 cm^3 = 0.3310 ÷ 332 = 9.96 × 10^{-4} mol

concentration of $AgNO_3$ =
9.96 × 10^{-4} × 2 ÷ (100/1000) = 1.99 × 10^{-2} mol dm^{-3}

Maths skills

1

If the rate determining step is first order for both Fe^{2+}(aq) and Co^{3+}(aq),
The rate equation will be:

Initial rate = k [Fe^{2+}(aq)]^1 [Co^{3+}(aq)]^1

and second order overall, i.e. 1 + 1 = 2 in the rate equation
substituting Initial rate = 10 × 0.05 × 0.05 = 0.025 mol dm^{-3} s^{-1}

2

$\ln k$ = $\ln A - E_a/RT$

E_a/RT = $\ln A - \ln k$

= 42.7 – (–7.2) = 49.9

E_a = 49.9 × 8.314 × 660

= 274000 J mol^{-1}

= 274 kJ mol^{-1}

3

(a) $\ln k' - \ln k = \text{Ea}/\text{R} \left(\dfrac{1}{T} - \dfrac{1}{T'} \right)$

∴ $-4.19 - (-5.64) = \text{Ea}/\text{R} \left(\dfrac{1}{283} - \dfrac{1}{293} \right)$

∴ $1.45 = E_a/8.314 (3.53 \times 10^{-3} - 3.41 \times 10^{-3})$

∴ $1.45 = E_a/8.314 (1.20 \times 10^{-4})$

∴ $E_a = \left(\dfrac{1.45 \times 8.314}{1.20 \times 10^{-4}} \right) = 100460$ J mol^{-1}

= 100 kJ mol^{-1}

(b) Substituting in
$\ln k = \ln A + \left(\dfrac{-E_a}{RT} \right)$

$= 36.86 - \dfrac{100000}{8.31 \times 288}$

$= 36.86 - 41.78$

$= -4.92$

$k = 7.30 \times 10^{-3}$ s^{-1}

4

Moles of monomeric benzoic acid = 1.29/122 = 1.06 × 10^{-2}

Mass of dimeric benzoic acid present = 12.20 – 1.29

= 10.91 g

Moles of dimeric benzoic acid = 10.91/244 = 4.47 × 10^{-2}

K_c = $[C_6H_5COOH]^2/[(C_6H_5COOH)_2]$

= $(1.06 \times 10^{-2}/1)^2/(4.47 \times 10^{-2}/1)$

= 1.12 × 10^{-4}/4.47 × 10^{-2} = 0.00251 mol dm^{-3}

5

(a) Using $pK_a = -\log K_a$
Fluoroethanoic acid
2.66 = $-\log K_a$
K_a = 2.19 × 10^{-3} mol dm^{-3}
Bromoethanoic acid
2.90 = $-\log K_a$
K_a = 1.26 × 10^{-3} mol dm^{-3}

(b) Fluoroethanoic acid – as it will dissociate to a greater extent, as it has the higher K_a / lower pK_a value. Hence the [H^+] will be higher and the pH relatively lower.

6

Slope = $-37888 = -E_a/R$

E_a = 37888 × 8.314 J mol^{-1}

= 315000 J mol^{-1}

= 315 kJ mol^{-1}

Test yourself answers

Unit 3

3.1

1 A species that oxidises another species and in the process it is reduced. [1]

2 Nb goes from Nb^{5+} to Nb^{3+} so it gains two electrons and has been reduced OR the oxidation state of Nb has gone from +5 to +3 and as it is less positive it has been reduced. [1]

   Hydrogen has been oxidised as its oxidation state has become more positive, going from 0 to +1 OR charge gone from neutral to H^+ so lose electrons. [1]

3 Oxidation: $Zn(s) \longrightarrow Zn^{2+}(aq) + 2e^-$ [1]
   Reduction: $UO_2^{2+}(aq) + 4H^+(aq) + 2e^- \longrightarrow U^{4+}(aq) + 2H_2O(l)$ [1]

4 A half-cell consisting of $H_2(g)$ at 1 atm pressure bubbled over a platinum electrode in a solution of $H^+(aq)$ at a concentration of 1 mol dm^{-3} at a temperature of 298 K. [2]

5 (a)

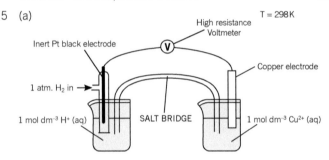

High resistance Voltmeter
T = 298 K
Inert Pt black electrode
Copper electrode
1 atm. H_2 in
1 mol dm^{-3} H^+ (aq)
SALT BRIDGE
1 mol dm^{-3} Cu^{2+} (aq)

   1 mark for diagram, 1 mark for each two complete correct labels.
   (b) Copper electrode. [1]
   (c) From hydrogen electrode to copper electrode through wire. [1]

6 Chlorine is a stronger oxidising agent that Fe^{3+} as it has a more positive standard electrode potential. This means that chlorine is able to oxidise Fe^{2+} to Fe^{3+} and form iron(III) chloride. [1] Iodine is a weaker oxidising agent than Fe^{3+} as it has a less positive standard electrode potential. This means that iodine is not able to oxidise Fe^{2+} to Fe^{3+} so it cannot form iron(III) iodide, and so it forms iron(II) iodide. [1]

7 For phosphinic acid to reduce a species it must have a more negative standard electrode potential than that of the other species. Phosphinic acid has a more negative $E^\ominus$ than that for dichromate (−0.50 V vs +1.33 V) so it is able to reduce this to form Cr^{3+}. Phosphinic acid has a more negative $E^\ominus$ than that for Cr^{3+} (−0.50 V vs −0.42 V) so it is able to reduce the Cr^{3+} produced in the first step to form Cr^{2+}. [1] Phosphinic acid has a less negative $E^\ominus$ than that for Cr^{2+} (−0.91 V vs −0.42 V) so it is unable to reduce this to form Cr. The main chromium-containing product is therefore Cr^{2+}. [1]

8 Do not release carbon dioxide/greenhouse gases when used. [1]
   More efficient release of energy / less energy is wasted. [1]

9 (a) 0.00 V [1]
   (b) K [1]
   (c) 1.20 V [1]
   (d) No reaction as EMF is negative (−0.27 V) [1]

3.2

1 $MnO_4^- + 4H^+ + 3e^- \longrightarrow MnO_2 + 2H_2O$ [1]

2 (a) $Fe^{2+} + Ce^{4+} \longrightarrow Fe^{3+} + Ce^{3+}$ [1]
   (b) Moles Ce^{4+} = Moles Fe^{2+} = $18.40 \times 10^{-3} \times 0.220$ = 4.048×10^{-3} [1]
   Concentration = $4.048 \times 10^{-3} \div (25/1000)$ = 0.162 mol dm^{-3} [1]

3 (a) $MnO_4^- + 8H^+ + 5e^- \longrightarrow Mn^{2+} + 4H_2O$ [1]
   (b) $2MnO_4^- + 5(COOH)_2 + 6H^+ \longrightarrow 2Mn^{2+} + 10CO_2 + 8H_2O$ [1]
   (c) (i) (29.85 + 29.75 + 29.65)/3 = 29.75 cm^3 [1]
       (ii) Moles manganate in 25 cm^3
       = $1 \times 10^{-2} \times 29.75 \times 10^{-3} = 2.975 \times 10^{-4}$ mol
       Moles oxalic acid in 250 cm^3
       = $2.975 \times 10^{-4} \times (5/2) \times 10 = 7.4375 \times 10^{-3}$ mol [1]
       Mass oxalic acid = $7.4375 \times 10^{-3} \times 90.02 = 0.670$ g [1]
       (iii) Lower concentration would need more than 50.0 cm^3 and this would not fit in a burette. [1] Higher concentration would give a lower volume and smaller measurements have greater percentage errors. [1]

4 (a) Moles of thiosulfate
   = $0.1 \times 17.85 \times 10^{-3} = 1.785 \times 10^{-3}$ mol = moles of Cu^{2+} [1]
   Concentration of Cu^{2+}
   = $1.785 \times 10^{-3} / 0.025 = 0.0714$ mol dm^{-3} [1]
   (b) Mass of $Cu(OH)_2 = 0.0714 \times (250/1000) \times 97.52$ = 1.741 g [1]
   Percentage by mass = 1.741 / 1.944 × 100 = 89.5% [1]

5 (a) NaOH turns the solution yellow. [1] The dichromate forms chromate (VI) ions (CrO_4^{2-}) as the sodium hydroxide removes H^+ [1] and forces the equilibrium below to the right. [1]
   $$Cr_2O_7^{2-} + H_2O \rightleftharpoons 2CrO_4^{2-} + 2H^+$$ [1]
   (b) Fe^{2+} turns the solution dark green [1] as the $Cr_2O_7^{2-}$ is reduced to dark green Cr^{3+}. [1]
   $$Cr_2O_7^{2-} + 6Fe^{2+} + 14H^+ \longrightarrow 2Cr^{3+} + 6Fe^{3+} + 7H_2O$$ [1]
   (c) In (a) the oxidation state of chromium stays at +6 so it is not a redox reaction. [1] In (b) the chromium is reduced from +6 to +3 so it is a redox reaction. [1]

3.3

1 Phosphorus has d-orbitals in its outer shell that are available to allow the octet to expand and form PCl_5 that contains 10 outer shell electrons. Nitrogen does not have d-orbitals available in its outer shell so cannot expand its octet and is limited to 8 outer shell electrons in NCl_3. [1]

2   The inert pair effect increases down the group, so the stable oxidation state of Pb is +2 and of Sn is +4. [1] PbO_2 therefore will be reduced easily to its more stable oxidation state and act as an oxidising agent. SnO_2 is in the stable oxidation state so will not be reduced. [1]

3   (a) Amphoteric means that it will react with both acids and bases. [1]
    (b) $Al(OH)_3$ reacting with acid:
    $Al(OH)_3 + 3HNO_3 \longrightarrow Al(NO_3)_3 + 3H_2O$ [1]
    $Al(OH)_3$ reacting with base: $Al(OH)_3 + NaOH \longrightarrow Na[Al(OH)_4]$ [1]

4   Al is electron deficient and Cl has lone pairs. [1] By forming coordinate bonds the Al is no longer electron deficient. [1]

5   CCl_4 – no reaction, forms a separate layer; $SiCl_4$ – vigorous, exothermic reaction releasing steamy fumes of HCl, bubbles and white solid form in water. [1]
    Explanation – Si has d-orbitals in outer shell available to bond with water to start the reaction, whilst C does not. [1]

6   Brick-red flame test = Ca^{2+} present; anion must be bromide as bromide and iodide both give coloured fumes but iodide will also produce a rotten egg smell. [1] The compound is $CaBr_2$. [1]

7   In the cold: $Cl_2 + 2NaOH \longrightarrow NaOCl + NaCl + H_2O$ [1]
    Cl has oxidation state zero at the start and +1 (in NaOCl) and −1 (in NaCl) at the end. As the same element has been oxidised (from 0 to +1) and reduced (from 0 to −1) this is a disproportionation reaction.
    In the hot: $3Cl_2 + 6NaOH \longrightarrow NaClO_3 + 5NaCl + 3H_2O$ [1]
    Cl has oxidation state zero at the start and +5 (in $NaClO_3$) and −1 (in NaCl) at the end. As the same element has been oxidised (from 0 to +5) and reduced (from 0 to −1) this is a disproportionation reaction. [1]

8   B contains K^+ as it gives a lilac flame test; C contains Cu^{2+} as it is pale blue. [1]
    A+B must form PbI_2 so A and B contain one each of Pb^{2+} and I^-. [1]
    B+C indicated Cu^{2+} mixed with I^- (see Chapter 3.2). As C contains Cu^{2+}, B must contain I^- and A must contain Pb^{2+}. [1]
    A+C white precipitate could be any of a range of lead compounds, as most are insoluble. Must be $PbSO_4$ as one solution contains sulfate.
    A = $Pb(NO_3)_2$ or $Pb(CH_3COO)_2$ [the only soluble lead compounds]; B = KI; C = $CuSO_4$. [1]

3.4

1   V: $1s^2 2s^2 2p^6 3s^2 3p^6 3d^3 4s^2$ Cr: $1s^2 2s^2 2p^6 3s^2 3p^6 3d^5 4s^1$
    Co: $1s^2 2s^2 2p^6 3s^2 3p^6 3d^7 4s^2$ [1] V^{2+}: $1s^2 2s^2 2p^6 3s^2 3p^6 3d^3$
    Mn^{2+}: $1s^2 2s^2 2p^6 3s^2 3p^6 3d^5$ [1]

2   Transition elements have partially filled d sub-shells in the atom or ions. Zn^{2+} and Zn both have filled d sub-shells, so Zn is not a transition element [1] but in Cu^{2+} the d sub-shell is partially filled so it is a transition element. [1]

3   The energies of the d-orbitals are similar to each other and to the s-orbitals. [1] It takes a similar amount of energy to remove each one and so the balance between the energy needed to ionise, and the energy released by forming bonds, is similar for different oxidation states, so more than one is possible. [1]

4   A: $[Cu(H_2O)_6]^{2+}$ B: $Cu(OH)_2$ C: $[Cu(NH_3)_4(H_2O)_2]^{2+}$ [1 each]

5   In solution $CuSO_4$ forms $[Cu(H_2O)_6]^{2+}$ and the water ligands cause the d sub-shell to split into 3 lower energy and 2 higher energy

orbitals. [1] Electrons from the lower shell can absorb light energy of a specific energy to be promoted to the higher energy level. This light absorbed is a specific colour and the remaining colours are what is seen – in this case red would be absorbed leaving blue to be reflected and seen. [1] The d sub-shell in Zn^{2+} is full so electrons cannot move to the higher energy level, so they do not absorb any frequencies of light. [1]

6   Fe in the Haber process / Ni in hydrogenation of alkenes or unsaturated oils / Pt in the oxidation of ammonia. [1]
    Heterogeneous catalysts provide a surface where the reactants are adsorbed by forming co-ordinate bonds with the metal atoms. This brings the reactants together and allows them to react. [1]

7   Homogeneous catalysts are in the same physical states as the reactants. [1] The transition metal atoms use their empty orbitals to bond to the reactants and then their variable oxidation states to oxidise/reduce the reactants making them much more reactive. [1]

8   (a) [1]

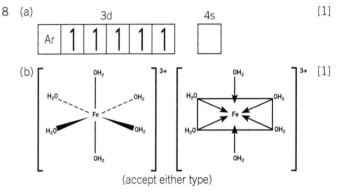

(accept either type)

    (c) The ligand splits the d-orbitals into 2 higher levels and 3 lower levels. [1] As an electron is promoted from a lower to a higher level, energy is absorbed from the visible spectrum. [1] The colour seen is the unabsorbed colours. [1]
    (d) For Fe^{3+}: Brown/yellow/red precipitate for Fe^{3+} that is insoluble in excess of the sodium hydroxide solution. [1]
    $Fe^{3+} + 3OH^- \longrightarrow Fe(OH)_3$
    For Cr^{3+}: Forms a grey-green precipitate. [1]
    $Cr^{3+} + 3OH^- \longrightarrow Cr(OH)_3$ [1]
    Dissolves in excess to form a green solution. [1]
    $Cr(OH)_3 + 3OH^- \longrightarrow [Cr(OH)_6]^{3-}$ [1]

3.5

1   Measuring volume of gas produced over time. [1]

2   Taking samples at set time intervals and adding these to ice water to stop the reaction. [1]

3   For a rate equation such as Rate = $k[A]^m[B]^n$, the order is the sum of the powers = $m+n$. [1]

4   (a) Colorimetry [1]
    (b) Order with respect to Br_2 = 0; [1]
        Order with respect to CH_3COCH_3 = 1. [1]
    (c) Concentration of H^+ is 10× greater at pH 0 compared to 1 [1] so rate will be × 10 = 7.0×10^{-6} mol dm^{-3} s^{-1} [1]
    (d) Rate = $k[CH_3COCH_3][H^+]$ [1]
    (e) Rate = $k[CH_3COCH_3][H^+]$ so k = Rate ÷ ($[CH_3COCH_3][H^+]$). [1]
        Using data set 1 = 2.8×10^{-5} mol^{-1} dm^3 s^{-1} [1]

5   (a) When concentration doubles, rate doubles
        Therefore first order or rate is proportional to concentration

(must give reason to obtain this mark.)

Alternative approaches:
Calculate k for each and show that all values are the same;
Calculate k for one concentration and use to calculate other values. [1]

(b) $k = $ Rate $\div [N_2O_5]$
e.g. $k = 3.00 \times 10^{-5} \div 4.00 \times 10^{-3}$
$= 7.50 \times 10^{-3}$
must be 3 significant figures; Unit $= s^{-1}$ [2]

(c) Rate-determining step must have one N_2O_5 molecule as reactant.
Mechanism A matches this rate equation.
Need reason to get this mark. [2]

6 (a) Second order as the units of the rate constant are those of a second order reaction [1]

(b) (i) Activation energy is the minimum amount of energy required for a collision to be successful/ for particles to react. [1]

(ii) $k = A \times e^{-(Ea/RT)}$ so $A = k / e^{-(Ea/RT)}$
Insert relevant values:
$A = 3 \times 10^7 / e^{-(23000/8.314 \times 350)}$ [1]
$= 8.124 \times 10^{10}$ mol^{-1} dm^3 s^{-1} [1]

(iii) $k = A \times e^{-(Ea/RT)} = 8.124 \times 10^{10} \times e^{-(23000/8.314 \times 400)}$ [1]
$= 8.06 \times 10^7$ (mol^{-1} dm^3 s^{-1}) [1]

(c) (i) $k = A \times e^{-(Ea/RT)} = 8.124 \times 10^{10} \times e^{-(11500/8.314 \times 350)}$
$= 1.56 \times 10^9$ (mol^{-1} dm^3 s^{-1}) [1]

(ii) Catalysts can increase the rate more than temperature;
Increasing temperature requires much more energy which is expensive/releases greenhouse gases;
Increasing the temperature of reversible reactions can decrease the yield. [1 mark for each reason]

7 Rate for a clock reaction $= 1/$time so rates are $1/46 = 0.022$, $1/23 = 0.044$, $1/23 = 0.044$, $1/23 = 0.044$; [1] Comparing 1 and 2 shows it is first order wrt H_2O_2, comparing 2 and 3 shows it is order zero wrt H^+, comparing 1 and 4 shows it is first order wrt I^-. [1]
Rate $= k[H_2O_2][I^-]$ [1]

3.6

1 The enthalpy change of a reaction is the same regardless of the route taken. [1]

2 (a) $K(s) \longrightarrow K(g)$ (b) $Na_2O(s) \longrightarrow 2 Na^+(g) + O^{2-}(g)$
(c) $Br^-(g) + aq \longrightarrow Br^-(aq)$ (d) $KI(s) + aq \longrightarrow KI(aq)$ [1 each]

3 Enthalpy of solution $= 2018 - 1360 - 2 \times 335 = -12$ kJ mol^{-1}. [1]
As this is exothermic the barium chloride will be soluble. [1]

4 $Mg^{2+}(g) + O^{2-}(g) \longrightarrow Mg^+(g) + O^{2-}(g) \longrightarrow Mg(g) + O^{2-}(g) \longrightarrow Mg(g) + O^-(g) \longrightarrow Mg(g) + O(g) \longrightarrow Mg(g) + \frac{1}{2} O_2(g) \longrightarrow Mg(s) + \frac{1}{2} O_2(g) \longrightarrow MgO(s)$
$- (1450) - (736) - (791) - (-141) - (\frac{1}{2} \times 496) - (149) + (-602) = $
so -3835 kJ mol^{-1} [3]

5 Standard enthalpy change of hydration of NaI $= - (-314) - (494) - (109) - (107) + (-296) = -692$ kJ mol^{-1} [2]
Standard enthalpy change of hydration of $I^- = -692 - (-406) = -286$ kJ mol^{-1} [1]

3.7

1 Magnesium, Mercury, Nitrogen, Air [1]

2 $-286 - (+44) = -242$ J K^{-1} mol^{-1} [1]

3 $\Delta G = \Delta H - T\Delta S$ [1]

4 Mg $= 669$ K; Ca $= 1106$ K; [1] Sr $= 1374$ K; Ba $= 1552$ K; [1]
The carbonates become more stable down the group as a higher temperature is needed to decompose them. [1]

5 (a) CO and CO_2 are gases and the other substances are solids. Gases have more freedom so a higher entropy than solids. [1]

(b) (i) Changes $= Fe + CO_2 - FeO - CO$; Enthalpy change $= -17$ kJ mol^{-1}; [1] Entropy change $= -11$ J K^{-1} mol^{-1} [1]

(ii) $\Delta G = -17 - (1100 \times -0.011)$ [1] $= -4.9$ kJ mol^{-1}. [1]
The reaction is feasible as the Gibbs free energy change is negative. [1]

(iii) The student is incorrect as Gibbs free energy change is negative at lower temperatures [1] – it only becomes positive at high temperatures, above 1545K. [1]

(c) Enthalpy change $= +319$ kJ mol^{-1}; [1] Entropy change $= +22$ J K^{-1} mol^{-1}; [1] Minimum temperature $= \Delta H/\Delta S = 14500$ K [1]

3.8

1 If a system in dynamic equilibrium experiences a change in conditions, then the equilibrium will shift to minimise the change. [1]

2 $K_c = \dfrac{[NH_3]^2}{[N_2][H_2]^3}$ [1]

3 $K_c = \dfrac{[NO]^2}{[N_2][O_2]}$ [1]

4 $K_p = \dfrac{P^2_{HI}}{P_{H_2} P_{I_2}}$ [1]

5 $K_p = (P_{NO_2})^2/(P_{N_2O_4})$ [1] so $8.33 \times 10^6 = (2.8 \times 10^5)^2 / (P_{N_2O_4})$ [1]
therefore $P_{N2O4} = 9.4 \times 10^3$ Pa [1]

6 $K_c = [H_2][I_2]/[HI]^2$ so $160 = 0.042 \times 0.042 / [HI]^2$ therefore $[HI]^2 = 1.1025 \times 10^{-5}$ and $[HI] = 3.32 \times 10^{-3}$ mol dm^{-3} [1]

7 Equilibrium concentrations are: $NH_3 = 0.24$ mol dm^{-3}; [1]
$N_2 = 0.28$ mol dm^{-3}; $H_2 = 0.44$ mol dm^{-3} [1]
so $K_c = 2.41$ mol^{-2} dm^6 [1]

8 (a) Equilibrium concentrations are: $PCl_5 = 0.32$ mol dm^{-3}; [1]
$PCl_3 = 0.38$ mol dm^{-3}; $Cl_2 = 0.38$ mol dm^{-3} [1] so
$K_c = 0.45$ mol dm^{-3} [1]

(b) Increasing temperature will increase the products for an endothermic reaction (according to Le Chatelier's principle). [1]
As the amount of products increase, and the amounts of reactants decrease, the value of K_c will increase. [1]

9 (a) $2SO_2 + O_2 \rightleftharpoons 2SO_3$ [1]

(b) (i) mol^{-1} dm^3 [1]

(ii) Concentration of SO_3 at equilibrium $= 4.0 \times 10^{-5}$ mol dm^{-3} [1]
$[O_2] = [SO_3]^2 / [SO_2]^2 \times K_c = (4.0 \times 10^{-5})^2 / (8.0 \times 10^{-5})^2 \times 4.5$ [1] $= 5.0 \times 10^{-2}$ mol dm^{-3} [1]

(iii) If K_c is smaller then less product at higher temperature [1] so the reaction must be exothermic. [1]

3.9

1 A proton donor that partially dissociates in solution / releases H^+ ions in a reversible reaction. [1]

2 pH $= - \log[H^+]$ [1]

3 $K_w = [H^+][OH^-]$ [1]

4 $[H^+] = 2 \times 0.20 = 0.40$; [1] pH $= - \log(0.40) = 0.40$ [1]

5 $[H^+] = 10^{-pH} = 10^{-2.45} = 3.55 \times 10^{-3}$ mol dm^{-3} [1]

6 (a) $K_a = \dfrac{[CH_3CH_2COO^-][H^+]}{[CH_3CH_2COOH]}$ [1]

   (b) $[H^+]^2 = K_a \times [CH_3CH_2COOH] = 1.34 \times 10^{-5} \times 0.1 = 1.34 \times 10^{-6}$ so $[H^+] = 1.16 \times 10^{-3}$ mol dm^{-3} and
pH $= -\log(1.16 \times 10^{-3}) = 2.94$ [1]

   (c) (i) A solution that keeps its pH value (almost) constant when small amounts of acid or base are added. [1]

      (ii) The acid dissociates in a reversible reaction and the salt dissociates completely:
      $CH_3CH_2COOH \rightleftharpoons CH_3CH_2COO^- + H^+$
      $CH_3CH_2COONa \rightarrow CH_3CH_2COO^- + Na^+$ [1]
      Addition of a small amount of acid causes the equilibrium to shift to the left to remove the H$^+$ and form more undissociated propanoic acid molecules. [1]
      Addition of a small amount of base causes the H$^+$ to neutralise the base, and the equilibrium shifts to the right to replace the H$^+$ lost. [1]

      (iii) $K_a = [H^+][SALT]/[ACID]$ [1] so $[H^+] = K_a[ACID]/[SALT] = 1.34 \times 10^{-5} \times 0.1/0.2 = 6.7 \times 10^{-6}$ [1]
      pH $= -\log(6.7 \times 10^{-6}) = 5.17$ [1]

7 $K_w = [H^+][OH^-] = 10^{-14}$ so $[H^+] = 10^{-14}/0.25 = 4 \times 10^{14}$ [1]
   and pH $= 13.4$ [1]

8 pH 7.5–10. [1] The sodium ethanoate dissociates fully releasing ethanoate ions.
   $CH_3COONa \rightarrow CH_3COO^- + Na^+$
   The ethanoate ions react with some of the H$^+$ ions produced by dissociation of water molecules, reducing the H$^+$ concentration and raising the pH.
   $CH_3CH_2COO^- + H^+ \rightleftharpoons CH_3CH_2COOH$ [1]

9 (a) 20.0 cm^3; [1] 0.080 mol dm^{-3} [1]
   (b) Cresol red and Phenolphthalein [1] as these change colour completely within the vertical region of the graph. [1]
   (c) Half neutralisation is 10 cm^3 of NaOH and the pH at this point is 3.2 (accept 2.9 to 3.5). [1]
   $pK_a =$ pH at half neutralisation so $K_a = 10^{-pH} = 6.3 \times 10^{-4}$ mol dm^{-3} (accept 3.2×10^{-4} to 1.3×10^{-3}). [1]

Unit 4

4.1

1 (a) not chiral
   (b) [6]
   (c)
   (d) not chiral
   (e)
   (f) $(H_3C)_3SiCl$ not chiral

2 [1]

3 M_r 165 [1]
   16.5g of phenylalanine in total [1]
   Therefore 8.25g of each enantiomer [1]

4 E-form [2]
   Z-form

5 (a) [2]
   (b) [1]

4.2

1 The delocalised ring of electrons is an area of high electron density and will repel nucleophiles such as OH$^-$. [1]

2 [2]

3 To polarise the Br–Br molecule so that attack by the ring electrons can occur. [1]

4 (a) Concentrated nitric and sulfuric acids [1]
   (b) NO_2^+ [1]
   (c) Fractional distillation [1]

5 [1]

6 $CH_3C^+HCH_3$ and NO^+ [1]

7 The process of addition would disrupt the stable delocalised electron system and the resulting product would be less stable. [1]

8 (a) $C_{10}H_8$ [1]
   (b) Number of moles of naphthalene $= 25.6/128 = 0.200$ [1]
      M_r of 1-nitronaphthalene $= 173$ [1]
      If a 90% yield, then number of moles of 1-nitronaphthalene
      $= 0.200 \times 90/100$
      $= 0.180$ [1]
      Therefore mass of 1-nitronaphthalene $= 0.180 \times 173 = 31.1$ g

4.3

1 (a) Primary [1]
   (b) $HOCH_2–CH_2OH$ [1]
   (c) Propanone [1]

2 (a) Pentan-1-ol   Primary   [1]

(b) (4-methylphenyl)methanol   Primary   [1]

(c) Hexan-2,4-diol   Both alcohol groups are secondary   [1]

3 Aqueous sodium / potassium hydroxide   [1]

4 $NaBH_4$ and $LiAlH_4$ [2]

5   [2]

6 The organic product of the chlorination (often a liquid) is easier to separate from the co-products SO_2 and HCl, which are gases.   [1]

$CH_3–CH_2–CH(CH_3)–CH_2OH + SOCl_2 \rightarrow$
$CH_3–CH_2–CH(CH_3)–CH_2Cl + SO_2 + HCl$ [1]

7 The formula for the ester is wrong. It should be $CH_3CH_2COOCH_2CH_3$ / there should be CH_3CH_2 on each side   [1]
The other product, HCl, is not given.

8 The equation can be represented as
$C_3H_6O_2 + R–OH \rightarrow C_6H_{12}O_2 + H_2O$, where R–OH represents the formula of the alcohol. The R alkyl group must be C_3H_7.   [1]

The alkyl group must be 1-propyl, $CH_3CH_2CH_2$ or 2-propyl, $(CH_3)_2CH$.   [1]

The alcohol is propan-1-ol or propan-2-ol.

9 Compound **B** must be a phenol, as these are acidic but not acidic enough to produce carbon dioxide with sodium hydrogencarbonate.   [1]

Compound **C** is an alcohol, as these are neutral to pH paper and do not react with $NaHCO_3$.   [1]

Compound **D** is a carboxylic acid as these are acidic and will produce bubbles of carbon dioxide with $NaHCO_3$.   [1]

10 M_r of 2,4,6-tribromophenol ($C_6H_3Br_3O$) is 331.   [2]
Mass of 2,4,6-tribromophenol produced is
$0.050 \times 331 = 16.6$ g   [2]

11

CH_2OH   and   OCH_3

4.4

1 Compound **A** is not a ketone as there is only one carbon atom bonded directly to the carbonyl carbon atom.   [1]

Compound **B** is not a ketone as a carbonyl group (C=O) is not present.   [1]

Compound **C** is a (cyclic) ketone as the carbonyl carbon atom is bonded directly to two other carbon atoms.   [1]

Compound **D** is not a ketone as there is only one carbon atom bonded directly to the carbonyl carbon atom.   [1]

2   [1]

3   [1]

4 $^-:C\equiv N:$ [1]

5   [3]

hydroxynitrile   hydroxyacid

6 Compound **E** 2-Hydroxy-1-phenylpropane   [1]
Compound **F** alkaline iodine   [1]
Compound **G** Triiodomethane   [1]

7 Butane-1,4-dial   [1]

8 Ethanol will be largely oxidised to ethanoic acid by the large excess of hot acidified dichromate.   [1]

4.5

1 (a)   (b)   [5]

(c) $HOOC—CH_2—CH_2—COOH$ (d) $CH_3(CH_2)_4 COOH$
(e) $HOOC—(CH_2)_4—COOH$

2 (a) Propanedioic acid
(b) phenylethanoic acid
(c) benzene-1,4-dioic acid
(d) (E)-butenedioic acid
(e) 2-phenylpropanoic acid   [5]

3   [1]

4 2 'O' atoms $\rightarrow$ 32 $\equiv$ 31.3%   [3]
Therefore 100% $\equiv$ $100 \times 32 / 31.3 = 102$
M_r CH_3 is 15 and M_r COOH is 45, so therefore R group has
M_r 102 – (15 + 45) = 42
R must be C_3H_6
Formula of the acid is $CH_3(CH_2)_2COOH$ or $(CH_3)_3C–COOH$

5 $CH_3CH(OH)COOH + NaOH \rightarrow CH_3CH(OH)COO^-Na^+ + H_2O$ [5]
No. of mol of lactic acid = No. of mol of
NaOH = $47.2 \times 0.500/1000 = 0.0236$
In 25.00 cm^3 $\rightarrow$ 0.0236 mol of diluted acid, and in 1 dm^3
$0.0236 \times 1000/25.00 = 0.944$ mol
Diluted 10 times, original concentration is 9.44 mol dm^{-3}
or $9.44 \times 90 = 850$ g dm^{-3}

6 $CH_3CH_2CH_2COO^-Na^+ + NaOH \rightarrow Na_2CO_3 + CH_3CH_2CH_3$
propane   [2]

7 $(CH_3CH_2CH_2COO)_2Ca \rightarrow CaCO_3 + (CH_3CH_2CH_2)_2C=O$
heptan-4-one   [2]

8 Reagent **R** $\rightarrow$ $SOCl_2$ / PCl_3 / PCl_5 Reagent **S** $\rightarrow$ Ammonia   [2]

9 1.36 g of ammonia ≡ 1.36/17 = 0.080 mol [3]
   1:1 ratio, therefore 0.080 mol of the amide
   M_r of the amide = 5.84/0.080 = 73
   M_r of $CONH_2$ is 44, therefore M_r of the R group is 73 – 44 = 29
   R must be CH_3CH_2 which has M_r 29

4.6

1 [2]

2 The nitrogen atom is directly bonded to more than one carbon atom [1]

3 Ammonia has a nitrogen atom that has a lone pair of electrons [1]
   The bromine atom is substituted by an NH_2 group. [1]

4 $H_2N–CH_2–CH_2–NH_2$ [1]

5 (a) Tin or iron, and hydrochloric acid [1]
   (b) (i) Phenylamine [1]

   (ii) 449 nm is in the blue region; it is absorbing blue and transmitting yellow. The dye is seen to have a yellow colour. [2]

6 [1]

7 [2]

   The ethyl group is bonded to the nitrogen atom and the name is based on ethanamide.

8 $C_6H_5CH_2NH_2 + HNO_2 \longrightarrow N_2$ + other products
   No. of mol of phenylmethylamine = No. of mol of nitrogen [1]
   0.015 mol of nitrogen formed
   Volume = 0.015 × 24.5 × 1000 = 368 cm³ [1]

9 (a) 5–10 °C [1]
   (b) (i) 4.81 ≡ 89% therefore 100% ≡ 4.81 × 100/89 = 5.40 g
       No. of mol of 2-methylphenol
       = 5.40 / M_r = 5.40 / 108 = 0.0500 [1]
       No. of mol of 2-methylphenylamine is also 0.0500 mol
       Mass of 2-methylphenylamine used
       = No. of mol × M_r = 0.0500 × 107 = 5.35 g [1]
   (ii) Volume used = Mass / Density
        = 5.35 / 1.01 [1] = 5.30g to 3 sf [1]

4.7

1 [1]

---

2 (a) 2-amino-3-methylbutanoic acid [1]
   (b) [1]

   (c) [1]

3 [1]

4 (a) a compound containing both NH_2 and $COOH$ groups with the NH_2 group bonded to the adjacent carbon atom to the $COOH$ group [1]
   (b) polypeptide containing many condensed amino acids [1]
   (c) a macromolecular biological catalyst [1]
   (d) decomposition by water (usually irreversibly) [1]
   (e) 'water hating' usually insoluble in water [1]

5 [1]

4.8

1 Reagent **L** potassium / sodium cyanide / KCN / NaCN [1]
   Reagent **M** lithium tetrahydridoaluminate(III) / $LiAlH_4$ [1]
   Reagent **N** aqueous sulfuric or hydrochloric acids [1]

2 Compound **O** is $C_6H_4(CH_3)CH(OH)CN$
   Compound **P** is $C_6H_4(CH_3)CH(OH)COOH$ [2]

3 [5]

4 The bulb of the thermometer is in the wrong place, it should be opposite the side arm [1]
   The water connections are the wrong way round water should enter from the bottom [1]

5 Compound **R** is triiodomethane / CHI_3
   Compound **S** is ethane / C_2H_6 [2]

6 4.25/118 = 0.036 mol of acid. At the end 0.006 is in the aqueous layer. Therefore in ethoxyethane there is 0.030 mol. It is 5 times more soluble in ethoxyethane. [3]

7 Equal volumes of solvents used – 1 unit in water and 1 unit in CCl_4. Total 91 units. In water there is 1/91 parts. Therefore mass in water is 1 × 0.03 / 91 = ~ 3.3×10^{-4}g [2]

8 (i) 1,1-dichloroethene [1]
   (ii) alkene, ester [2]

9

10 (a) HOOC—$(CH_2)_4$—COOH + $2NH_3$ $\rightarrow$
$H_4N^{+-}OOC$—$(CH_2)_4$—$COO^{-+}NH_4$ [2]

(b) $H_4N^{+-}OOC$—$(CH_2)_4$—$COOH^{-+}NH_4$ $\rightarrow$
$N{\equiv}C(CH_2)_4\,C{\equiv}N + 4H_2O$ [2]

(c) nickel [1]

11 [4]

C^aH_3—C—C—$C^bH_2CCl_3$ Singlet 3 singlet 2

—C^aH_2—C^bH_3 quartet 2 triplet 3

12 (a) 'Total' area = 20 + 36 + 16 = 72% of the largest peak
= 36 × 100/72 = 50 [2]

(b) (i) [1]

(ii) 3 signals [2]

Unit 5

1 (a) Appropriate axes (evenly spaced, horizontal from 0 to 30, vertical from 0 to 240 or 250) [1], Points plotted correctly [2], smooth curve [1]

(b) Tangent drawn at t = 0 [1], gradient = 15.5±1.0 [1], rate = 1.55×10^{-2} dm^3 min^{-1} [1]

(c) (i) Rate = $k[C_6H_5N_2Cl]$ [1]
(ii) $k = 1.55 \times 10^{-2} \div 0.10$ [1] = 0.155 min^{-1} [1]

2 Any 8 of:
- Add nitric acid to all solid samples to make solutions of each
- Identify any which fizz – those that do are $CuCO_3$ and Na_2CO_3
- $CO_3^{2-} + 2 H^+ \longrightarrow CO_2 + H_2O$
- Differentiate between $CuCO_3$ and Na_2CO_3 by adding NaOH(aq) – Cu^{2+} will give a pale blue precipitate, Na^+ does not give a precipitate.
- $Cu^{2+} + 2 OH^- \longrightarrow Cu(OH)_2$
- Add sodium hydroxide to the four other solutions – three give white precipitate but KI does not.
- $Mg^{2+} + 2 OH^- \longrightarrow Mg(OH)_2$
  $Pb^{2+} + 2 OH^- \longrightarrow Pb(OH)_2$
  $Al^{3+} + 3 OH^- \longrightarrow Al(OH)_3$
- Add excess sodium hydroxide and two precipitates redissolve, but $Mg(OH)_2$ does not.
- $Pb(OH)_2 + 2 OH^- \longrightarrow [Pb(OH)_4]^{2-}$
  $Al(OH)_3 + OH^- \longrightarrow [Al(OH)_4]^-$
- Add KI solution to solutions of $Pb(NO_3)_2$ and $Al(NO_3)_3$ and Pb^{2+} will give a bright yellow precipitate, which identifies the final two.
- $Pb^{2+} + 2 I^- \longrightarrow PbI_2$

3 Any 6 of:
- Measure 25 cm^3 or 50 cm^3 of HCl(aq) using a pipette / burette.
- Place in a polystyrene cup that insulates and prevents heat leaving the cup.
- Measure temperature every 30s for a few minutes to ensure temperature is constant.
- Measure mass of Mg powder OR cut ribbon into small pieces.
- Add Mg to acid at a known time and replace thermometer and lid.
- Mix acid and Mg.
- Measure temperature every 30s until the temperature starts to drop.
- Plot the temperature against time.
- Extrapolate data at start and end to find temperature rise.
- Use Energy = $mC\Delta T$ where m = mass of acid and divide by moles of magnesium (or acid if this is the limiting reagent)

Answers to other questions

Unit 3 Answering examination questions – answers

1   (a) Lead nitrate gives a white precipitate that dissolves in excess NaOH whilst the white precipitate formed from magnesium nitrate does not dissolve. [1] This is because lead is an amphoteric metal but magnesium is not. [1]

  (b) • Lead metal has metallic bonding.
     • Attraction between delocalised electrons and lattice of metal ions must be broken to melt so it has a (relatively) high melting point.
     • Delocalised electrons can move and so it is an electrical conductor.
     • Carbon forms graphite/diamond with (strong) covalent bonds between atoms.
     • Need to break covalent bonds to melt so have a very high melting temperature.
     • Graphite is a conductor as it has delocalised electrons / Diamond is an electrical conductor as it has no delocalised electrons.
     • Stable oxidation state of carbon is +4 and for lead it is +2.
     • Due to the inert pair effect increasing down the group.
     • CO_2 is an acidic oxide as it reacts with bases e.g. $CO_2 + 2\,NaOH \longrightarrow Na_2CO_3 + H_2O$
     • PbO is an amphoteric oxide as it reacts with both acids and bases.
     • $PbO + 2\,H^+ \longrightarrow Pb^{2+} + H_2O$
     • $PbO + 2\,OH^- \longrightarrow [PbO_2]^{2-} + H_2O$ / $PbO + 2\,OH^- + H_2O \longrightarrow [Pb(OH)_4]^{2-}$

5–6 marks: At least 6 points with at least one from each bullet point in the question.

3–4 marks: At least 5 points with reference to at least two bullet points

1–2 marks: At least 3 points.

  (c) $T = \Delta H / \Delta S$ $T = 97.0 \div (151 \times 10^{-3})$ [1] $= 576\,K$ (answer to 3 sig figs) [1]

2   (a) Ca^{2+} / Calcium ion [1]

  (b) CO_3^{2-} [1] as it fizzes with acid [1]; I^- [1] as it gives the smell of rotten eggs with concentrated sulfuric acid. [1]

  (c) Moles water = 2.16 / 18.02 = 0.1199 [1]
     $x = 0.01199 / 0.0200 = 6$ [1]

  (d) Formula containing all ions from (a) and (b) and correct number of water molecules [1]

     Balanced charges, e.g. $Ca_2CO_3I_2.6H_2O$
       or $Ca_3(CO_3)_2I_2.6H_2O$ [1]

Unit 4 Answering examination questions – answers

(a) (i) [1]

  (ii) CHO [1]
  (iii) Filter the solution.
      Wash the crystals with water (to remove soluble impurities).
      Dry the crystals at a temperature below its melting temperature. [3]

(b) (i) [2]

  (ii) [2]

(c) Relative molecular masses calculated (116 and 197). [1]
   A yield of 100% will give 5.22 × 197 / 116 = 8.87 g of 2-bromobutanedioic acid [1]
   The yield is 80%. Mass of 2-bromobutanedioic acid = 8.87 × 80 / 100 = 7.10 g [1]

Exam practice answers

Unit 3

The following mark schemes are based on that provided by WJEC, but WJEC bears no responsibility for the answers provided within this publication.

Question 1: WJEC Unit 3 2017 Question 9

Question			Marking details	Marks available					
				A01	A02	A03	Total	Maths	Prac
1	(a)		mass of $NO_2 = 25.0 \times 0.714 \div 100 = 0.1785$ (1) moles $NO_2 = 0.1785 \div 46 = 3.88 \times 10^{-3}$ (1) answer must be to 3 sig figs to gain full marks		2		2	2	
	(b)	(i)	$K_c = \dfrac{[NO_2]^2}{[N_2O_4]}$ (1) unit mol dm^{-3} (1) allow ecf for unit		2		2	1	
		(ii) I	concentration of $NO_2 = 2 \times (0.400 - 0.0581) = 0.6838$ (1) value of $K_c = 8.05$ (1) solvent is CCl_4 (1) (allow ecf)		2	1	3	1	
		II	CS_2 given with any valid reason (1) explanation in terms of position of equilibrium or amount of product formed (1)	1		1	2		
	(c)	(i)	catalyst allows a lower temperature to be used (1) this leads to an increased yield (1) (must get first mark)	1		1	2		
		(ii)	a higher pressure should be used (1) shifts equilibrium to the right as this has fewer gas molecules (1) a lower temperature should be used (1) shifts equilibrium to the right as this is the exothermic direction (1) (allow one explanation mark if reference to equilibrium shifting to oppose change with no further detail) accept alternative remove product as it forms (1) equilibrium moves further towards the right / product (1)		1 1 1	1 1	4		
			Question 1 total	0	10	5	15	4	0

Question 2: WJEC Unit 3 2019 Question 12

Question			Marking details	Marks available					
				A01	A02	A03	Total	Maths	Prac
2	(a)	(i)	all four carbonates (1) acids react with metal carbonates to form carbon dioxide gas (1)	2			2		2
		(ii)	(hydrochloric acid would form) <u>insoluble</u> compound with Pb^{2+} (1) should use nitric acid / ethanoic acid (1)			2	2		2
		(iii)	award (2) for all four correct award (1) for any two correct Mg^{2+}(aq) white (precipitate) Fe^{2+}(aq) (dark) green (precipitate) Cr^{3+}(aq) (grey) green (precipitate) Pb^{2+}(aq) white (precipitate)	2			2		2
		(iv)	add excess sodium hydroxide (1) award (1) for either of following • magnesium hydroxide white precipitate remains but lead hydroxide precipitate dissolves (giving a colourless solution) • iron(II) hydroxide green precipitate remains but chromium hydroxide precipitate dissolves (giving a dark green solution) lead and chromium are amphoteric (iron and magnesium are not) (1)			3	3		3

Question			Marking details	Marks available					
				A01	A02	A03	Total	Maths	Prac
2	(b)	(i)	Mg^{2+} $\dfrac{91}{24.3} = 3.7449$ Ca^{2+} $\dfrac{50}{40.1} = 1.2469$ (1) ratio is 3Mg : 1Ca (1) formula must contain $4 \times CO_3$ so it is $Mg_3Ca(CO_3)_4$ accept alternative correct representations, e.g. $CaCO_3.3MgCO_3$ $(MgCO_3)_3.CaCO_3$ (1)		1	2	3	2	
		(ii)	concentration $= 1.25 \times 10^{-6}$ mol dm^{-3} (from calcium or magnesium concentrations) (1) moles of solid used $= \dfrac{220 \times 10^{-6}}{353} = 0.6232 \times 10^{-6}$ volume $= 0.50$ dm^3 (1) allow any value in the range 0.498–0.501 dm^3		1 1	1	3	2	
			Question 2 total	4	3	8	15	4	9

Question 3: WJEC Unit 3 2018 Question 11

Question			Marking details	Marks available					
				A01	A02	A03	Total	Maths	Prac
3	(a)		profile with start and end at same energies and activation energy of 58 kJ mol^{-1} allow tolerance ± 1 small square each way for peak maximum		1		1		
	(b)	(i)	$k = A\,e^{-Ea/RT}$ accept $\ln k = \ln A - \dfrac{E_a}{RT}$	1			1	1	
		(ii)	$e^{-Ea/RT} = 1.19 \times 10^{-4}$ (1) ecf possible from part (i) $E_a = 22.5$ kJ mol^{-1} (1) activation energy is lower so this catalyst is more effective (1) ecf from activation energy calculated			3	3	2	
	(c)		$\Delta_r H^\ominus = 2 \times \Delta_r H^\ominus(H_2O) - 2 \times \Delta_r H^\ominus(H_2O_2)$ (1) $2\Delta_r H^\ominus(H_2O_2) = (2 \times -286) - (-98) = -474$ $\Delta_r H^\ominus(H_2O_2) = -237$ (1)		1 1		2	1	
	(d)		positive because the reaction forms a gas (and gases have a higher entropy than liquids)		1				
	(e)	(i) I	**A** HCl/H^+(aq) (1 mol dm^{-3}) **B** platinum/Pt (electrode) **both** required	1			1		1
		II	electrons flowing from standard hydrogen electrode to dichromate half cell		1		1		1
		III	salt bridge completes circuit without allowing solutions to mix **both** required	1			1		1
		IV	this makes the reading less positive/smaller (must give reason to gain this mark) (1) increase in concentration of chromium(III) ions will shift equilibrium to the left (1)			1 1	2		
		(ii) I	$2Cr^{3+} + 3H_2O_2 + H_2O \longrightarrow Cr_2O_7^{2-} + 8H^+$		1		1		
		II	reaction is feasible (must give reason to gain this mark) (1) as standard electrode potential for hydrogen peroxide is more positive than that for chromium/dichromate / EMF calculated as +0.44V and positive value means reaction is feasible (1)		1	1	2		
		III	electrochemical methods are better (must give reason for this mark) (1) because the reaction is in solution but standard enthalpies of formation use standard states			2	2		
			Question 3 total	3	7	8	18	4	3

Question 4: WJEC Unit 3 2018 Question 9

Question			Marking details	Marks available					
				A01	A02	A03	Total	Maths	Prac
4	(a)		(1) (1)	2			2		
	(b)		ligands cause d-orbitals to split into three lower and two higher energy levels (1) electrons absorb energy and are promoted to a higher energy level (1) colour seen is the colours not absorbed (1)	3			3		
	(c)		white precipitate (1) silver chloride forms / silver ions cause a precipitate with chloride (1) solution turns pink (1) removal of chloride causes equilibrium to shift to left (1)		1 1 	 1 1	4		1 1
	(d)	(i)	$f = 5.79 \times 10^{14}$ (1) $E = hf = 3.84 \times 10^{-19}$ (1) $E = 231$ (1)		3		3	3	
		(ii)	value above 600 nm as this is absorbed by chloro but not aqua / value below 560 nm as this is absorbed by aqua complex but not chloro		1		1		1
		(iii)	$\dfrac{[[CoCl_4]^{2-}] \times [H_2O]^6}{[[Co(H_2O)_6]^{2+}] \times [Cl^-]^4}$ (1) $mol^2\ dm^{-6}$ (1)		2		2	1	1
		(iv)	concentration of aqua complex $= 0.596$ concentration of chloride $= 0.224$ / concentration of chloro complex $= 0.124$ award (2) for all three concentrations correct award (1) for any two correct $K_c = 14.0$ (1)			3	3	3	
			Question 4 total	5	8	5	18	7	3

Question 5: Eduqas Component 1 2018 Question 13

Question			Marking details	Marks available					
				A01	A02	A03	Total	Maths	Prac
5	(a)		moles thiosulfate $= 2.809 \times 10^{-3}$ (1) moles Cu^{2+} in 25 cm³ $= 2.809 \times 10^{-3}$ moles Cu^{2+} in 250 cm³ $= 2.809 \times 10^{-2}$ (1) percentage by mass $= \dfrac{(2.809 \times 10^{-2} \times 63.5)}{2.877}$ (1) percentage by mass $= 62.0\%$ (1) must show method		2	2	4	4	3
	(b)		mass of precipitate in weighing 1 [$Cu(OH)_2$ and $Zn(OH)_2$] $= 1.78g$ mass of precipitate in weighing 2 [$Cu(OH)_2$ only] $= 1.11g$ (1) moles $Cu(OH)_2 = \dfrac{1.11}{97.52} = 0.01138$ moles $Zn(OH)_2 = \dfrac{0.67}{99.42} = 6.74 \times 10^{-3}$ (1) percentage by mass $= \dfrac{(0.01138 \times 63.5)}{(0.01138 \times 63.5) + (6.74 \times 10^{-3} \times 65.4)}$ (1) percentage by mass $= 62.1\%$ (1) must show method		2	2	4	4	2

Continued ▶

Question			Marking details	Marks available					
				A01	A02	A03	Total	Maths	Prac
	(c)		the evidence supports this as the percentages of copper are the same in both methods (within experimental error) – allow reverse argument if values do not match (1) could confirm this by analysis (or any named analytical method) of first solution for percentage by mass of zinc / undertake qualitative analysis to exclude the presence of other metals in alloy / appropriate alternative method (1)		2	2			2
	(d)	(i)	$[H^+] = 10^{-pH} = 10^{1.2}$ (1) $[HNO_3] = 15.8$ mol dm^{-3} (1)		2		2	2	
		(ii)	$[H^+] = \dfrac{10^{-14}}{2} = 5 \times 10^{-15}$ (1) pH $= -\log(5 \times 10^{-15}) = 14.3$ (1)		2		2	2	
			Question 5 total	0	8	6	14	12	7

Unit 4

Question 1: WJEC Unit 4 2017 Q13

Question			Marking details	Marks available					
				A01	A02	A03	Total	Maths	Prac
1	(a)	(i)	it is compound **N** (1) singlet for the —CH$_3$ group bonded to the carbon atom of the double bond that is also bonded to the bromine atom (2.2δ) (1) the CH proton is as a quartet – split by the adjacent CH$_3$ protons (5.7δ) (1) the other —CH$_3$ group is a doublet, split by the adjacent CH proton (1.7δ) (1)		4		4		
		(ii) I	add aqueous bromine / aqueous acidified manganate(VII) – this is decolourised by **L**, **M** and **N** but unaffected by bromocyclobutane		1		1		1
		II	bromocyclobutane would only give three ^{13}C signals (1) **L**, **M** and **N** would each give four signals as each carbon atom is in a different environment for these alkenes (1) accept answer based on C=C at δ 90 to 150 present in **L**, **M** and **N** but not in bromocyclobutane		2		2		
	(b)		Curly arrows (1) partial / full charges (1) regeneration of catalyst (1)	3			3		
	(c)	(i)	1:1 molar ratio $0.150 \times 159.8 = 23.97$ g (1)		2		2		
		(ii)	potassium cyanide / sodium cyanide	1					1
		(iii)	dilute sulfuric acid / hydrochloric acid	1					1
			Question 1 total	5	5	4	14	0	3

Question 2: Eduqas Component 2 2017 Q12

Question			Marking details	Marks available					
				A01	A02	A03	Total	Maths	Prac
2	(a)	(i)	warm 2-chloroethanol with aqueous sodium hydroxide (1) acidify the mixture with (aqueous) nitric acid (1) add aqueous silver nitrate (1) white precipitate (of silver chloride) confirms the presence of a C—Cl bond (1)	2	2		4		4
		(ii)	chloroethanol to chloroethanoic acid $\longrightarrow$ mole ratio = 1:1 (1) 83% conversion, therefore originally $100 \times 0.0600 / 83 = 0.072$ mol of chloroethanol (1) mass of 2-chloroethanol = $0.072 \times 80.6 = 5.80$g (1)		1 1	1	3	1	
	(b)	(i)	it does not contain polar groups (accept examples) that can hydrogen bond with water		1		1		
		(ii)	hydrogen and platinum / nickel catalyst		1		1		
		(iii)	total relative peak area = $48 + 13 + 10 + 9 = 80$ (1) 48 is equivalent to 0.018 mol kg^{-1} total concentration = $80 \times 0.018 / 48 = 0.030$ (mol kg^{-1}) (1)			2	2	1	
	(c)		Indicative content • in the upper atmosphere a C—Cl bond is broken by UV radiation • giving a chlorine atom / radical • $CFCl_3 \longrightarrow \cdot CFCl_2 + Cl\cdot$ • chorine radical attacks ozone • $Cl\cdot + O_3 \longrightarrow ClO\cdot + O_2$ • the chlorine radical is regenerated, therefore reaction is catalytic • $ClO\cdot + O_3 \longrightarrow Cl\cdot + 2O_2$ • environmental effects, for example ○ skin cancer ○ plant death ○ global warming	3	3		6		

5-6 marks

The response includes a good description of reactions leading to breakdown of ozone (could be by chemical equations) and two appropriate environmental effects

The candidate constructs a relevant, coherent and logically structured account including all key elements of the indicative content. A sustained and substantiated line of reasoning is evident and scientific conventions and vocabulary are used accurately throughout.

3-4 marks

The response includes a general description of breakdown of ozone and an environmental effect

The candidate constructs a coherent account including many of the key elements of the indicative content. Some reasoning is evident in the linking of key points and use of scientific conventions and vocabulary is generally sound.

1-2 marks

The response is more limited and includes reference to breakdown of ozone / environmental effect

The candidate attempts to link at least two relevant points from the indicative material. Coherence is limited by omission and/or inclusion of irrelevant materials. There is some evidence of appropriate use of scientific conventions and vocabulary.

0 marks

The candidate does not make any attempt or give an answer worthy of credit.

Question			Marking details	A01	A02	A03	Total	Maths	Prac
	(d)		**Mass spectrum** molecular ions at 154 and 156 in 3:1 ratio (1) reflecting only one chlorine atom present in the compound and isotopic ratio of 3:1 for $^{35}Cl : ^{37}Cl$ (1) **^{13}C NMR** two peaks (1) showing two distinct environments / CF_3 and CF_2Cl (1)		2 2		4		
			Question 2 total	5	12	4	21	2	4

Question 3: Eduqas Component 2 2018 Q13

Question			Marking details	A01	A02	A03	Total	Maths	Prac
3	(a)		they contain both an acidic and alkaline functional groups	1			1		
	(b)	(i)	award (1) for any of following • the burette had been rinsed with water and this was not replaced entirely with sodium hydroxide • inadequate shaking • rough titration / overshot end point	1			1		1
		(ii)	concordant titres chosen – 35.90, 36.00 and 36.10 cm^3 (1) mean titre = 36.00 cm^3 (1) $n(NaOH) = \dfrac{36.00 \times 0.105}{1000} = 0.00378$ (1) 1:1 ratio therefore number of moles of the amino acid is also 0.00378 250 cm^3 contain 0.0378 mol (1) M_r of the amino acid = $\dfrac{4.95}{0.0378} = 131$ (1)		5		5	1 1	

Continued ▶

Question			Marking details	Marks available					
				AO1	AO2	AO3	Total	Maths	Prac
(b)	(iii)		—CH(NH$_2$)COOH 'M_r' = 74 (1) 'M_r' of chain is 131 − 74 = 57 so must be C$_4$H$_9$ ecf possible from part (ii) formula must be CH$_3$CH$_2$CH$_2$CH$_2$CH(NH$_2$)COOH (1)			2	2		
(c)	(i)		compound **T** as this is the only one that contains a chiral centre / asymmetric carbon atom		1		1		
	(ii)		only compound **T** would show an N—H stretching frequency at 3300-3500 cm^{-1}		1		1		
	(iii)		compound **S** could only form one dipeptide via its COOH group, as it does not contain an N—H bond			1	1		
(d)	(i)		the reaction proceeds via secondary carbocations which are more stable / have lower activation energies accept explanation using Markovnikov's rule			1	1		
	(ii)	I	bromine is more electronegative than carbon / has greater electron attracting power than carbon (so is δ−) accept converse argument	1			1		
		II	it acts as a base / nucleophile	1			1		
	(iii)		e.g. 		1		1		
			Question 3 total	4	8	4	16	2	1

Unit 5

Question 1: Eduqas Chemistry in Practice 2017 Q2

Question			Marking details	Marks available					
				AO1	AO2	AO3	Total	Maths	Prac
1	(a)		**X** is magnesium Mg + H$_2$O ⟶ MgO + H$_2$ (1)	1			1		1
	(b)	(i)	n[**Y**(OH)$_2$] in 600cm^3 = 0.0431 × 600/1000 = 0.02586 (1) 1:1 ratio of **Y**:**Y**(OH)$_2$ therefore n(**Y**) = 0.02586 (1) A_r = 2.27/0.02586 = 87.8 therefore **Y** is strontium / Sr (1)		1 1 1		3	2	1
	(ii)	I	**Y**(OH)$_2$(aq) + Na$_2$CO$_3$(aq) ⟶ **Y**CO$_3$(s) + 2NaOH(aq) **Y**$^{2+}$(aq) + CO$_3$$^{2-}$(aq) ⟶ **Y**CO$_3$(s) accept either equation but must be balanced and include state symbols; accept Sr in place of **Y** ecf possible from part (i) if incorrect Group 2 metal identified	1			1		
		II	200cm^3 of solution = 0.0431 × 200/1000 = 0.00862 mol of **Y**CO$_3$ (1) mass of **Y**CO$_3$ = 0.00862 × M_r(**Y**CO$_3$) = 0.00862 × 147.6 (1) = 1.27 g **must** be given to 3 sig figs (1)		3		3	2	
	(c)		**Indicative content** **carbonates more thermally stable down the group** description of sensible method such as following examples: • carbonate taken in clamped test tube connected to delivery tube dipping into second tube containing limewater • carbonate heated using Bunsen flame **until limewater turns milky** • control variables ✓ **same number of moles of carbonate** ✓ same volume + concentration of limewater ✓ same distance of test tube to flame + same flame temperature • **time** taken for limewater to turn milky (cloudy) **increases** for carbonates on **going down** the group • carbonate taken in a crucible and heated to determine the **loss in mass of the carbonate** • control variables	1					

Continued ▶

Question			Marking details	Marks available					
				AO1	AO2	AO3	Total	Maths	Prac
	(c)		✓ **same number of moles of carbonate** ✓ same distance of crucible to flame + same temperature of flame ✓ carbonates heated for the same amount of time • **mass loss** on heating for the **same set time** is **less** on **going down** the group no trend or incorrect trend in thermal stability and no description of how to control variables used should be considered as 'significant omissions' and credit must be limited to the 1-2 marks band			5	6		6

5-6 marks

All aspects of question covered and key details given, including those in bold print

The candidate constructs a relevant, coherent and logically structured account including all key elements of the indicative content. A sustained and substantiated line of reasoning is evident and scientific conventions and vocabulary are used accurately throughout.

3-4 marks

Correct description of trend in thermal stability; basic description of apparatus / method used; reference to one control variable; observation and expected results

The candidate constructs a coherent account including many of the key elements of the indicative content. Some reasoning is evident in the linking of key points and use of scientific conventions and vocabulary is generally sound.

1-2 marks

Description of apparatus or method used / simple observation

The candidate attempts to link at least two relevant points from the indicative material. Coherence is limited by omission and/or inclusion of irrelevant materials. There is some evidence of appropriate use of scientific conventions and vocabulary.

0 marks

The candidate does not make any attempt or give an answer worthy of credit.

				AO1	AO2	AO3	Total	Maths	Prac
			Question 1 total	3	5	6	14	4	7

Question 2: WJEC Practical Methods and Analysis Task 2019 Q2

Question			Marking details	AO1	AO2	AO3	Total	Maths	Prac
2			outlines a suitable plan that would allow the identification of **all** four species (2) outlines a suitable plan that would allow the identification of **two/three** species (1) award (1) each for up to **three** observations linked to the reagent(s) used and species identified award (1) each for up to **three** ionic equations linked to the reagent(s) used and species identified **carbonate ion** addition of dilute H_2SO_4 fizzing and CO_3^{2-}(aq) identified $CO_3^{2-} + 2H^+ \longrightarrow H_2O + CO_2$ **iodide ion** addition of $AgNO_3$(aq) – yellow precipitate and I^-(aq) identified $Ag^+ + I^- \longrightarrow AgI$ **chlorine** addition of aqueous iodide - yellow / brown coloured solution formed and Cl_2(aq) identified $Cl_2 + 2I^- \longrightarrow 2Cl^- + I_2$ **or** addition of $AgNO_3$(aq) – white precipitate and Cl^-(aq) identified explanation of formation of Cl^-(aq) required for both marks to be awarded e.g. $Cl_2 + H_2O \longrightarrow HCl + HClO$ or Cl_2(aq) contains Cl^-(aq) $Ag^+ + Cl^- \longrightarrow AgCl$ **thiosulfate ion** dropwise addition of I_2 solution formed – yellow / brown solution of I_2 becomes colourless and $S_2O_3^{2-}$(aq) identified $I_2 + 2S_2O_3^{2-} \longrightarrow 2I^- + S_4O_6^{2-}$ **or** addition of dilute sulfuric acid – off-white precipitate / cloudy white solution and $S_2O_3^{2-}$(aq) identified $S_2O_3^{2-} + 2H^+ \longrightarrow SO_2 + S + H_2O$	3	2 3	8			2 3
			Question 2 total	3	0	5	8	0	5

Question 3: WJEC Practical Methods and Analysis Task 2018 Q3

Question			Marking details	AO1	AO2	AO3	Total	Maths	Prac
3	(a)		award (1) for correct number of moles of **both** reactants $n(C_6H_8O_7) = 0.050$ mol $n(NaHCO_3) = 0.190$ mol award (1) for statement that compares the ratio of [1] moles of both reactants with reference to the 3:1 stoichiometry		2		2	2	
	(b)		energy change $= 78800 \times 0.050 = 3940$ J (1)		1			1	
			energy change $= m \times c \times \Delta T$ $3940 = 50 \times 4.18 \times \Delta T$ $\Delta T = 3940/(50 \times 4.18) = 18.9$ °C (1)	1				1	
			final temperature $= 24.4 - 18.9 = 5.5$ °C (1) ecf possible throughout		1	3			3
			Question 3 total	1	2	2	5	4	3

Glossary

Activation energy is the minimum amount of energy required for a collision to be successful.

Oxidising agent A substance that takes electrons from another substance and so it is reduced.

Reducing agent A substance that donates electrons to another substance and so it is oxidised.

Amphoteric A material that reacts with both acids and bases.

Aromatic primary amine Has the $-NH_2$ group directly bonded to the benzene ring.

Azo dye Contains the grouping Ar–N=N–Ar where Ar is an aromatic ring system.

Buffer Resists changes in pH as small amounts of acid and alkali are added.

Catalysts Substances which increase the rate of a chemical reaction by providing an alternative pathway with a lower activation energy.

Chiral centre An atom in a molecule that is bonded to four different atoms or groups.

A **chromophore** is a structural unit in a molecule that is primarily responsible for the absorption of radiation of a certain wavelength, generally in the visible or ultraviolet region of the electromagnetic spectrum.

Concentrated In a concentrated solution there is a large amount of an acid or base dissolved in a set volume of water.

Condensation Loss of a small molecule, often water or hydrogen chloride, as two or more organic molecules combine to give a larger molecule.

d-block The set of elements whose outer electrons are found in d-orbitals.

Decarboxylation Loss of carbon dioxide from the organic molecule usually giving calcium carbonate or sodium carbonate as the inorganic product.

Dehydration Loss of two hydrogen atoms and an oxygen atom as water.

Dilute In a dilute solution there is a small amount of acid or base dissolved in a set volume of water.

Disproportionation reaction One in which some atoms of an element are oxidised and other atoms of the same element are reduced. It forms products containing the element in two different oxidation states.

Dynamic equilibrium A reversible reaction where the rates of the forward and reverse reactions are equal.

Electron affinity This is the enthalpy change when one mole of gaseous negative ions are formed from gaseous atoms of a substance by gaining an electron.

Electrophile An electron-deficient species that can accept a lone pair of electrons.

Enantiomers Non-superimposable mirror image forms of each other that rotate the plane of plane polarised light in opposite directions.

In NMR spectroscopy the **environment** means the nature of the surrounding atoms or groups in the molecule.

An **equimolar mixture** is one that has equal amounts of moles (and hence equal concentrations) of each substance.

Essential amino acids Those α-amino acids that cannot be synthesised in the body and must be supplied through the diet.

Ester Always contains a

 group.

Heterogeneous catalysts Catalysts that are in a different physical state from the reactants in the reactions that they catalyse.

Homogeneous catalysts Catalysts that are in the same physical state as the reactants in the reactions that they catalyse.

Inert pair effect The tendency of the s pair of electrons in an atom to stay paired leading to a lower oxidation state.

Indicator A substance that has different colours in low and high pH solutions. This can be used to differentiate between an acid and an alkali but cannot distinguish between acids of different pH or alkalis of different pH.

Ionisation This is the enthalpy change when one mole of gaseous positive ions are formed from gaseous atoms of an element by losing an electron.

Miscible Liquids that are completely soluble in each other at all concentrations. An example of this is ethanol and water.

Monomer A starting compound from which a polymer is produced. For example, ethene is the monomer of poly(ethene).

Nucleophiles are ions or compounds possessing a lone pair of electrons that can seek out a relatively positive site (often a δ+ carbon atom). Common nucleophiles include ^-OH, ^-CN and NH_3. Nucleophiles can be negatively charged, or neutral molecules.

Optical activity This occurs in molecules that possess chiral centre(s). These molecules rotate the plane of plane polarised light.

The **order of reaction** with respect to a particular reactant is the power to which the concentration is raised in the rate equation.

Oxidation is the loss of electrons.

Polyester A compound made up of repeating units that each contain the

$$-\overset{\overset{\displaystyle O}{\displaystyle \|}}{C}-O \text{ group.}$$

Quenching The sudden stopping of a chemical reaction to allow for analysis to occur without the reaction proceeding further. It is usually undertaken by adding the sample to ice water as this both cools and dilutes the reactants.

Racemic mixture An equimolar mixture of both enantiomers that produces no overall rotation of the plane of plane polarised light.

Rate is the rate of change of the concentration, or of the amount, of a particular reactant or product.

Rate constant is a constant in the rate equation. It is not affected by changing the concentrations of the reactants at a particular temperature.

Rate-determining step The slowest step in a reaction mechanism.

Reduction is the gain of electrons.

Salt bridge A piece of apparatus that connects the solutions in two half-cells so that the circuit is complete and the current can flow without the solutions mixing.

Sampling Taking samples from the reaction mixture at regular time intervals.

Standard conditions are:
A temperature of 298 K (25ºC).
A concentration of 1 mol dm^{-3} for solutions.
A pressure of 101 kPa, or one atmosphere (1 atm), for gases.

Standard electrode potential The potential difference between any half-cell under standard conditions and the standard hydrogen electrode. Standard conditions are a temperature of 298 K, a pressure of 1 atm and concentrations of 1 mol dm^{-3}.

Standard enthalpy change of hydration, $\Delta_{hyd}H^{\ominus}$ This is the enthalpy change when one mole of an ionic compound in solution is formed from ions of the elements in the gas phase.

Standard enthalpy change of lattice formation, $\Delta_{latt}H^{\ominus}$ This is the enthalpy change when one mole of an ionic compound is formed from ions of the elements in the gas phase.

Standard enthalpy change of solution, $\Delta_{sol}H^{\ominus}$ This is the enthalpy change that occurs when one mole of a substance dissolves completely in a solvent under standard conditions to form a solution.

Standard state is the physical state of a substance under standard conditions, such as oxygen gas, liquid water or sodium chloride solid.

A **strong acid** is one that dissociated fully to release H$^+$ ions.

Transition element A metal that possesses a partially filled d sub-shell in its atom or stable ions.

A **weak acid** is one that dissociates partially to release H$^+$ ions in a reversible reaction.

Zwitterion A dipolar form of an amino acid where the carboxylic acid group loses a proton, becoming COO$^-$ and the amino group gains the proton, becoming $^+$NH$_3$.

Index

THE PERIODIC TABLE

Group

Key

A_r	relative atomic mass
Symbol	
Name	
Z	atomic number

Period	1	2													3	4	5	6	7	0
1	1.01 H Hydrogen 1																			4.00 He Helium 2
2	6.94 Li Lithium 3	9.01 Be Beryllium 4													10.8 B Boron 5	12.0 C Carbon 6	14.0 N Nitrogen 7	16.0 O Oxygen 8	19.0 F Fluorine 9	20.2 Ne Neon 10
3	23.0 Na Sodium 11	24.3 Mg Magnesium 12													27.0 Al Aluminium 13	28.1 Si Silicon 14	31.0 P Phosphorus 15	32.1 S Sulfur 16	35.5 Cl Chlorine 17	40.0 Ar Argon 18
4	39.1 K Potassium 19	40.1 Ca Calcium 20	45.0 Sc Scandium 21	47.9 Ti Titanium 22	50.9 V Vanadium 23	52.0 Cr Chromium 24	54.9 Mn Manganese 25	55.8 Fe Iron 26	58.9 Co Cobalt 27	58.7 Ni Nickel 28	63.5 Cu Copper 29	65.4 Zn Zinc 30			69.7 Ga Gallium 31	72.6 Ge Germanium 32	74.9 As Arsenic 33	79.0 Se Selenium 34	79.9 Br Bromine 35	83.8 Kr Krypton 36
5	85.5 Rb Rubidium 37	87.6 Sr Strontium 38	88.9 Y Yttrium 39	91.2 Zr Zirconium 40	92.9 Nb Niobium 41	95.9 Mo Molybdenum 42	98.9 Tc Technetium 43	101 Ru Ruthenium 44	103 Rh Rhodium 45	106 Pd Palladium 46	108 Ag Silver 47	112 Cd Cadmium 48			115 In Indium 49	119 Sn Tin 50	122 Sb Antimony 51	128 Te Tellurium 52	127 I Iodine 53	131 Xe Xenon 54
6	133 Cs Caesium 55	137 Ba Barium 56	139 La Lanthanum 57 ▲	179 Hf Hafnium 72	181 Ta Tantalum 73	184 W Tungsten 74	186 Re Rhenium 75	190 Os Osmium 76	192 Ir Iridium 77	195 Pt Platinum 78	197 Au Gold 79	201 Hg Mercury 80			204 Tl Thallium 81	207 Pb Lead 82	209 Bi Bismuth 83	(210) Po Polonium 84	(210) At Astatine 85	(222) Rn Radon 86
7	(223) Fr Francium 87	(226) Ra Radium 88	(227) Ac Actinium 89 ▲▲																	

s block · **d block** · **p block** · **f block**

▲ Lanthanoid elements

140 Ce Cerium 58	141 Pr Praseodymium 59	144 Nd Neodymium 60	(147) Pm Promethium 61	150 Sm Samarium 62	(153) Eu Europium 63	157 Gd Gadolinium 64	159 Tb Terbium 65	163 Dy Dysprosium 66	165 Ho Holmium 67	167 Er Erbium 68	169 Tm Thulium 69	173 Yb Ytterbium 70	175 Lu Lutetium 71

▲▲ Actinoid elements

232 Th Thorium 90	(231) Pa Protactinium 91	238 U Uranium 92	(237) Np Neptunium 93	(242) Pu Plutonium 94	(243) Am Americium 95	(247) Cm Curium 96	(245) Bk Berkelium 97	(251) Cf Californium 98	(254) Es Einsteinium 99	(253) Fm Fermium 100	(256) Md Mendelevium 101	(254) No Nobelium 102	(257) Lr Lawrencium 103